IRMA Lectures in Mathematics and Theoretical Pi

Edited by Vladimir G. Turaev

Institut de Recherche Mathématique Avancée
Université Louis Pasteur et CNRS
7 rue René Descartes
67084 Strasbourg Cedex
France

IRMA Lectures in Mathematics and Theoretical Physics

Infinite Dimensional Groups and Manifolds

Editor

Tilmann Wurzbacher

W DE G

Walter de Gruyter · Berlin · New York

Editor

Tilmann Wurzbacher
Laboratoire de Mathématiques et Applications de Metz
UMR 7122, Université de Metz et C.N.R.S., Ile du Saulcy, 57045 Metz cedex 01, France
e-mail: wurzbacher@poncelet.univ-metz.fr

Series Editor

Vladimir G. Turaev
Institut de Recherche Mathématique Avancée (IRMA), Université Louis Pasteur − C.N.R.S.,
7, rue René Descartes, 67084 Strasbourg Cedex, France, e-mail: turaev@math.u-strasbg.fr

Mathematics Subject Classification 2000:
22E65, 58Bxx, 81-xx, 53-xx, 57Rxx

Key words:
Infinite dimensional Lie groups, infinite dimensional manifolds, quantum physics, differential
geometry, differential topology

⊚ Printed on acid-free paper which falls within the guidelines of the ANSI
to ensure permanence and durability.

Library of Congress Cataloging-in-Publication Data

Infinite dimensional groups and manifolds / edited by Tilmann
[sic] Wurzbacher.
     p.   cm. − (IRMA lectures in mathematics and theo-
retical physics ; 5)
    Includes bibliographical references.
    ISBN 3-11-018186-X (pbk. : acid-free paper)
    1. Infinite-dimensional manifolds.   2. Infinite dimen-
sional Lie algebras.   3. Differential equations, Partial.
4. Quantum field theory.   I. Wurzbacher, Tilmann, 1961 −
II. Series.
QA613.2.I54   2004
512′.55−dc22

                              2004011405

ISBN 3-11-018186-X

Bibliographic information published by Die Deutsche Bibliothek

Die Deutsche Bibliothek lists this publication in the Deutsche Nationalbibliografie;
detailed bibliographic data is available in the Internet at <http://dnb.ddb.de>.

Printed in Germany.
Cover design: I. Zimmermann, Freiburg.
Depicted on the cover is the Strasbourg Cathedral.
Typeset using the authors' TₑX files: I. Zimmermann, Freiburg.
Printing and binding: Hubert & Co. GmbH & Co. KG, Göttingen.

Preface

Infinite dimensional groups and manifolds are inevitably present in physics once the passage is made from point particle mechanics to classical field theory and, a fortiori, from quantum mechanics to quantum field theory. Though mathematicians time and again considered this subject during the 20th century, it seems fair to say that the renewed communication of the last two decades between the mathematical and the theoretical physics communities, centered around string theory and quantum fields, gave strong new stimuli to the study of infinite dimensional geometry, and related algebra and analysis.

The present collection of articles is an outgrowth of the 70th meeting of theoretical physicists and mathematicians at the Institut de Recherche Mathématique Avancée (IRMA) in Strasbourg in May 2002, organized by Vladimir Turaev and myself on *Groupes et variétés de dimension infinie en mathématiques et physique quantique*. Since, together with Claude Roger, I had the opportunity to organize a colloquium in the Centre International de Rencontres Mathématiques (CIRM) in Marseille-Luminy on the related subject *Géométrie de dimension infinie et applications à la théorie des champs* in November 2002, some of the speakers of the latter colloquium also contributed to this collection of articles.

Since the subject of infinite dimensional geometry is rapidly developing and touching by its very nature many areas, a general "tour d'horizon" seems very ambitious. Instead, I would like to point out some central topics of current research that will show up in these proceedings:

- the intimate relation between flows on infinite dimensional manifolds and partial differential equations, ranging from integrable equations and solitons to fluid dynamics;
- the search for both general and sensible classes of examples of manifolds and groups and the study of their structure, notably groups of maps and gauge groups, and groups of operators;
- the extension of fundamental algebraic and geometric constructions from finite to infinite dimensions and the developement of the necessary analytic tools;
- rigorous geometric and algebraic approaches to (parts of) quantum field theory, especially in the presence of symmetries;
- the study of "large N limits" or "$1/N$ expansions" (as, e.g., but not only for $SU(N)$ gauge theories) in quantum physics and in mathematics.

I now sketch the content of the individual contributions.

In "Lie groups of germs of analytic mappings", Helge Glöckner gives some foundational material on the theory of this class of infinite dimensional groups. More precisely, he considers, for a non-empty compact set K of a metrizable topological vector space over $\mathbb{R}$ or $\mathbb{C}$, and a Banach–Lie group G, the group $\Gamma(K, G)$ of germs around K of G-valued analytic mappings. In particular, he establishes a useful "push-forward property" and obtains, as the main result of this article, that $\Gamma(K, G)$ is an analytic Baker–Campbell–Hausdorff Lie group, i.e. an analytic Lie group having a locally diffeomorphic exponential map and having its multiplication being locally given by the Baker–Campbell–Hausdorff series.

In "The flow completion of the Burgers equation", Boris A. Khesin and Peter W. Michor study certain partial differential equations from the point of view of vector fields on infinite dimensional manifolds. More precisely, given a smooth map $f : \mathbb{R}^n \to \mathbb{R}^k$, the equation

$$u_t + (f(u) \cdot \nabla)u = 0$$

in the unknown u in $C^\infty(\mathbb{R}^k, \mathbb{R}^n)$ is equivalent to $u_t = X(u)$, where $X(u) = -(f(u) \cdot \nabla)u$ is re-interpreted as a vector field on the mapping manifold $C^\infty(\mathbb{R}^k, \mathbb{R}^n)$. Upon generalizing Richard Palais' construction of the flow completion of a manifold-with-vector field to a (possibly non-separated) manifold-with-complete vector field to infinite dimensions, they describe the flow completion in the above case explicitly. This yields, e.g., a geometric framework for the study of the question how solutions of this class of partial differential equations, including notably the Burgers equation, develop shocks.

Marcos Mariño reviews in "Enumerative geometry and knot invariants" the recently emerged idea of a large N duality between links in three-dimensional manifolds and strings in associated six-dimensional manifolds, N specifying here $\mathrm{SU}(N)$ or $\mathrm{U}(N)$ resp. the number of certain D-branes, i.e. submanifolds specifying boundary conditions for open strings. This string/gauge theory duality conjecture predicts a close relation between knot invariants provided by Chern–Simons gauge theory on a three-manifold M, and enumerative invariants, counting holomorphic curves on a Calabi–Yau manifold associated to M. In his contribution, Marcos Mariño recalls first some background on closed and open strings and on Chern–Simons theory, reviews then "large N transitions", and finally tests this duality by considering several of its numerical predictions in detail.

In "Gerbes, (twisted) K-theory, and the supersymmetric WZW model", Jouko Mickelsson discusses the important issue of non-preservation of classical symmetries upon passage to a quantum theory. In many cases, this phenomenom is mathematically described by a projective representation (as opposed to a linear representation) of the group of classical symmetries, on a Hilbert space of quantum states. Such "anomalies" can often be expressed in terms of Dixmier–Douady classes (in the integer-valued third cohomology group) or in terms of gerbes, or via twisted K-theory. Jouko Mickelsson explains in his contribution these different appearances of anomalies, and he gives explicitly a very interesting class of examples related to loop groups and, via restriction to the zero modes, to the Kostant–Dirac operator on finite dimensional homogeneous spaces, coined "supersymmetric Wess–Zumino–Witten models".

Karl-Hermann Neeb considers in "Current groups for non-compact manifolds and their central extensions" groups of maps from a manifold into a Lie group. These current groups, first studied by physicists because of their rôle in classical and quantum field theory, form an important class of infinite dimensional Lie groups since they are a natural and non-trivial generalization of the well-studied class of loop groups (having as source the circle S^1 and as target typically a finite dimensional compact Lie group). Though the representation theory of current groups is only poorly understood by now, the importance of projective representations and thus of central extensions should persist upon going from loop groups to these more general groups. Karl-Hermann Neeb studies these extensions here for two different classes of current groups of smooth maps from a non-compact manifold M to a possibly infinite dimensional Lie group K. For Lie algebra two-cocycles of "product type" on the corresponding current algebras, known since the work of Andrew Pressley and Graeme Segal, he gives a very detailed answer to the question whether one can "integrate" a given Lie algebra extension to a Lie group extension. A principal result is that a certain delicate condition, the discreteness of a period group associated to the Lie algebra cocycle, is – in the case of product type cocycles – fulfilled for all finite dimensional manifolds M if and only if it is fulfilled for M being the circle.

In "Traces and characteristic classes on loop spaces", Sylvie Paycha and Steven Rosenberg pursuit the Chern–Weil approach to the construction of characteristic classes of infinite dimensional vector bundles. Given a (topological non-trivial) structure group, e.g. for the tangent bundle of a manifold, the theory of classifying spaces yields of course automatically characteristic classes in the topological sense. Sylvie Paycha and Steven Rosenberg are working on the explicit construction of these classes via connections and "traces", i.e. certain continuous linear functionals on algebras containing the Lie algebra of the structure group. In particular they give a detailed analysis of several traces and the differential forms obtained from mimickimg the finite dimensional Chern–Weil theory in the case that the structure group is $Cl_0^*(M, E)$, the group of zeroth-order invertible classical pseudo-differential operators acting on a finite dimensional vector bundle over a closed finite dimensional manifold M.

S. G. Rajeev advocates in "New classical limits of quantum theories" the idea that not only the usual limit $\hbar \to 0$ but also other limits, as, e.g., $N \to \infty$ for a SU(N) Yang–Mills theory might yield a "underlying classical theory." He illustrates this by recalling the classical Hartree–Fock theory and the Thomas–Fermi approximation, as well as by considering atoms in the limit of large spatial dimension. Then S. G. Rajeev explains the beautiful idea of treating the space of modular forms of weight k on the Hecke congruence subgroup $\Gamma_0(n)$ of SL$(2, \mathbb{Z})$ as converging to a (semi-)classical limit via $k \to \infty$ and to a "neo-classical limit" as $n \to \infty$.

Let me take the opportunity to thank Vladimir Turaev for his efficient and pleasant co-organisation of the conference in Strasbourg and his unfailing support for these proceedings, as well as the whole staff of the IRMA, notably Josiane Moreau and Claudine Orphanides, for providing excellent conditions in May 2002. My thanks are due also to Claude Roger and the staff of the CIRM in Luminy, especially to

Anna Zeller-Meier. It is of course a pleasure to acknowledge the financial support of the IRMA and the CNRS for the Strasbourg meeting, and of the CIRM and the GDR SG-MAT for the Luminy meeting. Finally, I would like to express my gratitude to all referees for their effort for this volume.

Metz, May 2004   *Tilmann Wurzbacher*

Table of Contents

Lie groups of germs of analytic mappings

Helge Glöckner

TU Darmstadt, FB Mathematik AG 5
Schlossgartenstr. 7, 64289 Darmstadt, Germany
email: `gloeckner@mathematik.tu-darmstadt.de`

Abstract. Let X be a metrizable topological vector space over $\mathbb{K} \in \{\mathbb{R}, \mathbb{C}\}$, $K \subseteq X$ be a non-empty compact subset, and G be a Banach-Lie group over $\mathbb{K}$. In this paper, we turn the group $\Gamma(K, G)$ of germs around K of $\mathbb{K}$-analytic G-valued mappings into a $\mathbb{K}$-analytic Baker–Campbell–Hausdorff Lie group.

2000 Mathematics Subject Classification: 22E65; 22E67, 46H05, 46T25.

Introduction

Besides groups of real analytic diffeomorphisms of compact manifolds ([19], [20]) it is important in connection with Lie pseudogroups associated with involutive systems of analytic partial differential equations to consider also Lie groups of germs of $\mathbb{K}$-analytic local diffeomorphisms around $0 \in \mathbb{K}^n$ fixing the origin, and generalizations of such groups (see [23] for $\mathbb{K} = \mathbb{C}$, [18] for $\mathbb{K} = \mathbb{R}$). Similarly, replacing globally defined mappings with germs, it is our goal here to consider not groups of smooth or real analytic Lie group-valued mappings (as usually done), but groups of germs of analytic mappings with values in Lie groups.

Throughout the introduction, let X be a metrizable topological $\mathbb{K}$-vector space over $\mathbb{K} \in \{\mathbb{R}, \mathbb{C}\}$, and $K \subseteq X$ be a non-empty compact subset. We are mostly interested in the case where X is locally convex, but the constructions work just as well for general X.

Groups of germs. If G is a Banach–Lie group over $\mathbb{K}$, we consider the group $\Gamma(K, G)$ of germs $[\gamma]$ around K of $\mathbb{K}$-analytic mappings $\gamma : U \to G$ defined on open neighbourhoods $U \subseteq X$ of K. As our main result, we show that $\Gamma(K, G)$ can be made a $\mathbb{K}$-analytic Lie group modelled on the space of germs $\Gamma(K, L(G))$, equipped with its natural locally convex direct limit topology (Theorem 5.10). By construction, $\Gamma(K, G)$ will be a so-called Baker–Campbell–Hausdorff (BCH-) Lie group, *i.e.*, it has an exponential function inducing a local isomorphism of $\mathbb{K}$-analytic manifolds on some zero-neighbourhood, and the group multiplication close to $(1, 1)$ is given by the BCH-series, in exponential coordinates. BCH-Lie groups are particularly well-behaved infinite-dimensional Lie groups, whose basic Lie theory closely resembles the familiar finite-dimensional case (see [24], [15]).

Mappings between spaces of germs. To facilitate our constructions, we establish analyticity properties for typical mappings between spaces of germs. Given Banach spaces E and F over $\mathbb{K}$, an open neighbourhood U of K in X, an open subset $V \subseteq E$, and a $\mathbb{K}$-analytic mapping $f : U \times V \to F$, we show that the push-forward

$$f_* : \Gamma(K, V) \to \Gamma(K, F), \quad f_*([\gamma]) := [x \mapsto f(x, \gamma(x))]$$

is $\mathbb{K}$-analytic on the open set $\Gamma(K, V) := \{[\gamma] \in \Gamma(K, E) : \gamma(K) \subseteq V\}$ (Propositions 3.3 and 4.6). Once this main technical tool is established, standard ideas from the theory of mapping groups $C^r(M, G)$ on compact manifolds (see [21] or [15]) can be used to create the Lie group structure on $\Gamma(K, G)$.

Intricacies of the discussion of f_*. If both X and E are finite-dimensional, then $\Gamma(K, E)$ is a Silva space (as is well-known), *i.e.*, a locally convex direct limit of an ascending sequence of Banach spaces, with compact bonding maps. In this special case, analyticity of f_* (in the complex case) might also be proved in the naïve way, by a mere verification that f_* is complex analytic on each step of the directed system (which is the case here by standard arguments). This simpler approach is correct because the locally convex direct limit topology on a Silva space $Y = \varinjlim Y_n$ coincides with the topology of direct limit topological space [12, §7.1]; as also finite powers of Y are Silva, this entails that a map $g : W \to Z$ on an open subset $W \subseteq Y$ is C^r in the Michal–Bastiani sense (as in [21] or [13]), resp., complex analytic if and only if so is each restriction $g|_{W \cap Y_n} : W \cap Y_n \to Z$. More generally, for locally convex X and finite-dimensional E it is known that $\Gamma(K, E)$ is Silva if and only if X is a Schwartz space (cf. [3], Thm. 7 and Prop. 9, where $\mathbb{K} = E = \mathbb{C}$). If E is an infinite-dimensional Banach space, then $\Gamma(K, E)$ never is a Silva space.[1] In the case of locally convex direct limits which are not Silva spaces, analyticity of a map g on the steps does not entail analyticity of g : counter-examples are well known (e.g., the algebra multiplication map in [14, §10.9] or analogous examples in [10]). Therefore the naïve argument just described cannot be used anymore. Instead, we have to prove analyticity of f_* on a more technical level, working directly with the locally convex direct limit topology.

Algebras of germs and their unit groups. If A is a unital Banach algebra over $\mathbb{K}$, it is known that $\Gamma(K, A)$ is a so-called "continuous inverse algebra," *viz.* a locally convex, unital, associative topological $\mathbb{K}$-algebra whose group of units is open and whose inversion map is continuous (see [26, Thm. 1] for the important special case

---

[1] If $\Gamma(K, E)$ is Silva, the unit ball $\mathrm{ev}_x(\mathrm{Hol}_b(U, B_1(0))) = B_1(0) \subseteq E$ is relatively compact, whence $\dim(E) < \infty$. Here U is an open neighbourhood of K in X, $x \in K$, and $\mathrm{ev}_x : \Gamma(K, E) \to E$ is evaluation.

where $X = \mathbb{C}^n$, $A = \mathbb{C}$; [17, 1.15] when $K \subseteq \mathbb{C}^n$ (without proof); cf. also [27, §6.6]). Indeed, the unit group is open as it contains the open convex identity neighbourhood $\Gamma(K, B_1(1))$ of germs taking values in the unit ball $B_1(1) \subseteq A^\times$ around $1 \in A$; and by [10], $\Gamma(K, A)$ is locally m-convex as a countable inductive limit of Banach algebras (whence inversion is continuous). Using that pushforwards f_* are $\mathbb{K}$-analytic, we present here a self-contained alternative proof which provides additional information: the unit group $\Gamma(K, A)^\times$ is isomorphic to $\Gamma(K, A^\times)$ and therefore is a $\mathbb{K}$-analytic BCH-Lie group (regardless of completeness properties).

For general information concerning continuous inverse algebras, we refer to [25]–[27], [17], [14]. The algebras $\Gamma(K, \mathbb{C})$, $K \subseteq \mathbb{C}^n$, play an important role in the theory of such algebras, in connection with multi-variable holomorphic functional calculus [26], [4]. In [14], continuous inverse algebras B were inspected from the point of view of infinite-dimensional Lie theory. It turned out that $B^\times$ is a $\mathbb{K}$-analytic BCH-Lie group provided B is Mackey complete, but there exist non-Mackey complete continuous inverse algebras whose unit groups are not BCH and do not even have a globally defined exponential map [14]. Completeness (and related) properties of spaces of germs have been investigated in [2], [7], [11], [22] and further works by these authors. For example, $\Gamma(K, E)$ is complete if E is finite-dimensional and X is locally convex ([11], [22]).

1 Preliminaries

In this section, we describe preliminaries concerning analytic mappings. For basic definitions and facts in infinite-dimensional Lie theory, we refer to [21] (devoted to Lie groups modelled on sequentially complete, locally convex spaces) and [13]. As before, $\mathbb{K} \in \{\mathbb{R}, \mathbb{C}\}$.

Definition 1.1 ([6, Defn. 5.6]). Let E be a complex topological vector space, F be a locally convex complex topological vector space, $U \subseteq E$ be an open, non-empty subset, and $f : U \to F$ a map. Then f is called *complex analytic* or $\mathbb{C}$-*analytic* if it is continuous and for every $x \in U$, there exists a zero-neighbourhood $V \subseteq E$ such that $x + V \subseteq U$ and

$$f(x + h) = \sum_{n=0}^{\infty} \beta_n(h) \quad \text{for all } h \in V \text{ as a pointwise limit,}$$

for suitable continuous homogeneous polynomials $\beta_n : E \to F$ of degree $n \in \mathbb{N}_0$.

1.2. Recall that a mapping $\beta : E \to F$ is called a *homogeneous polynomial of degree $n \in \mathbb{N}_0$* if there exists a complex n-linear map $\mu : E^n \to F$ such that $\beta(x) = \mu(x, \ldots, x)$ for all $x \in E$ (resp., $\beta(x) = \mu(0)$ if $n = 0$). Here μ can always be chosen as a symmetric n-linear map, and this symmetric choice is unique; it is denoted $\overline{\beta} := \mu$.

The homogeneous polynomial β is continuous if and only if so is $\overline{\beta}$ (see [5] for these facts). Also recall that each β_n in Definition 1.1 is uniquely determined; it is given by $\beta_n = \frac{1}{n!}\delta_x^n f$ where $\delta_x^n f : E \to F$ is the n-th Gateaux-differential of f at x defined via

$$\delta_x^n f(h) := \frac{d^n}{dt^n}\Big|_{t=0} f(x+th) = d^n f(x, h, \ldots, h) \quad \text{for } h \in E.$$

Here $d^n f : U \times E^n \to F$ is the nth differential of f defined via $d^n f(x, h_1, \ldots, h_n) := (D_{h_1} \ldots D_{h_n} f)(x)$ using directional derivatives in the directions h_j.

We remark that, for mappings from normed spaces to locally convex spaces, the preceding definition of complex analytic mappings is equivalent to the one in [8], due to [6, Prop. 5.1].

Definition 1.3. Let E be a real topological vector space, F be a locally convex real topological vector space, $\emptyset \neq U \subseteq E$ be open, and $f : U \to F$ be a map. Then f is called *real analytic* or *$\mathbb{R}$-analytic* if f extends to a complex analytic mapping $V \to F_{\mathbb{C}}$, defined on some open neighbourhood V of U in the complexification $E_{\mathbb{C}} = E \oplus iE$ of E.

This notion of a real analytic mapping is stronger than the one used in [6] (which mimics Definition 1.1 above), enabling us to use some of their results. Unfortunately, many results in [6] are only formulated for sequentially complete ranges F. Part of the results (for instance, the Identity Theorems) remain valid for trivial reasons also for general F, since analytic mappings into F are also analytic as mappings into the completion $\overline{F}$ of F. Less trivial generalizations can be looked up in [13] (for locally convex spaces), as well as [1] and [16] (for not necessarily locally convex domains). Essentially, we only need: (a) Compositions of composable $\mathbb{K}$-analytic mappings are $\mathbb{K}$-analytic. (b) Real analyticity is a local property: If $(U_i)_{i \in I}$ is an open cover of U and $f|_{U_i} : U_i \to F$ is real analytic for each $i \in I$, then f is real analytic [16, Rem. 1.12].

1.4. For later use, we define $\|\beta\| := \sup\{\|\beta(v)\| : v \in E \text{ such that } \|v\| \leq 1\}$ for a continuous homogeneous polynomial $\beta : E \to F$ of degree n between Banach spaces, and $\|\mu\| := \sup\{\|\mu(v_1, \ldots, v_n)\| : v_1, \ldots, v_n \in E \text{ such that } \|v_1\|, \ldots, \|v_n\| \leq 1\}$ for a continuous n-linear map $\mu : E^n \to F$.

2 Spaces of germs of complex analytic mappings

Given a metrizable, complex topological vector space X, a non-empty, compact subset $K \subseteq X$ and a complex Banach space E, we let $\Gamma_{\mathbb{C}}(K, E)$ ($\Gamma(K, E)$ for short) be the complex vector space of germs of complex analytic E-valued maps around K (defined in the obvious way). Since every complex analytic map $\gamma : U \to E$ on an open neighbourhood U of K is bounded on some smaller open neighbourhood of K, we have $\Gamma(K, E) = \varinjlim \mathrm{Hol}_b(U, E)$ as a vector space, where U ranges through the directed set of open neighbourhoods of K and $\mathrm{Hol}_b(U, E)$ denotes the Banach space of bounded complex analytic mappings from U to E, equipped with the supremum norm.[2] The bonding maps are the restriction maps $\mathrm{Hol}_b(U, E) \to \mathrm{Hol}_b(V, E)$, $\gamma \mapsto \gamma|_V$ (where $K \subseteq V \subseteq U$), which apparently are linear maps of operator norm ≤ 1, and the limit maps are $\mathrm{Hol}_b(U, E) \to \Gamma(K, E)$, $\gamma \mapsto [\gamma]$. We give $\Gamma(K, E)$ the (*a priori* not necessarily Hausdorff) locally convex direct limit topology. Since K is compact and X is metrizable, there exists a descending sequence $U_1 \supseteq U_2 \supseteq \cdots$ of open neighbourhoods of K in X such that every neighbourhood of K contains some U_n (a "fundamental sequence of open neighbourhoods of K"), which can (and will) always be chosen such that every connected component of each U_n meets K. This entails that the bonding maps $\rho_{m,n} : \mathrm{Hol}_b(U_n, E) \to \mathrm{Hol}_b(U_m, E)$ are injective for all $n \leq m$ and so are the limit maps $\lambda_n : \mathrm{Hol}_b(U_n, E) \to \Gamma(K, E)$. Then $\Gamma(K, E) = \varinjlim \mathrm{Hol}_b(U_n, E)$ by cofinality.

As in the well-known case where X is locally convex, also in the general case the locally convex space $\Gamma(K, E)$ is Hausdorff. [Let $g \in \overline{\{0\}} \subseteq \Gamma(K, E)$, say $g = [\gamma]$, where $\gamma : U \to E$. Given $x \in K$, $k \in \mathbb{N}_0$, $\Lambda \in E'$ and $v_1, \ldots, v_k \in X$, using the direct limit universal property the linear map $\phi : \Gamma(K, E) \to \mathbb{C}$, $\phi([\eta]) := \Lambda(d^k \eta(x; v_1, \ldots, v_k))$ is easily seen to be continuous (cf. [5, Thm. A] and [6, Prop. 6.5]). Now $\mathbb{C}$ being Hausdorff, we deduce from the preceding that $d^k \gamma(x; v_1, \ldots, v_k) = 0$. Hence γ vanishes on a neighbourhood of the arbitrary element $x \in K$ and thus $[\gamma] = 0$].

Clearly $\Gamma(K, E_1 \times \cdots \times E_n) \cong \Gamma(K, E_1) \times \cdots \times \Gamma(K, E_n)$ canonically as locally convex spaces, if $E_1, \ldots, E_n$ are complex Banach spaces.

3 Pushforwards of germs of complex analytic maps

3.1. Let X be a metrizable complex topological vector space, $K \subseteq X$ a non-empty, compact subset, E and F complex Banach spaces, $U \subseteq X$ an open neighbourhood of K, $V \subseteq E$ a non-empty, open subset, and $f : U \times V \to F$ be a complex analytic map. We define

$$\Gamma(K, V) := \{[\gamma] \in \Gamma(K, E) : \gamma(K) \subseteq V\}.$$

Apparently, if $[\gamma] \in \Gamma(K, V)$, then $\gamma^{-1}(V)$ is an open neighbourhood of K. We consider the "push-forward"

$$f_* : \Gamma(K, V) \to \Gamma(K, F), \quad f_*([\gamma]) := [f_*(\gamma)],$$

where the representative $\gamma : W \to E$ is chosen such that $W \subseteq U$ and $\mathrm{im}(\gamma) \subseteq V$, and $f_*(\gamma)$ is defined via $f_*(\gamma)(x) := f(x, \gamma(x))$ for $x \in W$. If X is locally convex, then

---

[2] $\mathrm{Hol}_b(U, E)$ is closed in the Banach space $C_b(U, E)$ of bounded continuous E-valued maps as the limit $\gamma \in C_b(U, E)$ of any uniformly convergent sequence in $\mathrm{Hol}_b(U, E)$ is $\mathbb{K}$-analytic by [6, Prop. 6.5].

$f_*(\gamma) = f \circ (\mathrm{id}_W, \gamma) \colon W \to F$ is complex analytic as a composition of complex analytic maps. If X is not locally convex, then $(\mathrm{id}_W, \gamma) \colon W \to X \times E$ is *not* complex analytic (as its range is not locally convex). However, being a composition of $C_{\mathbb{C}}^{\infty}$ maps, $f_*(\gamma)$ is $C_{\mathbb{C}}^{\infty}$ and hence complex analytic as its range is locally convex [1, §7].

If V is an open, convex zero-neighbourhood, then $\Gamma(K, V)$ is open in $\Gamma(K, E)$, as it is convex and apparently its inverse image in each $\mathrm{Hol}_b(U_n, E)$ is open. This entails that $\Gamma(K, V)$ is open for any open subset $V \subseteq E$.

3.2 Notation. Abbreviate $A_n = \mathrm{Hol}_b(U_n, E)$ for $n \in \mathbb{N}$, where $U_1 \supseteq U_2 \supseteq \cdots$ is a fundamental sequence of open neighbourhoods of K. It is easy to see that the sets

$$\mathcal{V}(\varepsilon) := \mathrm{conv}\left(\bigcup_{n \in \mathbb{N}} \lambda_n(B_{\varepsilon_n}^{A_n}(0)) \right) \subseteq \Gamma(K, E)$$

are open and form a basis for the filter of zero-neighbourhoods in $\Gamma(K, E)$ when $\varepsilon = (\varepsilon_n)_{n \in \mathbb{N}}$ ranges through $(\mathbb{R}^+)^{\mathbb{N}}$; here $B_{\varepsilon_n}^{A_n}(0)$ is the open ε_n-ball around 0 in A_n, with respect to the supremum norm $\|.\|_{A_n}$. Similarly, writing $B_n := \mathrm{Hol}_b(U_n, F)$, we define analogous sets $\mathcal{W}(\varepsilon)$ for $\varepsilon \in (\mathbb{R}^+)^{\mathbb{N}}$ forming a basis of open zero-neighbourhoods for $\Gamma(K, F)$.

We are now in the position to prove our main technical tool.

Proposition 3.3. *In the situation of 3.1, the map* $f_* \colon \Gamma(K, V) \to \Gamma(K, F)$ *is complex analytic. If V is a balanced open zero-neighbourhood in particular, then*

$$f_*([\eta]) = \sum_{n=0}^{\infty} \frac{\delta_0^n(f_*)([\eta])}{n!} \quad \textit{for all } [\eta] \in \Gamma(K, V), \tag{1}$$

where the n-th Gateaux differential $\delta_0^n(f_*) \colon \Gamma(K, E) \to \Gamma(K, F)$ is the continuous homogeneous polynomial of degree n given by $\delta_0^n(f_*)([\eta]) = [x \mapsto \delta_0^n f_x(\eta(x))]$, where $f_x := f(x, \bullet) \colon V \to F$ for $x \in U$.

Proof. Given $[\gamma] \in \Gamma(K, V)$, we now show that f_* is continuous at $[\gamma]$, and given by a convergent power series on some neighbourhood of $[\gamma]$ (whence f is complex analytic).

3.4. Let $[\gamma] \in \Gamma(K, V)$, where $\gamma \colon W \to E$; we may assume that $W \subseteq U$. There is $\delta > 0$ such that $\gamma(K) + B_{2\delta}(0) \subseteq V$. Replacing W with $\gamma^{-1}(\gamma(K) + B_{\delta}(0))$, we achieve that $\mathrm{im}(\gamma) + B_{\delta}(0) \subseteq V$ for some $\delta > 0$. Then $g \colon W \times B_{\delta}(0) \to F$, $g(x, y) := f(x, \gamma(x) + y) - f(x, \gamma(x))$ is complex analytic, and apparently $f_*([\gamma] + [\eta]) = g_*([\eta]) + f_*([\gamma])$ for all $[\eta] \in \Gamma(K, B_{\delta}(0))$. Thus f_* will be continuous at $[\gamma]$ and given by a power series around $[\gamma]$ if we can show that g_* is continuous at 0 and given by a power series around 0. Replacing f with g and V with $B_{\delta}(0)$, we may therefore assume that $0 \in V$, $[\gamma] = 0$, and $f(x, 0) = 0$ for all $x \in U$.

3.5. As $K \times \{0\} \subseteq X \times E$ is compact and f continuous, there exists an open neighbourhood $U_0 \subseteq U$ of K in X and $\rho > 0$ such that $B_\rho(0) \subseteq V$ and $f(U_0 \times B_\rho(0)) \subseteq f(K \times \{0\}) + B_1(0) = B_1(0)$. Thus $f(U_0 \times B_\rho(0))$ is bounded in F. Hence, shrinking U and V if necessary, we may assume that f is bounded, of supremum norm $M < \infty$, say, and $V = B_\rho(0)$ for some $\rho > 0$.

3.6. Given $x \in U$, consider the complex analytic mapping $f_x := f(x, \bullet) : V = B_\rho(0) \to F$. Let $\mathbb{D} \subseteq \mathbb{C}$ be the open unit disk around 0. Given $w \in B_\rho(0)$, we have $\delta_0^n f_x(w) = d^n f_x(0, w, \dots, w) = h^{(n)}(0)$ for each $n \in \mathbb{N}_0$, where $h : \mathbb{D} \to F, h(z) := f_x(zw)$. The Cauchy Estimates [6, Cor. 3.2] entail that $\frac{\|\delta_0^n f_x(w)\|}{n!} = \frac{\|h^{(n)}(0)\|}{n!} \leq \frac{M}{1^n} = M$. Hence

$$\left\| \frac{\delta_0^n f_x}{n!} \right\| \leq M\rho^{-n}, \quad \text{for all } x \in U \text{ and } n \in \mathbb{N}_0. \tag{2}$$

With (2) and [5], proof of Prop. 1, Part 1^0, we deduce that

$$R_n := \sup\left\{ \left\| \frac{d^n f_x(0, \bullet)}{n!} \right\| : x \in U \right\} \leq \frac{2^n n^n}{n!} \sup\left\{ \left\| \frac{\delta_0^n f_x}{n!} \right\| : x \in U \right\} \leq \frac{2^n n^n}{n!} M\rho^{-n}. \tag{3}$$

Since $\lim_{n \to \infty} \frac{n}{\sqrt[n]{n!}} = e$ as a consequence of Stirling's formula (where e is Euler's constant), (3) entails that

$$R := \limsup_{n \to \infty} \sqrt[n]{R_n} \leq 2e\rho^{-1} < \infty.$$

3.7. We may assume that the fundamental sequence $U_1 \supseteq U_2 \supseteq \cdots$ of open neighbourhoods of K in X is chosen such that $U_1 \subseteq U$. We define complex analytic maps $h_n : U \times E^n \to F$ and $p_n : U \times E \to F$ via

$$h_n(x, v_1, \dots, v_n) := \frac{d^n f_x(0, v_1, \dots, v_n)}{n!} = \frac{d^n f((x, 0), (0, v_1), \dots, (0, v_n))}{n!} \quad \text{and}$$

$$p_n(x, v) := \frac{\delta_0^n f_x(v)}{n!} = h_n(x, v, \dots, v),$$

for $n \in \mathbb{N}$. For $n, j \in \mathbb{N}$, we get a symmetric n-linear map $\beta_{n,j} : \mathrm{Hol}_b(U_j, E)^n \to \mathrm{Hol}_b(U_j, F)$ via $\beta_{n,j}(\eta_1, \dots, \eta_n)(x) := h_n(x, \eta_1(x), \dots, \eta_n(x))$. We let

$$\beta_n : \Gamma(K, E)^n \to \Gamma(K, F)$$

be the n-linear mapping determined by

$$\beta_n([\eta_1], \dots, [\eta_n]) := [\beta_{n,j}(\eta_1, \dots, \eta_n)] \quad \text{if } \eta_1, \dots, \eta_n \in \mathrm{Hol}_b(U_j, E), j \in \mathbb{N}.$$

Then $(p_n)_* : \Gamma(K, E) \to \Gamma(K, F)$ apparently satisfies $(p_n)_*([\eta]) = \beta_n([\eta], \dots, [\eta])$, and thus $(p_n)_*$ is a homogeneous polynomial of degree n.

3.8. Choose $r \in]0, \rho]$ such that $\frac{1}{r} > R$. Then $\sum_{n=1}^{\infty} R_n r^n < \infty$ and hence, for each $k \in \mathbb{N}$ and $\eta \in \mathrm{Hol}_b(U_k, E) =: A_k$ such that $\|\eta\|_{A_k} < r$,

$$f(x, \eta(x)) = \sum_{n=1}^{\infty} p_n(x, \eta(x)),$$

with convergence uniform in $x \in U_k$. Hence

$$f_*(\eta) = \sum_{n=1}^{\infty} (p_n)_*(\eta)$$

in $\mathrm{Hol}_b(U_k, F) =: B_k$ and thus

$$f_*([\eta]) = \mu_k \left(\sum_{n=1}^{\infty} (p_n)_*(\eta) \right) = \sum_{n=1}^{\infty} (p_n)_*([\eta])$$

in $\Gamma(K, F)$, the map $\mu_k : \mathrm{Hol}_b(U_k, F) \to \Gamma(K, F), \zeta \mapsto [\zeta]$ being continuous linear.

3.9. Suppose that $\varepsilon = (\varepsilon_j)_{j \in \mathbb{N}} \in (\mathbb{R}^+)^{\mathbb{N}}$ is given. For $j \in \mathbb{N}$, we define

$$\delta_j := \min \left\{ r, \frac{r \varepsilon_j}{2^j (1 + \sum_{n=1}^{\infty} R_n r^n)} \right\}.$$

Let $\delta := (\delta_j)_{j \in \mathbb{N}}$. We claim that

$$f_*([\eta]) = \sum_{n=1}^{\infty} (p_n)_*([\eta]) \in \mathcal{W}(\varepsilon) \tag{4}$$

for all $[\eta] \in \mathcal{V}(\delta)$. In fact, given $[\eta]$, there exists $m \in \mathbb{N}$, real numbers $t_1, \ldots, t_m \geq 0$ and elements $\eta_j \in A_j = \mathrm{Hol}_b(U_j, E)$ such that $\sum_{j=1}^{m} t_j = 1$, $\|\eta_j\|_{A_j} < \delta_j$ for $j = 1, \ldots, m$, and

$$[\eta] = \sum_{j=1}^{m} t_j [\eta_j]. \tag{5}$$

Then $\|\eta_j|_{U_m}\|_{A_m} < \delta_j \leq r$ for each $j = 1, \ldots, m$, entailing that $\left\| \sum_{j=1}^{m} t_j \eta_j|_{U_m} \right\|_{A_m} < r$ (where $\sum_{j=1}^{m} t_j \eta_j|_{U_m}$ has the same germ as η), and thus $f_*([\eta]) = \sum_{n=1}^{\infty} (p_n)_*([\eta])$ by 3.8, establishing the first half of (4). Next, using (5), we obtain for each $N \in \mathbb{N}$:

$$\sum_{n=1}^{N} (p_n)_*([\eta]) = \sum_{n=1}^{N} \beta_n([\eta], \ldots, [\eta])$$

$$= \sum_{n=1}^{N} \sum_{j_1, \ldots, j_n = 1}^{m} \left(\prod_{i=1}^{n} t_{j_i} \right) \beta_n([\eta_{j_1}], \ldots, [\eta_{j_n}]) = \sum_{j=1}^{m} 2^{-j} \mu_j(\xi_j),$$

abbreviating $\xi_j := 2^j \sum_{n=1}^N \sum_{\alpha \in I_{n,j}} \left(\prod_{i=1}^n t_{\alpha_i} \right) \cdot \beta_{n,j}(\rho_{j,\alpha_1}(\eta_{\alpha_1}), \ldots, \rho_{j,\alpha_n}(\eta_{\alpha_n}))$
where $I_{n,j} := \{\alpha = (\alpha_i) \in \{1, \ldots, m\}^n : \|\alpha\|_\infty = j\}$, and $\rho_{j,\alpha_i} : \mathrm{Hol}_b(U_{\alpha_i}, E) \to$
$\mathrm{Hol}_b(U_j, E)$ is the restriction map. Note that $\|\beta_{n,j}(\rho_{j,\alpha_1}(\eta_{\alpha_1}), \ldots, \rho_{j,\alpha_n}(\eta_{\alpha_n}))\|_{B_j} \leq$
$R_n \cdot \delta_j \cdot r^{n-1}$. Hence

$$\|\xi_j\|_{B_j} \leq \frac{2^j \delta_j}{r} \sum_{n=1}^N R_n r^n \underbrace{\sum_{\alpha \in I_{n,j}} \left(\prod_{i=1}^n t_{\alpha_i} \right)}_{\leq \sum_{j_1, \ldots, j_n = 1}^m (\prod_{i=1}^n t_{j_i}) = 1} \leq \frac{\delta_j 2^j}{r} \sum_{n=1}^N R_n r^n < \varepsilon_j.$$

Set $s := \sum_{j=1}^m 2^{-j} < 1$. By the preceding, $\sum_{n=1}^N (p_n)_*([\eta]) = s \sum_{j=1}^m \frac{2^{-j}}{s} \mu_j(\xi_j) \in$
$s \, \mathcal{W}(\varepsilon)$, for all $N \in \mathbb{N}$. Letting $N \to \infty$, we deduce that

$$f_*([\eta]) = \sum_{n=1}^\infty (p_n)_*([\eta]) \in \overline{s \, \mathcal{W}(\varepsilon)} = s \, \overline{\mathcal{W}(\varepsilon)} \subseteq \mathcal{W}(\varepsilon),$$

as asserted. Consequently, f_* is continuous at 0.

3.10. Note that, since p_n might play the role of f, by what has already been shown the map $(p_n)_*$ is continuous at 0 and thus continuous, being a homogeneous polynomial [5, Thm. 1]. Since $f_*([\eta]) = \sum_{n=1}^\infty (p_n)_*([\eta])$ for all $[\eta] \in \Gamma(K, B_r(0))$ by 3.8, we see that f_* is given on the zero-neighbourhood $\Gamma(K, B_r(0))$ by a convergent series of continuous homogeneous polynomials. This completes the proof of the complex analyticity of f_*.

3.11. To prove the final assertion, note that the complex analytic mapping f_* is given on the balanced zero-neighbourhood $\Gamma(K, V)$ by its Taylor series $\sum_{n=1}^\infty \frac{\delta_0^n(f_*)}{n!}$ (cf. [6, Prop. 5.5]). On the other hand, by 3.8, f_* is given by $f_*(0) + \sum_{n=1}^\infty (p_n)_*$ on some zero-neighbourhood. Thus $\delta_0^0(f_*) = f_*(0)$ and $\frac{\delta_0^n(f_*)}{n!} = (p_n)_*$ for $n \in \mathbb{N}$, which entails the final assertion. $\qquad\square$

4 Spaces of germs of real analytic mappings

In this section, we transfer the definitions and results of Sections 2 and 3 to the real case.

4.1. Let X be a metrizable, real topological vector space, $K \subseteq X$ be a non-empty, compact subset, and E be a real Banach space. Let $X_\mathbb{C} = X \oplus i X$ and $E_\mathbb{C} = E \oplus i E$ be the complexifications of X and E, respectively. Then the real vector space $\Gamma_\mathbb{R}(K, E)$ of germs $[\gamma]$ about K of real analytic E-valued mappings $\gamma : U \to E$ on open neighbourhoods $U \subseteq X$ of K is defined as in the complex case, replacing the word "complex" with "real" there.

4.2. If $\gamma : U \to E$ is a real analytic E-valued map on an open neighbourhood U of K in X, then there is a complex analytic map $\tilde{\gamma} : \tilde{U} \to E_{\mathbb{C}}$ on an open neighbourhood $\tilde{U}$ of U in $X_{\mathbb{C}}$ which extends γ, by definition of real analyticity. It is easy to see that the map

$$\Lambda : \Gamma_{\mathbb{R}}(K, E) \to \Gamma_{\mathbb{C}}(K, E_{\mathbb{C}}), \qquad \Lambda([\gamma]) := [\tilde{\gamma}] \tag{6}$$

is well defined, injective, $\mathbb{R}$-linear, and that $\operatorname{im} \Lambda \subseteq \Gamma_{\mathbb{C}}(K, E_{\mathbb{C}})$ consists precisely of those germs possessing a representative $\eta : W \to E_{\mathbb{C}}$ defined on an open neighbourhood W of K in $X_{\mathbb{C}}$, such that $\eta(W \cap X) \subseteq E$. (To perform the simple proofs, the following observation is useful: Given a germ in $\Gamma_{\mathbb{R}}(K, E)$, we can always find a representative $\gamma : U \to E$ such that $\tilde{\gamma}$ as above is defined on $\tilde{U} := U + iV$ for some balanced open zero-neighbourhood V in X; then $\tilde{\gamma}$ is the unique complex analytic map on $\tilde{U}$ extending γ).

4.3. Let $L := \operatorname{im} \Lambda$. Then $\Gamma_{\mathbb{C}}(K, E_{\mathbb{C}}) = L \oplus iL$ internally as a real topological vector space, since L and iL are the $(+1)$-eigenspace and (-1)-eigenspace, respectively, of the continuous antilinear involution σ of $\Gamma_{\mathbb{C}}(K, E_{\mathbb{C}})$ defined via $\sigma([\gamma]) := [u + iv \mapsto \overline{\gamma(u - iv)}]$, where the bar indicates complex conjugation in $E_{\mathbb{C}}$, and $u \in U$, $v \in V$, where the domain of definition of the representative γ of the germ is chosen as $U + iV$ with U an open neighbourhood of K in X and V an open balanced zero-neighbourhood in X.

4.4. We give $\Gamma_{\mathbb{R}}(K, E)$ the real locally convex vector topology induced by Λ. Then apparently $\Gamma_{\mathbb{C}}(K, E_{\mathbb{C}}) = \Gamma_{\mathbb{R}}(K, E)_{\mathbb{C}}$ is the complexification of $\Gamma_{\mathbb{R}}(K, E)$ (via Λ).

Remark 4.5. $\Gamma_{\mathbb{R}}(K, E)$ can be described as the locally convex direct limit of the spaces $\{[\gamma|_U] : \gamma \in \operatorname{Hol}_b(U + iV, E_{\mathbb{C}}), \gamma(U) \subseteq E\} \subseteq \Gamma_{\mathbb{R}}(K, E)$, equipped with the topology induced by $\operatorname{Hol}_b(U + iV, E_{\mathbb{C}})$. Furthermore, intrinsic descriptions of the topology can be given.

If V is an open subset in E, we set $\Gamma_{\mathbb{R}}(K, V) := \{[\gamma] \in \Gamma_{\mathbb{R}}(K, E) : \gamma(K) \subseteq V\}$. Then $\Gamma_{\mathbb{R}}(K, V) = \Lambda^{-1}(\Gamma_{\mathbb{C}}(K, V + iE))$, showing that $\Gamma_{\mathbb{R}}(K, V)$ is open in $\Gamma_{\mathbb{R}}(K, E)$.

Proposition 4.6. *Proposition* 3.3 *remains valid if* $\mathbb{C}$ *is replaced with* $\mathbb{R}$, *except that it may be necessary to shrink the zero-neighbourhood* U *in order that* (1) *holds.*

Proof. Being real analytic, f extends to a complex analytic mapping $\tilde{f} : \tilde{U} \to F_{\mathbb{C}}$ on some open neighbourhood $\tilde{U}$ of $K \times V$ in $X_{\mathbb{C}} \times E_{\mathbb{C}}$. Let $[\gamma] \in \Gamma_{\mathbb{R}}(K, V)$, say $\gamma : W \to E$. Then $\gamma(K) \subseteq V$. Using the compactness of K and $\gamma(K)$, we find an open neighbourhood $U_1 \subseteq U$ of K in X, open zero-neighbourhoods U_2 in X and V_2 in E, and an open neighbourhood $V_1 \subseteq V$ of $\gamma(K)$ in E such that $(U_1 + iU_2) \times (V_1 + iV_2) \subseteq \tilde{U}$. Then $g := f|_{U_1 \times V_1}$ is real analytic, with complex analytic extension $\tilde{g} := \tilde{f}|_{(U_1+iU_2) \times (V_1+iV_2)}$. There is a complex analytic extension $\tilde{\gamma} : \tilde{W} \to E_{\mathbb{C}}$ of γ to an open neighbourhood $\tilde{W}$ of W in $X_{\mathbb{C}}$. After replacing γ with its restriction to $U_1 \cap \gamma^{-1}(V_1)$ and $\tilde{\gamma}$ with its restriction to $(U_1 + iU_2) \cap \tilde{\gamma}^{-1}(V_1 + iV_2)$, we see that $g_*(\tilde{\gamma})$ is a complex analytic extension of $f_*(\gamma)$, and thus $f_*(\gamma)$ is real analytic, whence $f_*([\gamma]) = [f_*(\gamma)]$ defines an element of $\Gamma_{\mathbb{R}}(K, F)$ indeed. Next, by Proposition 3.3, the map $(\tilde{g})_* : \Gamma_{\mathbb{C}}(K, V_1 + iV_2) \to \Gamma_{\mathbb{C}}(K, F_{\mathbb{C}})$ is complex analytic. Now

$$(\tilde{g})_* \circ \Lambda_1 |_{\Gamma_{\mathbb{R}}(K, V_1)}^{\Gamma_{\mathbb{C}}(K, V_1 + iV_2)} = \Lambda_2 \circ g_*, \tag{7}$$

where $\Lambda_1 : \Gamma_{\mathbb{R}}(K, E) \to \Gamma_{\mathbb{C}}(K, E_{\mathbb{C}})$ and $\Lambda_2 : \Gamma_{\mathbb{R}}(K, F) \to \Gamma_{\mathbb{C}}(K, F_{\mathbb{C}})$ are as in (6). Thus g_* has a complex analytic extension and hence is real analytic. Since $f_*|_{\Gamma_{\mathbb{R}}(K, V_1)} = g_*$, we see that f_* is real analytic on the open neighbourhood $\Gamma_{\mathbb{R}}(K, V_1)$ of the given germ $[\gamma]$. Thus f_* is locally real analytic and hence real analytic. Taking $[\gamma] = 0$ here, in view of (7) the final assertion of the proposition follows from the corresponding assertion in Proposition 3.3. $\qquad\square$

Corollary 4.7. *Let X be a metrizable topological vector space over $\mathbb{K} \in \{\mathbb{R}, \mathbb{C}\}$, $K \subseteq X$ be a non-empty, compact subset, E and F Banach spaces over $\mathbb{K}$, $V \subseteq E$ be a non-empty, open subset, and $f : V \to F$ be a $\mathbb{K}$-analytic map. Then*

$$\Gamma_{\mathbb{K}}(K, f) : \Gamma_{\mathbb{K}}(K, V) \to \Gamma_{\mathbb{K}}(K, F), \quad [\gamma] \mapsto [x \mapsto f(\gamma(x))]$$

is a $\mathbb{K}$-analytic mapping. If V is an open zero-neighbourhood, then there exists an open zero-neighbourhood $Q \subseteq V$ in E such that

$$\Gamma_{\mathbb{K}}(K, f)([\eta]) = \sum_{n=0}^{\infty} \frac{\Gamma_{\mathbb{K}}(K, \delta_0^n f)([\eta])}{n!} \quad \text{for all } [\eta] \in \Gamma_{\mathbb{R}}(K, Q). \tag{8}$$

Proof. The map $g : X \times V \to F$, $g(x, y) := f(y)$ is $\mathbb{K}$-analytic, and $\Gamma_{\mathbb{K}}(K, f) = g_*$. The assertions now follow immediately from Proposition 3.3 (if $\mathbb{K} = \mathbb{C}$), resp. 4.6 ($\mathbb{K} = \mathbb{R}$). $\qquad\square$

5 Lie groups of germs of analytic mappings

Let $\mathbb{K} \in \{\mathbb{R}, \mathbb{C}\}$, X be a metrizable topological $\mathbb{K}$-vector space, $K \subseteq X$ be a non-empty compact subset, and G be a Banach–Lie group over $\mathbb{K}$. In this section, we turn the group $\Gamma_{\mathbb{K}}(K, G)$ of germs of G-valued $\mathbb{K}$-analytic mappings around K into a $\mathbb{K}$-analytic Lie group, modelled on the locally convex topological $\mathbb{K}$-vector space $\Gamma_{\mathbb{K}}(K, L(G))$.

5.1. The group $\Gamma_{\mathbb{K}}(K, G)$ of germs of G-valued $\mathbb{K}$-analytic maps around K is defined in the obvious way. The group multiplication is defined via $[\gamma] \cdot [\eta] := [W_1 \cap W_2 \ni x \mapsto \gamma(x)\eta(x)]$ for $\mathbb{K}$-analytic mappings $\gamma : W_1 \to G$, $\eta : W_2 \to G$ on open neighbourhoods W_1, W_2 of K. The identity element is $[x \mapsto e_G]$.

5.2. We choose a norm $\|.\|$ on $L(G)$ defining its topology and making $L(G)$ a normed Lie algebra, viz. $\|[u, v]\| \leq \|u\| \cdot \|v\|$ for all $u, v \in L(G)$. There is an open balanced zero-neighbourhood $U \subseteq L(G)$ such that $U_1 := \exp_G(U)$ is open in G, $\psi := \exp_G |_U^{U_1}$ a diffeomorphism, and such that the BCH-series $\sum_{n=1}^{\infty} \beta_n$ converges on $U \times U$ to a $\mathbb{K}$-analytic map $* = \mu : U \times U \to L(G)$ [9, §II.7.2, Prop. 1]. Here $\beta_n : L(G)^2 \to L(G)$ is the homogeneous polynomial of degree n in the BCH-series.

5.3. The map $\Gamma_{\mathbb{K}}(K, \mu) : \Gamma_{\mathbb{K}}(K, U \times U) \to \Gamma_{\mathbb{K}}(K, L(G))$ is $\mathbb{K}$-analytic by Corollary 4.7, and there is an open zero-neighbourhood $U_0 \subseteq U$ such that

$$\Gamma_{\mathbb{K}}(K, \mu)([(\gamma, \eta)]) = \sum_{n=1}^{\infty} [\beta_n \circ (\gamma, \eta)] \quad \text{for all } [(\gamma, \eta)] \in \Gamma_{\mathbb{K}}(K, U_0 \times U_0). \quad (9)$$

5.4. Since $\Gamma(K, U^2) \cong \Gamma(K, U)^2$ as $\mathbb{K}$-analytic manifolds, we deduce from 5.3 that $m : \Gamma(K, U)^2 \to \Gamma(K, L(G))$, $m([\gamma], [\eta]) := [x \mapsto \gamma(x) * \eta(x)]$ is $\mathbb{K}$-analytic. We let $V \subseteq U$ be any open, balanced zero-neighbourhood in $L(G)$ such that $V * V \subseteq U$, and set $V_1 := \exp_G(V)$. We give $\Gamma_{\mathbb{K}}(K, U_1) := \{[\gamma] \in \Gamma_{\mathbb{K}}(K, G) : \gamma(K) \subseteq U_1\}$ the $\mathbb{K}$-analytic manifold structure which makes the bijection $\Psi : \Gamma_{\mathbb{K}}(K, \psi) : \Gamma_{\mathbb{K}}(K, U) \to \Gamma_{\mathbb{K}}(K, U_1), [\gamma] \mapsto [\psi \circ \gamma]$ a diffeomorphism of $\mathbb{K}$-analytic manifolds, where $\Gamma_{\mathbb{K}}(K, U)$ is considered as an open submanifold of $\Gamma_{\mathbb{K}}(K, L(G))$.

5.5. U being connected, the Identity Theorem [6, Prop. 6.6 I] shows that $\exp_G(u*v) = \exp_G(u) \exp_G(v)$ for all $(u, v) \in U \times U$, as this holds in some zero-neighbourhood. Hence $\Psi(m([\gamma], [\eta])) = \Psi([\gamma])\Psi([\eta])$ for all $[\gamma], [\eta] \in \Gamma_{\mathbb{K}}(K, V)$, entailing that the group multiplication of $\Gamma_{\mathbb{K}}(K, G)$ restricts to a $\mathbb{K}$-analytic map $\Gamma_{\mathbb{K}}(K, V_1)^2 \to \Gamma_{\mathbb{K}}(K, U_1)$. Similarly, the inversion map $\Gamma_{\mathbb{K}}(K, U_1) \to \Gamma_{\mathbb{K}}(K, U_1)$ corresponds to $\Gamma_{\mathbb{K}}(K, U) \to \Gamma_{\mathbb{K}}(K, U), [\gamma] \mapsto -[\gamma]$ under Ψ, and is therefore $\mathbb{K}$-analytic.

5.6. Let $[\gamma] \in \Gamma_{\mathbb{K}}(K, G)$, where $\gamma : W \to G$. Since the map $h : W \times L(G) \to L(G)$, $h(x, u) := \mathrm{Ad}_{\gamma(x)}(u)$ is $\mathbb{K}$-analytic and hence continuous, there exists an open neighbourhood W_0 of K in W and an open zero-neighbourhood $P \subseteq U$ in $L(G)$ such that

$$\mathrm{Ad}_{\gamma(x)}.u \in U \quad \text{for all } x \in W_0 \text{ and all } u \in P. \quad (10)$$

After replacing the representative γ with $\gamma|_{W_0}$, we may assume that $W = W_0$. Note that $h(x, \bullet)$ is linear for each $x \in W$. Therefore Proposition 3.3 (resp., Proposition 4.6) shows that $h_* : \Gamma_{\mathbb{K}}(K, L(G)) \to \Gamma_{\mathbb{K}}(K, L(G))$, $h_*([\eta]) := [x \mapsto h(x, \eta(x))]$ is a $\mathbb{K}$-analytic and thus continuous linear map. Consider now the inner automorphisms $I_g : G \to G, k \mapsto gkg^{-1}$ for $g \in G$, and $I_{[\gamma]} : \Gamma_{\mathbb{K}}(K, G) \to \Gamma_{\mathbb{K}}(K, G)$, $[\eta] \mapsto [\gamma][\eta][\gamma]^{-1}$. Let $[\eta] \in \Gamma_{\mathbb{K}}(K, P)$, where $\eta : W_1 \to L(G)$ is chosen such that $\mathrm{im}\, \eta \subseteq P$ and $W_1 \subseteq W$. For each $x \in W_1$, we have $\exp_G(\mathrm{Ad}_{\gamma(x)}(\eta(x))) = I_{\gamma(x)}(\exp_G(\eta(x)))$, entailing that

$$\Psi(h_*([\eta])) = I_{[\gamma]}(\Psi([\eta])). \quad (11)$$

Since $\Psi|_{\Gamma_{\mathbb{K}}(K,P)}$ is a diffeomorphism of $\mathbb{K}$-analytic manifolds onto the open identity neighbourhood $Q := \Psi(\Gamma_{\mathbb{K}}(K,P))$, we deduce from (11) and the $\mathbb{K}$-analyticity of h_* that $I_{[\gamma]}|_Q$ is $\mathbb{K}$-analytic.

5.7. In view of 5.5 and 5.6, [15, Prop. 1.13] provides a unique $\mathbb{K}$-analytic manifold structure on $\Gamma_{\mathbb{K}}(K,G)$ making it a $\mathbb{K}$-analytic Lie group, and making $\Gamma_{\mathbb{K}}(K,V_1)$, with its given $\mathbb{K}$-analytic manifold structure induced by $\Gamma_{\mathbb{K}}(K,U_1)$, an open submanifold of $\Gamma_{\mathbb{K}}(K,G)$.

5.8. The map $\operatorname{Exp} := \Gamma_{\mathbb{K}}(K,\exp_G)\colon \Gamma_{\mathbb{K}}(K,L(G)) \to \Gamma_{\mathbb{K}}(K,G), [\gamma] \mapsto [\exp_G \circ \gamma]$ is $\mathbb{K}$-analytic on $\Gamma_{\mathbb{K}}(K,V)$, by definition of the manifold structure on $\Gamma_{\mathbb{K}}(K,G)$. For each $n \in \mathbb{N}$ and $[\gamma] \in \Gamma_{\mathbb{K}}(K,nV)$, we have $\operatorname{Exp}([\gamma]) = \operatorname{Exp}([\frac{1}{n}\gamma])^n$, whence also $\operatorname{Exp}|_{\Gamma_{\mathbb{K}}(K,nV)}$ is $\mathbb{K}$-analytic. Hence Exp is $\mathbb{K}$-analytic on all of $\Gamma_{\mathbb{K}}(K,L(G))$ and thus smooth. In particular, for each $[\gamma] \in \Gamma_{\mathbb{K}}(K,L(G))$, say $\gamma: W \to L(G)$, the map $\xi: \mathbb{R} \to \Gamma_{\mathbb{K}}(K,G)$, $\xi(t) := \operatorname{Exp}(t[\gamma])$ is smooth. Given $s,t \in \mathbb{R}$, we have $\exp_G(s\gamma(x)+t\gamma(x)) = \exp_G(s\gamma(x))\exp_G(t\gamma(x))$ for all $x \in W$ and thus $\xi(s+t) = \xi(s)\xi(t)$. Furthermore, identifying $T_1(\Gamma_{\mathbb{K}}(K,G))$ with $\Gamma_{\mathbb{K}}(K,L(G))$ by means of the isomorphism of topological vector spaces $d\Psi^{-1}(1,\bullet)$ now and throughout the following, we have $\xi'(0) = [\gamma]$. Hence $\Gamma_{\mathbb{K}}(X,G)$ has a globally defined exponential map (see [21] or [13] for this concept), given by $\operatorname{Exp}: \Gamma_{\mathbb{K}}(K,L(G)) \to \Gamma_{\mathbb{K}}(K,G)$.

5.9. Note that by (9), $\Psi^{-1}(\Psi([\gamma])\Psi([\eta]))$ is given by the BCH-series of $\Gamma_{\mathbb{K}}(K,L(G))$, with respect to the continuous Lie bracket

$$[.,.]_1 := \Gamma_{\mathbb{K}}(K,[.,.])\colon \Gamma_{\mathbb{K}}(K,L(G)^2) \cong \Gamma_{\mathbb{K}}(K,L(G))^2 \to \Gamma_{\mathbb{K}}(K,L(G)).$$

The second order term in the BCH-series being the antisymmetric bilinear map $\frac{1}{2}[.,.]_1$, we deduce from [21, (5.2)] that the Lie bracket on $\Gamma_{\mathbb{K}}(K,L(G))$ as the Lie algebra of $\Gamma_{\mathbb{K}}(K,G)$ coincides with $[.,.]_1$. This in turn implies that (9) is the BCH-series of $\Gamma_{\mathbb{K}}(K,L(G))$ also when $\Gamma_{\mathbb{K}}(K,L(G))$ is identified with the Lie algebra of $\Gamma_{\mathbb{K}}(K,G)$. Summing up:

Theorem 5.10. *Let $\mathbb{K} \in \{\mathbb{R},\mathbb{C}\}$, X be a metrizable topological $\mathbb{K}$-vector space, $\emptyset \neq K \subseteq X$ a compact subset, and G be a Banach–Lie group over $\mathbb{K}$. Then there is a uniquely determined $\mathbb{K}$-analytic BCH-Lie group structure on $\Gamma_{\mathbb{K}}(K,G)$ modelled on $\Gamma_{\mathbb{K}}(K,L(G))$ such that $\Gamma_{\mathbb{K}}(K,\exp_G): \Gamma_{\mathbb{K}}(K,L(G)) \to \Gamma_{\mathbb{K}}(K,G)$ is a local diffeomorphism of $\mathbb{K}$-analytic manifolds on some zero-neighbourhood.* □

6 Algebras of germs of analytic mappings

The results are now applied to the case of algebras.

Theorem 6.1. *Let $\mathbb{K} \in \{\mathbb{R},\mathbb{C}\}$, X be a metrizable topological $\mathbb{K}$-vector space, $K \subseteq X$ a non-empty, compact subset, and A be a unital, associative Banach algebra* over $\mathbb{K}$. Then $\Gamma_{\mathbb{K}}(K, A)$ *is a continuous inverse algebra over* $\mathbb{K}$, *whose group of units* $\Gamma_{\mathbb{K}}(K, A)^{\times}$ *is a* $\mathbb{K}$-*analytic BCH-Lie group, with Lie algebra* $\Gamma_{\mathbb{K}}(K, A)$ *and exponential function* $\mathrm{Exp} \colon \Gamma_{\mathbb{K}}(K, A) \to \Gamma_{\mathbb{K}}(K, A)^{\times}$, $\mathrm{Exp}([\gamma]) = \sum_{n=0}^{\infty} \frac{1}{n!} [\gamma]^{n}$. *The map* $\Psi \colon \Gamma_{\mathbb{K}}(K, A^{\times}) \to \Gamma_{\mathbb{K}}(K, A)^{\times}$, $[\gamma] \mapsto [\gamma]$ *is an isomorphism of Lie groups.*

Proof. The algebra multiplication $\mu \colon A \times A \to A$ being a continuous $\mathbb{K}$-bilinear map and thus $\mathbb{K}$-analytic, Corollary 4.7 shows that the bilinear map

$$\Gamma_{\mathbb{K}}(K, A)^{2} \cong \Gamma_{\mathbb{K}}(K, A^{2}) \to \Gamma_{\mathbb{K}}(K, A), \quad ([\gamma], [\eta]) \mapsto \Gamma_{\mathbb{K}}(K, \mu)(\gamma, \eta) = [\gamma] \cdot [\eta]$$

is $\mathbb{K}$-analytic and thus continuous. Thus $\Gamma_{\mathbb{K}}(K, A)$ is a topological algebra.

The map $q \colon B_{1}(0) \to A$, $a \mapsto (1 - a)^{-1}$ is $\mathbb{K}$-analytic, where $B_{1}(0)$ is the unit ball in A. By Corollary 4.7, also $Q := \Gamma_{\mathbb{K}}(K, q) \colon \Gamma_{\mathbb{K}}(K, B_{1}(0)) \to \Gamma_{\mathbb{K}}(K, A)$ is $\mathbb{K}$-analytic. We have

$$(1 - [\gamma]) \cdot Q([\gamma]) = [x \mapsto (1 - \gamma(x))(1 - \gamma(x))^{-1}] = [1] = Q([\gamma]) \cdot (1 - [\gamma])$$

for each $[\gamma] \in \Gamma_{\mathbb{K}}(K, B_{1}(0))$. It follows that the open identity neighbourhood $1 - \Gamma_{\mathbb{K}}(K, B_{1}(0))$ is contained in $\Gamma_{\mathbb{K}}(K, A)^{\times}$, and $[\gamma]^{-1} = Q(1 - [\gamma])$ for each $[\gamma] \in 1 - \Gamma_{\mathbb{K}}(K, B_{1}(0))$, which is a $\mathbb{K}$-analytic function of $[\gamma]$. Hence $\Gamma_{\mathbb{K}}(K, A)^{\times}$ is open and that inversion is continuous on all of $\Gamma_{\mathbb{K}}(K, A)^{\times}$ (see [14, §2]). Thus $\Gamma_{\mathbb{K}}(K, A)$ is a continuous inverse algebra.

The exponential map $\exp_{A} \colon A \to A$, $\exp_{A}(x) := \sum_{n=0}^{\infty} \frac{1}{n!} x^{n}$ restricts to a $\mathbb{K}$-analytic diffeomorphism $\exp_{A} |_{U}^{V}$ from some open zero-neighbourhood U in A onto some open identity-neighbourhood V in $A^{\times}$ (cf. [9, Ch. I, §7, no. 3]). Corollary 4.7 (applied to $A_{\mathbb{C}}$ in the real case) entails that the map $\mathrm{Exp} := \Gamma_{\mathbb{K}}(K, \exp_{A}) \colon \Gamma_{\mathbb{K}}(K, A) \to \Gamma_{\mathbb{K}}(K, A)^{\times}$ is $\mathbb{K}$-analytic, and given for all $[\gamma] \in \Gamma_{\mathbb{K}}(K, A)$ by the convergent series $\mathrm{Exp}([\gamma]) = \sum_{n=0}^{\infty} \frac{1}{n!} [\gamma]^{n}$. Corollary 4.7 also entails that $\mathrm{Exp} |_{\Gamma_{\mathbb{K}}(K, U)}^{\Gamma_{\mathbb{K}}(K, V)} = \Gamma_{\mathbb{K}}(K, \exp_{A} |_{U}^{V}) \colon \Gamma_{\mathbb{K}}(K, U) \to \Gamma_{\mathbb{K}}(K, V)$ is a $\mathbb{K}$-analytic bijection with $\mathbb{K}$-analytic inverse $\Gamma_{\mathbb{K}}(K, (\exp |_{U}^{V})^{-1})$. Thus Exp induces an isomorphism of $\mathbb{K}$-analytic manifolds on some zero-neighbourhood.

It is clear that the map Ψ described in the theorem is well-defined, a homomorphism of groups, and bijective. Since $\Psi \circ \Gamma_{\mathbb{K}}(K, \exp_{A^{\times}}) = \Gamma_{\mathbb{K}}(K, \exp_{A}) = \mathrm{Exp}$, where $\mathrm{Exp} \colon \Gamma_{\mathbb{K}}(K, A) \to \Gamma_{\mathbb{K}}(K, A)^{\times}$ induces a local diffeomorphism of $\mathbb{K}$-analytic manifolds at 0 and so does $\Gamma_{\mathbb{K}}(K, \exp_{A^{\times}}) \colon \Gamma_{\mathbb{K}}(K, A) \to \Gamma_{\mathbb{K}}(K, A^{\times})$ (see Theorem 5.10, where this map is called Exp), we deduce that the isomorphism of groups Ψ induces an isomorphism of $\mathbb{K}$-analytic manifolds on some identity neighbourhood, and hence is an isomorphism of Lie groups. $\qquad\square$

References

[1]   Bertram, W., H. Glöckner and K.-H. Neeb, Differential calculus over general base fields and rings, to appear in *Expo. Math.* (see also arXiv:math.GM/0303300).

[2] Bierstedt, K.-D., J. Bonet and A. Peris, Vector-valued holomorphic germs on Fréchet-Schwartz spaces, *Proc. Roy. Irish Acad Sect.* A 94 (1994), 31–46.

[3] Bierstedt, K.-D. and R. Meise, Nuclearity and the Schwartz property in the theory of holomorphic functions on metrizable locally convex spaces, in *Infinite Dimensional Holomorphy and Applications* (Matos M. C., ed.), North-Holland, 1977, 93–129 .

[4] Biller, H., Analyticity and naturality of the multi-variable functional calculus, TU Darmstadt, Preprint 2332, April 2004.

[5] Bochnak, J. and J. Siciak, Polynomials and multilinear mappings in topological vector spaces, *Studia Math.* 39 (1971), 59–76.

[6] —, Analytic functions in topological vector spaces, *Studia Math.* 39 (1971), 77-112.

[7] Bonet, J., P. Domanski and J. Mujica, Complete spaces of vector-valued holomorphic germs, *Math. Scand.* 75 (1994), 150–160.

[8] Bourbaki, N., *Variétés différentielles et analytiques. Fascicule de résultats,* Hermann, Paris 1967.

[9] —, *Lie Groups and Lie Algebras,* Vol. 1–3, Springer-Verlag, 1989.

[10] Dierolf S. and J. Wengenroth, Inductive limits of topological algebras, *Linear Topol. Spaces Complex Anal.* 3 (1997), 45–49.

[11] Dineen, S., Holomorphic germs on compact subsets of locally convex spaces, in *Functional Analysis, Holomorphy, and Approximation Theory* (Rio de Janeiro, 1978), Springer-Verlag, 1981, 247–263.

[12] Floret, K., Lokalkonvexe Sequenzen mit kompakten Abbildungen, *J. Reine Angew. Math.* 247 (1971), 155–195.

[13] Glöckner, H., Infinite-dimensional Lie groups without completeness restrictions, in *Geometry and Analysis on Finite- and Infinite-Dimensional Lie Groups* (Strasburger, A. et al., eds.), Banach Center Publications 55, Warsaw 2002, 43–59.

[14] —, Algebras whose groups of units are Lie groups, *Studia Math.* 153 (2002), 147–177.

[15] —, Lie group structures on quotient groups and universal complexifications for infinite-dimensional Lie groups, *J. Funct. Anal.* 194 (2002), 347–409.

[16] —, Implicit functions from topological vector spaces to Banach spaces, TU Darmstadt Preprint 2271, March 2003; also arXiv:math.GM/0303320

[17] Gramsch, B., Relative Inversion in der Störungstheorie von Operatoren und Ψ-Algebren, *Math. Ann.* 269(1984), 27–71.

[18] Kamran, N., and T. Robart, A manifold structure for analytic Lie pseudogroups of infinite type, *J. Lie Theory* 11(2001), 57–80.

[19] Kriegl, A. and P. W. Michor, *The convenient setting of global analysis*, Math. Surveys Monogr. 53, Amer. Math. Soc., Providence, RI, 1997.

[20] Leslie, J., On the group of real analytic diffeomorphisms of a compact real analytic manifold, *Trans. Amer. Math. Soc.* 274 (1982), 651–669.

[21] Milnor, J., Remarks on infinite-dimensional Lie groups, in *Relativity, Groups and Topology II* (DeWitt, B., and R. Stora, eds.), North Holland, 1983, 1008–1057.

[22] Mujica, J., A completeness criterion for inductive limits of Banach spaces, in *Functional Analysis, Holomorphy and Approximation Theory, II* (Rio de Janeiro, 1981), 319–329, North Holland, 1984, 319–329.

[23] Pisanelli, D., An example of an infinite Lie group, *Proc. Amer. Math. Soc.* 62 (1977), 156–160.

[24] Robart, T., Sur l'intégrabilité des sous-algèbres de Lie en dimension infinie, *Canad. J. Math.* 49 (1997), 820–839.

[25] Waelbroeck, L., Les algèbres à inverse continu, *C.R. Acad. Sci. Paris* 238 (1954), 640–641.

[26] —, Le calcul symbolique dans les algèbres commutatives, *J. Math. Pures Appl.* 33 (1954), 147–186.

[27] —, The holomorphic functional calculus as an operational calculus, in *Spectral Theory*, Banach Center Publications 8, Warsaw 1982, 512–552.

The flow completion of the Burgers equation

Boris A. Khesin and Peter W. Michor*

Department of Mathematics, University of Toronto
100 St.George Street, Toronto, ON M5S 3G3, Canada
email: khesin@math.toronto.edu

Institut für Mathematik, Universität Wien
Strudlhofgasse 4, A-1090 Wien, Austria
and
Erwin Schrödinger International Institute of Mathematical Physics
Pasteurgasse 6/7, A-1090 Wien, Austria
email: peter.michor@esi.ac.at

Abstract. For a manifold equipped with vector field there exists the universal completion consisting of a (possibly non-Hausdorff) manifold with a complete vector field on it. We describe the universal completion of the partial differential equations $u_t + F(u)u_x = 0$ viewed as vector fields on infinite dimensional manifolds.

2000 Mathematics Subject Classification: 58D05, 58F07; 35Q53.

Keywords: Flow completion, Burgers equation, manifolds of mappings, diffeomorphism group.

1 Introduction

For a pair (M, X) consisting of a smooth manifold M and a vector field X on it there exists the universal completion $(\bar{M}, \bar{X})$, a possibly non-Hausdorff manifold $\bar{M}$ with a complete vector field $\bar{X}$, where (M, X) is embedded equivariantly as an open subspace. In this note we describe the universal completion of some partial differential equations viewed as vector fields on infinite dimensional manifolds. The equations are $u_t + f(u)u_x = 0$ where $u = u(x, t) : \mathbb{R}^n \times \mathbb{R} \to \mathbb{R}^n$, and $f : \mathbb{R}^n \to \mathbb{R}^n$ is some smooth map. A special case is the inviscid Burgers equation $u_t + 3uu_x = 0$ (also called the Hopf equation). The universal completion gives some insight at how solutions of these equations develop shocks. Namely, in the universal completion the

---
*B. A. Khesin and P. W. Michor were both supported by 'Fonds zur Förderung der wissenschaftlichen Forschung, Projekt P 14195 MAT'.

solutions are uniquely extended beyond the shocks, and become multivalued functions with infinite derivatives.

Recall that the inviscid Burgers equation can be regarded as the geodesic equation on the infinite dimensional group of diffeomorphisms of $\mathbb{R}^n$ (cf. e.g. [1, 9]). Such a derivation of the geodesic equation in the one-dimensional case, on the manifold of all embeddings $\mathrm{Emb}(\mathbb{R}, \mathbb{R})$, is reminded below, following [7]. The universal completion described in this note requires the consideration of multivalued velocity fields. These fields are solutions in the phase space of the system. In the configuration space, the completion corresponds to an extension of the diffeomorphism group to the semigroup of polymorphisms.

2 The universal flow completion

2.1 Vector fields on infinite dimensional manifolds. Let M be a connected smooth manifold of possibly infinite dimension, modeled on convenient vector spaces (see [5, section 27] for necessary definitions). Let us assume that M is smoothly Hausdorff, i.e., the global smooth functions on M separate points. Let X be a smooth (kinematic) vector field on M. We say that X *admits a local flow*, if there exists a smooth mapping

$$M \times \mathbb{R} \supset \mathcal{D}(X) \xrightarrow{\quad \mathrm{Fl}^X \quad} M$$

defined on a C^∞-open neighborhood $\mathcal{D}(X)$ of $M \times 0$ such that

(1) $\mathcal{D}(X) \cap (\{x\} \times \mathbb{R})$ is a connected open interval.

(2) If $\mathrm{Fl}^X_s(x)$ exists then $\mathrm{Fl}^X_{t+s}(x)$ and $\mathrm{Fl}^X_t(\mathrm{Fl}^X_s(x))$ exist simultaneously and are equal to each other.

(3) $\mathrm{Fl}^X_0(x) = x$ for all $x \in M$.

(4) $\frac{d}{dt} \mathrm{Fl}^X_t(x) = X(\mathrm{Fl}^X_t(x))$.

It is shown in [5, 32.14], that then for each integral curve c of X we have $c(t) = \mathrm{Fl}^X_t(c(0))$ (see [5, 32.14] for the proof, as well as for counterexamples against existence, uniqueness, etc. of integral curves for more general X). Thus there exists a unique maximal flow. Furthermore, X is Fl^X_t-related to itself, i.e., $T(\mathrm{Fl}^X_t) \circ X = X \circ \mathrm{Fl}^X_t$.

2.2 Theorem. *Let $X \in \mathfrak{X}(M)$ be a smooth vector field on a (connected) smooth, possibly infinite-dimensional, manifold M modeled on convenient vector spaces. Let us assume that the vector field X admits a local flow.*

Then there exists a universal flow completion $j : (M, X) \to (\bar{M}, \bar{X})$ of (M, X). Namely, there exists a (connected) smooth not necessarily Hausdorff manifold $\bar{M}$, *a complete vector field $\bar{X} \in \mathfrak{X}(\bar{M})$, and an embedding $j : M \to \bar{M}$ onto an open submanifold such that X and $\bar{X}$ are j-related: $Tj \circ X = \bar{X} \circ j$. Moreover, for any other equivariant morphism $f : (M, X) \to (N, Y)$ for a manifold N and a complete vector field $Y \in \mathfrak{X}(N)$ there exists a unique equivariant morphism $\bar{f} : (\bar{M}, \bar{X}) \to (N, Y)$ with $\bar{f} \circ j = f$. The leaf spaces M/X and $\bar{M}/\bar{X}$ are homeomorphic.*

An equivariant morphism $f : (M, X) \to (N, Y)$ is a smooth mapping $f : M \to N$ satisfying $Tf \circ X = Y \circ f$. It follows that then $f \circ \mathrm{Fl}_t^X = \mathrm{Fl}_t^Y \circ f$ wherever Fl_t^X is defined.

Sketch of Proof. The finite dimensional version of this theorem is due to Palais [8]. The formulation here is from [4] and the proof given in [4] goes through in the infinite-dimensional case as well.

Since we shall need the construction, we sketch it here: Consider the manifold $\mathbb{R} \times M$ with a coordinate function s on $\mathbb{R}$, the vector field $\tilde{X} := \partial_s \times X \in \mathfrak{X}(\mathbb{R} \times M)$, and let $\bar{M} := \mathbb{R} \times_{\tilde{X}} M$ be the orbit space (or leaf space) of the vector field $\tilde{X}$. We consider the flow mapping $\mathrm{Fl}^{\tilde{X}} : \mathcal{D}(\tilde{X}) \to \mathbb{R} \times M$ given by $\mathrm{Fl}_t^{\tilde{X}}(s, x) = (s + t, \mathrm{Fl}_t^X(x))$.

For each $s \in \mathbb{R}$ we have the injective mapping

$$j_s : M \xrightarrow{\mathrm{ins}_t} \{s\} \times M \subset \mathbb{R} \times M \xrightarrow{\pi} \mathbb{R} \times_{\tilde{X}} M = \bar{M}$$

which is a homeomorphism on its open image $j_s(M)$ in $\bar{M}$ in the quotient topology. We use the mappings $j_s : M \to \bar{M}$ as charts. The chart change for $r < s$ are then $(j_s)^{-1} \circ j_r = \mathrm{Fl}_{s-r}^X$ restricted to $(j_s)^{-1}(j_r(M)) \subset M$.

The flow $(t, (s, x)) \mapsto (s + t, x)$ on $\mathbb{R} \times M$ commutes with the flow of $\tilde{X}$ and thus induces a flow on the leaf space $\bar{M} = \mathbb{R} \times_{\tilde{X}} M$. Differentiating this flow we get a vector field $\bar{X}$ on $\bar{M}$.

The construction $(M, X) \mapsto (\bar{M}, \bar{X})$ is a functor from the category of smooth convenient smoothly Hausdorff manifolds with vector fields admitting local flows and smooth mappings intertwining the vector fields into the category of possibly non-Hausdorff manifolds with smooth vector fields with global flows and smooth mappings intertwining these fields. For a pair (M, X) with a complete vector field X the flow completion $(\bar{M}, \bar{X})$ is equivariantly diffeomorphic to (M, X) since then any of the charts $j_s : M \to \bar{M}$ is also surjective. From this the universal property follows. $\square$

2.3 Example. Consider $M = \mathbb{R}$, $X = -x^2 \partial_x$. The solutions of the ordinary differential equation $\dot{x} = -x^2$ are $x(t) = 1/(t + 1/x(0))$ which are all incomplete, and 0. The foliation in $\mathbb{R} \times M$ is given by the graphs of the functions $x(t) = 1/(t + 1/x(0))$. Consider the following illustration of $\mathbb{R} \times M$ and its foliation. Note that this incompleteness of a quadratic field is similar to the incompleteness of the Burgers equation described below. Examples leading to non-Hausdorff completions can be found in [4].

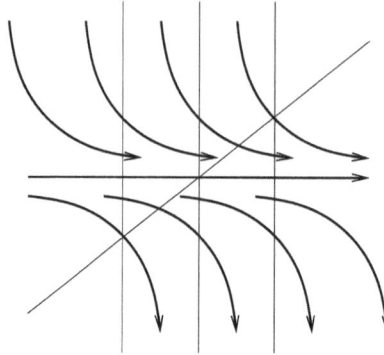

Figure 1. The flow of the field $\dot{x} = -x^2$. The embeddings j_t are induced by the vertical slices. The completion is $\bar{M} = \mathbb{R}$, the identification is given by the inclined line, for example.

2.4 Remark on Hamiltonian systems. Suppose that M is a symplectic or Poisson manifold and that X_f is the Hamiltonian vector field of a smooth function f. Then there exists a unique symplectic or Poisson structure on the flow completion $\bar{M}$ and a unique smooth function $\bar{f}$ such that $\bar{X}$ is again the Hamiltonian vector field of $\bar{f}$. Moreover, if $f = f_1, \ldots, f_n$ is a maximal Poisson commuting set of smooth function such that (M, X_f) is a completely integrable system, then there are unique extensions $\bar{f}_1, \ldots, \bar{f}_n$ to $\bar{M}$ such that the flow completion $(\bar{M}, \overline{X_f})$ is again a completely integrable system.

In the infinite-dimensional symplectic case (M, ω) should be a weak symplectic manifold and all (possibly, infinitely many) functions f_i have to be taken in the space $C_\omega^\infty(M, \mathbb{R})$ of smooth functions with a smooth ω-gradient, see [5], section 48. All this is an easy consequence of the fact that the symplectic or Poisson structures and the conservation laws f_i are invariant under the flow of X_f, and that restrictions of this flow are the chart transfer mappings for the atlas used to define the flow completion.

3 The Burgers equation as a geodesic equation

3.1 The principal bundle of embeddings. Let M and N be smooth connected finite-dimensional manifolds without boundary, such that $\dim M \leq \dim N$. The space $\mathrm{Emb}(M, N)$ of all embeddings (immersions which are homeomorphisms on their images) from M into N is an open submanifold of $C^\infty(M, N)$ which is stable under the right action of the diffeomorphism group of M. Here $C^\infty(M, N)$ is a smooth manifold modeled on spaces of sections $\Gamma_c(f^*TN)$ with compact support. In particular, the tangent space at f is canonically isomorphic to the space of vector fields along f with compact support in M. If f and g differ on a non-compact set then they belong to different connected components of $C^\infty(M, N)$.

Then $\mathrm{Emb}(M, N)$ is the total space of a smooth principal fiber bundle whose structure group is the diffeomorphism group of M. Its base, denoted by $B(M, N)$, is a Hausdorff smooth manifold modeled on nuclear (LF)-spaces. It can be thought of as the "nonlinear Grassmannian" of all submanifolds of N which are of type M. If we take a Hilbert space H instead of N, then $B(M, H)$ is the classifying space for $\mathrm{Diff}(M)$ if M is compact, and the classifying bundle $\mathrm{Emb}(M, H)$ carries also a universal connection, see details in [5, sections 42–44].

3.2 A geodesic equation. Consider the convenient manifold $\mathrm{Emb}(\mathbb{R}, \mathbb{R})$ of all embeddings of the real line into itself, which contains the diffeomorphism group $\mathrm{Diff}(\mathbb{R})$ as an open subset. Each connected component is a free orbit of the diffeomorphism group $\mathrm{Diff}(\mathbb{R})$ for the action of composition from the right. The tangent bundle is trivial, $T\,\mathrm{Emb}(\mathbb{R}, \mathbb{R}) = \mathrm{Emb}(\mathbb{R}, \mathbb{R}) \times C_c^\infty(\mathbb{R}, \mathbb{R})$, tangent vectors are smooth functions with compact support. For our purposes, we may restrict attention to the space of orientation-preserving embeddings, denoted by $\mathrm{Emb}^+(\mathbb{R}, \mathbb{R})$. The case S^1 is treated in a similar fashion and the results are also valid in this situation, where $\mathrm{Emb}(S^1, S^1) = \mathrm{Diff}(S^1)$.

Following V. Arnold's approach to Euler equations on diffeomorphism groups, we define the weak Riemannian metric on $\mathrm{Emb}^+(\mathbb{R}, \mathbb{R})$ by the formula:

$$G_f(h, k) = \int_\mathbb{R} h(x)k(x)|f_x(x)|\, dx, \quad f \in \mathrm{Emb}(\mathbb{R}, \mathbb{R}), \quad h, k \in C_c^\infty(\mathbb{R}, \mathbb{R}).$$

It is invariant under the right action of the diffeomorphism group. The energy of a curve f of embeddings is

$$E(f) = \tfrac{1}{2}\int_a^b G_f(f_t, f_t)dt = \tfrac{1}{2}\int_a^b \int_\mathbb{R} f_t^2 f_x\, dx dt.$$

Consider smooth variations of $f(x, t)$ with fixed endpoints. Then variational calculus provides the following form of the geodesic equation with its corresponding initial data:

$$f_{tt} = -2\frac{f_t f_{tx}}{f_x},$$

where

$$f(.\,, 0) \in \mathrm{Emb}^+(\mathbb{R}, \mathbb{R}), \quad f_t(.\,, 0) \in C_c^\infty(\mathbb{R}, \mathbb{R}).$$

The geodesic equation has the following conservation law: if instead of the obvious framing we change variables to $T\,\mathrm{Emb} = \mathrm{Emb} \times C_c^\infty \ni (f, h) \mapsto (f, hf_x^2) =: (f, H)$ then the geodesic equation becomes $H_t = \frac{\partial}{\partial t}(f_t f_x^2) = f_x^2(f_{tt} + 2\frac{f_t f_{tx}}{f_x}) = 0$, so that $H = f_t f_x^2$ is constant in t.

3.3 The geodesic property of the Burgers equation. We restrict our attention from the whole space $\mathrm{Emb}(\mathbb{R}, \mathbb{R})$ to the open subset $\mathrm{Diff}(\mathbb{R})$. Consider the trivialization of $T\,\mathrm{Diff}(\mathbb{R})$ by right translation. The derivative of the inversion $\mathrm{Inv} : g \mapsto g^{-1}$ is given by

$$T_g(\mathrm{Inv})h = -T(g^{-1}) \circ h \circ g^{-1} = \frac{h \circ g^{-1}}{g_x \circ g^{-1}} \qquad \text{for } g \in \mathrm{Diff}(\mathbb{R}),\ h \in C_c^\infty(\mathbb{R}, \mathbb{R}).$$

Defining $u := f_t \circ f^{-1}$, or, in more detail, $u(x, t) = f_t(f(\quad, t)^{-1}(x), t)$, we have

$$u_x = (f_t \circ f^{-1})_x = (f_{tx} \circ f^{-1})\frac{1}{f_x \circ f^{-1}} = \frac{f_{tx}}{f_x} \circ f^{-1},$$

$$u_t = (f_t \circ f^{-1})_t = f_{tt} \circ f^{-1} + (f_{tx} \circ f^{-1})(f^{-1})_t$$

$$= f_{tt} \circ f^{-1} + (f_{tx} \circ f^{-1})\frac{1}{f_x \circ f^{-1}}(f_t \circ f^{-1})$$

which, by the geodesic equation of 3.2 becomes

$$u_t = f_{tt} \circ f^{-1} - \left(\frac{f_{tx} f_t}{f_x}\right) \circ f^{-1} = -3\left(\frac{f_{tx} f_t}{f_x}\right) \circ f^{-1} = -3u_x u.$$

The geodesic equation on $\mathrm{Diff}(\mathbb{R})$ in right trivialization, that is, in Eulerian formulation, is hence

$$u_t = -3u_x u$$

which is just the inviscid Burgers equation. Similarly, one obtains the derivation in the n-dimensional case.

4 The flow completion of some hyperbolic systems

4.1 A partial differential equation. Let $f = (f_1, \ldots, f_k) : \mathbb{R}^n \to \mathbb{R}^k$ be smooth and consider the partial differential equation

$$u_t + (f(u) \cdot \nabla)u = 0$$

or, which is the same,

$$u_t + f_1(u)u_{x_1} + \cdots + f_k(u)u_{x_k} = 0, \qquad \mathbb{R}^k \times \mathbb{R} \supseteq U \xrightarrow{u} \mathbb{R}^n,$$

where U is an open neighborhood of $\mathbb{R}^k \times 0$ in $\mathbb{R}^k \times \mathbb{R}^n$, and u is a smooth $\mathbb{R}^n$-valued function on U. This type of equations are called hyperbolic conservation laws in physics, see [2].

   We consider now the manifold $C^\infty(\mathbb{R}^k, \mathbb{R}^n)$ of all smooth $\mathbb{R}^n$-valued functions on $\mathbb{R}^k$ with the manifold structure described in [5, section 42]. The tangent bundle is trivial, and the space $C_c^\infty(\mathbb{R}^k, \mathbb{R}^n)$ of functions with compact support serves as the fiber. Note that $C_c^\infty(\mathbb{R}^k, \mathbb{R}^n)$ is an open connected component in $C^\infty(\mathbb{R}^k, \mathbb{R}^n)$. We consider the characteristic vector field

$$X(u) = (f(u) \cdot \nabla)u.$$

It is a vector field on $C^\infty(\mathbb{R}^k, \mathbb{R}^n)$ if $X(u)$ has compact support for each u. This is the case if u has compact support. In the general case one has to leave the realm of manifolds with charts. (See [6] for a setting for infinite dimensional manifolds based on curves instead of charts, which is applicable in this situation.)

For the sake of simplicity, let us restrict attention to $C_c^\infty(\mathbb{R}^k, \mathbb{R}^n)$. There, flow lines of the vector field X are given by solutions of the above partial differential equation (where one has to adapt the domain of definition). We may thus consider $(C_c^\infty(\mathbb{R}^k, \mathbb{R}^n), X)$ as smooth convenient manifolds with vector fields admitting local flows.

4.2 Characteristics and solutions. To describe the universal completion of the quasi-linear equation $u_t + (f(u) \cdot \nabla)u = 0$ we apply the characteristic method (see e.g. [3] or [1], where in particular, the case of the Burgers equation is treated).

In the space $\mathbb{R}^{k+n}$ with coordinates (x, y) consider the vector field $Y(x, y) = (f(y), 0) = f_1(y)\partial_{x^1} + \cdots + f_k(y)\partial_{x^k}$ with differential equation $\dot{x} = f(y), \dot{y} = 0$. It has the complete flow $\mathrm{Fl}_t^Y(x, y) = (x + tf(y), y)$.

Let now $u(x, t)$ be a curve of functions. We ask when the graph of u can be reparametrized in such a way that it becomes a solution curve of the push forward vector field $Y_* : f \mapsto Y \circ f$ on the space of embeddings $\mathrm{Emb}(\mathbb{R}^k, \mathbb{R}^{k+n})$. Thus consider a time dependent reparametrization $z \mapsto x(z, t)$, i.e., $x \in C^\infty(\mathbb{R}^{k+1}, \mathbb{R}^k)$. The curve $t \mapsto (x(z, t), u(x(z, t), t))$ in $\mathbb{R}^{k+n}$ is an integral curve of Y if and only if

$$\begin{pmatrix} f \circ u \circ x \\ 0 \end{pmatrix} = \partial_t \begin{pmatrix} x \\ u \circ x \end{pmatrix} = \begin{pmatrix} x_t \\ u_t \circ x + (\nabla u \circ x) \cdot x_t \end{pmatrix} \tag{1}$$

$$\Longleftrightarrow \begin{cases} x_t = f \circ u \circ x \\ 0 = (u_t + (f \circ u) \cdot \nabla u) \circ x \end{cases} \tag{2}$$

This implies that the graph of $u(\cdot, t)$, namely the curve $t \mapsto (x \mapsto (x, u(x, t)))$, may be parameterized as a solution curve of the vector field Y_* on the space of embeddings $\mathrm{Emb}(\mathbb{R}^k, \mathbb{R}^{k+n})$ starting at $x \mapsto (x, u(x, 0))$ if and only if u is a solution of the partial differential equation $u_t + (f(u) \cdot \nabla)u = 0$. The parameterization $z \mapsto x(z, t)$ is then given by $x_t(z, t) = f(u(x(z, t), t))$ with $x(z, 0) = z \in \mathbb{R}^k$.

For $k = n$ the characteristics have a simple physical meaning. Consider freely flying particles in $\mathbb{R}^n$, and trace a trajectory $x(t)$ of one of the particles. Denote the velocity of a particle at the position x at the moment t by $u(t, x)$, or rather, by $f(u(x, t)) := \dot{x}(t)$. (For the inviscid Burgers equation, $u(x, t) := \dot{x}(t)$.) Due to the absence of interaction, the Newton equation of any particle is $\ddot{x}(t) = 0$.

Example. The inviscid 1D Burgers equation (see [1]). Consider the equation $u_t + 3uu_x = 0$ with $k = n = 1$ and $f(u) = 3u$. There the flow of the vector field $Y = 3u\partial_x$

is tilting the plane to the right with constant speed. The illustration shows how a graph of an honest function is moved through a shock (when the derivatives become infinite) towards the graph of a multivalued function; each piece of it is still a local solution. We

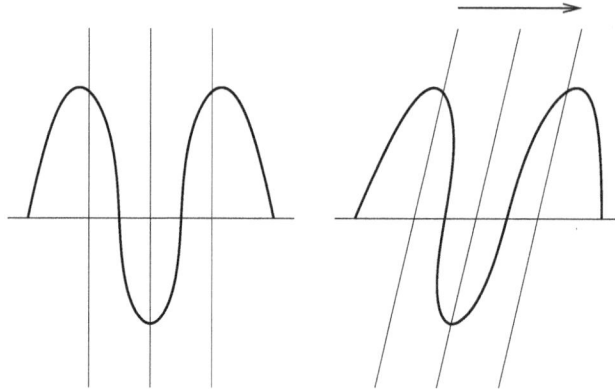

Figure 2. The characteristic flow of the inviscid Burgers equations tilts the plane.

also refer to [2] for a treatment of more general equations $u_t + A(u)u_x = 0$ (where A is matrix valued with all eigenvalues distinct) as the limits of equations with "viscous" right hand side $\epsilon \Delta u$.

We now interpret the characteristics in the space of graphs of functions. Given a function $u_0 \in C_c^\infty(\mathbb{R}^k, \mathbb{R}^n)$ with compact support we consider the graph of u_0 as the submanifold $\Gamma(u_0) = \{(x, u_0(x)) : x \in \mathbb{R}^k\}$ of $\mathbb{R}^{k+n}$. Let $\mathrm{pr}_1 : \mathbb{R}^{k+n} \to \mathbb{R}^k$ and $\mathrm{pr}_2 : \mathbb{R}^{k+n} \to \mathbb{R}^n$ be the projections. Consider the interval of all $t \in \mathbb{R}$ such that $\mathrm{pr}_1 | Fl_t^Y (\Gamma(u_0)) : Fl_t^Y(\Gamma(u_0)) \to \mathbb{R}^k$ is a diffeomorphism for all $t' \in [0, t]$ or $t' \in [t, 0]$, respectively. Then

$$u(x, t) = \mathrm{pr}_2(\mathrm{pr}_1 | Fl_t^Y (\Gamma(u_0)))^{-1}(x)$$

is a solution of equation 4.1 with initial value $u(x, 0) = u_0(x)$. Thus the vector field X on $C_c^\infty(\mathbb{R}^k, \mathbb{R}^n)$ admits a local flow.

4.3 The flow completion. Now one can see that after some time graphs of functions become graphs of multivalued functions. This explains the following construction of the completion.

We consider the principal bundle of all proper smooth embedded k-surfaces in $\mathbb{R}^{k+n}$ which deviate from $\mathbb{R}^k \times 0$ only in a compact set, with projection

$$\pi : \mathrm{Emb}_{\mathbb{R}^k}(\mathbb{R}^k, \mathbb{R}^{k+n}) \to B_{\mathbb{R}^k}(\mathbb{R}^k, \mathbb{R}^{k+n})$$

onto the convenient manifold of k-dimensional submanifolds of $\mathbb{R}^{k+n}$ which deviate from $\mathbb{R}^k \times 0$ only in a compact set. The structure group is the group of $\mathrm{Diff}_c(\mathbb{R}^k)$ of diffeomorphisms with compact support. We have the graph embedding, a smooth mapping

$$\gamma : C_c^\infty(\mathbb{R}^k, \mathbb{R}^n) \to \mathrm{Emb}_{\mathbb{R}^k}(\mathbb{R}^k, \mathbb{R}^{k+n}), \qquad \gamma(u)(x) = (x, u(x)).$$

Let us assume now that $f(0) = 0$. Then the flow $\mathrm{Fl}_t^Y(x, y) = (x + tf(y), y)$ of the vector field $Y(x, y) = f_1(y)\partial_{x^1} + \ldots f_k(y)\partial_{x^k}$ on $\mathbb{R}^{k+n}$ acts on parameterized k-surfaces in $\mathrm{Emb}_{\mathbb{R}^k}(\mathbb{R}^k, \mathbb{R}^{k+n})$ by $(\mathrm{Fl}_t^Y \circ (c_1, c_2))(x) = (c_1(x) + tf(c_2(x)), c_2(x))$ and is the flow of the vector field Y_* on $\mathrm{Emb}_{\mathbb{R}^k}(\mathbb{R}^k, \mathbb{R}^{k+n})$ given by $Y_*(c_1, c_2) = (f_1 \circ c_2)\partial_{x^1} + \cdots + (f_k \circ c_2)\partial_{x^k} = (f \circ c_2, 0)$. The vector field Y_* is invariant under the principal right action of $g \in \mathrm{Diff}_c(\mathbb{R}^k)$ which is given by $(c_1, c_2) \mapsto (c_1 \circ g, c_2 \circ g)$. Thus Y_* induces a smooth vector field Z on the base manifold $B_{\mathbb{R}^k}(\mathbb{R}^k, \mathbb{R}^{k+n})$ whose flow is again Fl_t^Y applied to closed submanifolds of $\mathbb{R}^{k+n}$.

We consider now the space $\mathcal{G}$ of all closed non-compact k-dimensional submanifolds $N \in B_{\mathbb{R}^k}(\mathbb{R}^k, \mathbb{R}^{k+n})$ such that for some $t \in \mathbb{R}$ the mapping $\mathrm{pr}_1 \circ \mathrm{Fl}_t^Y \,|\, N : N \to \mathbb{R}^k$ is a diffeomorphism. By the choice of topology on $B_{\mathbb{R}^k}(\mathbb{R}^k, \mathbb{R}^{k+n})$ the space $\mathcal{G}$ is open, and obviously invariant under the flow of the vector field Z.

Proposition. *Let $f(0) = 0$. Then the flow completion $\overline{C_c^\infty(\mathbb{R}^k, \mathbb{R}^n)}$ of the infinite dimensional manifold with vector field $(C_c^\infty(\mathbb{R}^k, \mathbb{R}^n), X)$ is diffeomorphic to $(\mathcal{G}, Z)$. The mapping $j_t : (C_c^\infty(\mathbb{R}, \mathbb{R}^n), X) \to \mathcal{G}$ is given by $j_t = \mathrm{Fl}_t^Z \circ \pi \circ \gamma$.*

Proof. In the proof of theorem 2.2 we have seen that the completion $\overline{C_c^\infty(\mathbb{R}^k, \mathbb{R}^n)}$ can be described by taking the pieces $j_t(C_c^\infty(\mathbb{R}, \mathbb{R}^n))$ which are all diffeomorphic to $C_c^\infty(\mathbb{R}, \mathbb{R}^n)$ and gluing them via the smooth mappings $(j_s)^{-1} \circ j_r = \mathrm{Fl}_{s-r}^X$ for $r < s$. But this is realized in the open subset $\mathcal{G} \subset B_{\mathbb{R}^k}(\mathbb{R}^k, \mathbb{R}^{k+n})$ by the global flow Fl^Z. Thus we reconstructed the atlas describing the completion in 2.2 as a smooth manifold. $\square$

References

[1] Arnold, V. I. *Geometrical methods in the theory of ordinary differential equations*, Grunglehren Math. Wiss. 250, Springer-Verlag, New York 1983.

[2] Bianchini, Stefano, Bressan, Alberto, Vanishing Viscosity Solutions of Nonlinear Hyperbolic Systems, 2002, 1–99; arXiv:math.AP/0111321.

[3] Eilenberger, G., *Solitons*, Springer Ser. Solid-State Sci. 19, Springer-Verlag, 1981.

[4] Kamber, Franz W.; Michor, Peter W. The flow completion of a manifold with vector field *Electron. Res. Announc. Amer. Math. Soc.* 6 (2000), 95–97; arXiv:math.DG/0007173.

[5] Kriegl, A.; Michor, P. W. *The Convenient Setting for Global Analysis*, Surveys and Monographs 53, Amer. Math. Soc., Providence, R.I., 1997.

[6] Michor, Peter W., A convenient setting for differential geometry and global analysis, I, II, *Cahiers Topologie Géom. Différentielle Catég.* 25 (1984), 63–109, 113–178.

[7] Michor, Peter W.; Ratiu, Tudor, Geometry of the Virasoro-Bott group, *J. Lie Theory* 8 (1998), 293–309.

[8] Palais, Richard S., *A global formulation of the Lie theory of transformation groups*, Mem. Amer. Math. Soc. 22, Amer. Math. Soc., Providence, R.I., 1957.

[9] Ovsienko, V. Yu.; Khesin, B. A. The super Korteweg-de Vries equation as an Euler equation. *Funct. Anal. Appl.* 21 (4) (1987), 81–82.

Enumerative geometry and knot invariants

Marcos Mariño

Jefferson Physical Laboratory, Harvard University
Cambridge MA 02138, U.S.A.
email: `marcos@mail.cern.ch`

Abstract. We review the string/gauge theory duality relating Chern–Simons theory and topological strings on noncompact Calabi–Yau manifolds, as well as its mathematical implications for knot invariants and enumerative geometry.

2000 Mathematics Subject Classification: 81T30; 57M27, 14N35.

Contents

1 Introduction

Enumerative geometry and knot theory have benefitted considerably from the insights and results in string theory and topological field theory. The theory of Gromov–Witten invariants has emerged mostly from the consideration of topological sigma models and topological strings, and mirror symmetry has provided a surprising point of view with powerful techniques and deep implications for the theory of enumerative invariants. On the other hand, the new invariants of knots and links that emerged in the eighties turned out to be deeply related to Chern–Simons theory, a topological gauge theory introduced by Witten in [92], which also provided a new family of invariants of three-manifolds. It is safe to say that these two topics, enumerative geometry and knot theory, have been deeply transformed through the emergence of these connections to physics.

A more recent surprise, however, is that, in many situations, knot invariants are related to enumerative invariants. The reason is that Chern–Simons gauge theory has a string description in the sense envisaged by 't Hooft [86], and this description turns out to involve topological strings, *i.e.* the physical counterparts of Gromov–Witten invariants. This relation between two seemingly unrelated areas of geometry is therefore based on a beautiful realization of the large N string/gauge theory duality. The connection between Chern–Simons theory and topological strings was first pointed out by Witten in [95], and the current picture emerged in the works of Gopakumar and Vafa [37] and Ooguri and Vafa [76].

In this paper we have tried to review these developments. We have focused mostly in presenting results, general ideas and examples. Some of the physical arguments leading to these results are not covered in detail, mostly for reasons of space, but also with the hope that mathematicians will find this review more readable. Important related developments, like the interplay with mirror symmetry and the relation with M-theory on manifolds of G_2 holonomy, are only mentioned in the text. Other reviews of the topics covered here include [88, 57], and more recently [40], which provides extensive mathematical background.

The plan of this paper is the following. In section 2 we review some basic facts about open and closed topological strings and their structure in terms of integer invariants. In section 3, we give a quick review of Chern–Simons theory and knot and link invariants. In section 4, we state the basic ideas of string/gauge theory duality in the $1/N$ expansion, and we show, following Gopakumar and Vafa, that Chern–Simons theory has a description in terms of closed strings on the resolved conifold. In section 5 we show in detail how to incorporate Wilson loops in the duality. It turns out that the Chern–Simons/string duality can be extended to closed strings propagating in more complicated toric geometries, and we summarize some of the results in section 6. Finally, some conclusions and open problems are collected in section 7.

2 Topological strings

2.1 Topological sigma models

The starting point to construct topological strings is an $\mathcal{N} = (2, 2)$ superconformal field theory, the $\mathcal{N} = (2, 2)$ nonlinear sigma model. This model can be twisted in two ways in order to produce a topological field theory [91, 58, 93], which are usually called the A and the B model. We will focus here on the A-model.

The field content of this model is the following. First, we have a map $x : \Sigma_g \to X$ from a Riemann surface of genus g to a target space X, that will be a Kähler manifold of complex dimension d. We also have fermions $\chi \in x^*(TX)$, which are scalars on Σ_g, and a fermionic one form ψ_α with values in $x^*(TX)$. This last field satisfies a selfduality condition which implies that its only nonzero components are $\psi_{\bar{z}}^I \in x^*(T^{(1,0)}X)$ and $\psi_z^{\bar{I}} \in x^*(T^{(0,1)}X)$, where $T^{(1,0)}X$, $T^{(0,1)}X$ denote, respectively, the holomorphic and the antiholomorphic tangent bundles, and $I, \bar{I}$ are the corresponding indices. The theory also has a BRST, or topological, charge Q which acts on the fields according to

$$
\begin{aligned}
\{Q, x\} &= i\chi, \\
\{Q, \chi\} &= 0, \\
\{Q, \psi_{\bar{z}}^I\} &= -\partial_{\bar{z}} x^I - i\chi^J \Gamma^I_{JK} \psi_{\bar{z}}^K, \\
\{Q, \psi_z^{\bar{I}}\} &= -\partial_z x^{\bar{I}} - i\chi^{\bar{J}} \Gamma^{\bar{I}}_{\overline{JK}} \psi_z^{\bar{K}}.
\end{aligned}
\tag{2.1}
$$

The twisted Lagrangian turns out to be Q-exact, up to a topological term:

$$
\mathcal{L} = i\{Q, V\} + \int_{\Sigma_g} x^*(\omega),
\tag{2.2}
$$

where $\omega = J + iB$ is the complexified Kähler class of X, and V (sometimes called the gauge fermion) is given by

$$
V = \int_{\Sigma_g} d^2 z \, G_{I\bar{J}} (\psi_{\bar{z}}^{\bar{I}} \partial_{\bar{z}} x^J + \partial_x x^{\bar{I}} \psi_{\bar{z}}^J).
\tag{2.3}
$$

In this equation, $G_{I\bar{J}}$ is the Kähler metric of X. Notice that the last term in (2.2) is a topological invariant characterizing the homotopy type of the map $x : \Sigma_g \to X$, therefore the energy-momentum tensor of this theory is given by:

$$T_{\alpha\beta} = \{Q, b_{\alpha\beta}\}, \tag{2.4}$$

where $b_{\alpha\beta} = \delta V/\delta g^{\alpha\beta}$. The fact that the energy-momentum tensor is Q-exact means that the theory is topological, and the fact that the Lagrangian is Q-exact up to a topological term means that the semiclassical approximation is exact. The classical solutions of the sigma model action are holomorphic maps $x : \Sigma_g \to X$, which are also known as worldsheet instantons, and the functional integral localizes to these configurations. The relevant operators in this theory, as in any topological theory of cohomological type, are the Q-cohomology classes. In this case they are given by operators of the form,

$$\mathcal{O}_\phi = \phi_{i_1 \ldots i_p} \chi^{i_1} \cdots \chi^{i_p}, \tag{2.5}$$

where $\phi = \phi_{i_1 \ldots i_p} dx^{i_1} \wedge \cdots \wedge dx^{i_p}$ is a closed p-form representing a nontrivial class in $H^p(X)$. Moreover, one can derive a selection rule for correlation functions of such operators: the vacuum expectation value $\langle \mathcal{O}_{\phi_1} \ldots \mathcal{O}_{\phi_\ell} \rangle$ vanishes unless

$$\sum_{k=1}^{\ell} \deg(\mathcal{O}_{\phi_k}) = 2d(1-g) + 2\int_{\Sigma_g} x^*(c_1(X)), \tag{2.6}$$

where $\deg(\mathcal{O}_{\phi_k}) = \deg(\phi_k)$. The right hand of this equation is nothing but the virtual dimension of the moduli space of holomorphic maps, $\mathcal{M}^{\text{hol}}_{\Sigma_g \to X}$. Since the operators (2.5) can be interpreted as differential forms on this moduli space, the above selection rule just says that we have to integrate top forms.

In the case of a Calabi–Yau manifold of complex dimension 3, we have $c_1(X) = 0$, and the selection rule says that at genus $g = 0$ (i.e. when the Riemann surface is a sphere $\mathbb{S}^2$) we have to insert three operators associated to 2-forms. The correlation functions can be evaluated by summing over the different topological sectors of holomorphic maps. These sectors can be labelled by "instanton numbers." Let Σ_i denote a basis of $H_2(X)$, with $i = 1, \ldots, b_2$. If the image of $x(\mathbb{S}^2)$ is in the homology class $\beta = \sum_i n_i \Sigma_i$, then we will say that the worldsheet instanton is in the sector specified by β, or equivalently, by the integers n_i. The trivial sector corresponds to $\beta = 0$, i.e. the image of the sphere is a point in the target, and in this case the correlation function is just the classical intersection number $D_1 \cap D_2 \cap D_3$ of the three divisors $D_i, i = 1, 2, 3$, associated to the 2-forms, while the nontrivial instanton sectors give an infinite series. The final answer looks, schematically,

$$\langle \mathcal{O}_{\phi_1} \mathcal{O}_{\phi_2} \mathcal{O}_{\phi_3} \rangle = (D_1 \cap D_2 \cap D_3) + \sum_\beta I_{0,3,\beta}(\phi_1, \phi_2, \phi_3) q^\beta \tag{2.7}$$

The notation is as follows: let $\omega = \sum_{i=1}^{b_2} t_i \omega_i$, be the complexified Kähler form of X, where ω_i is a basis for $H^2(X)$ dual to Σ_i, and t_i are the complexified Kähler parameters. Set $q_i = e^{-t_i}$. If $\beta = \sum_i n_i \Sigma_i$, then q^β denotes $\prod_i q_i^{n_i}$. The coefficients $I_{0,3,\beta}(\phi_1, \phi_2, \phi_3)$ "count" in some appropriate way the number of holomorphic maps from the sphere to the Calabi–Yau, in the topological sector specified by β, and in such a way that the point of insertion of $\mathcal{O}_{\phi_i}$ gets mapped to the divisor D_i. This is an example of a Gromov–Witten invariant, although to get the general picture we have to couple the model to gravity, as we will see very soon.

When $c_1(X) > 0$, correlation functions also have the structure of (2.7): the trivial sector gives just the classical intersection number of the cohomology ring, and then there are quantum corrections associated to the worldsheet instantons. One important aspect of the case $c_1(X) > 0$ is that the right hand side of (2.6) contains the positive integer $\sum_i n_i \int_{\Sigma_i} c_1(X)$, where n_i are the instanton numbers labelling the topological sector of the holomorphic map. As the n_i increase, it won't be possible to satisfy the selection rule for the insertions. Therefore, only a finite number of topological sectors contribute to the correlation function, which will be given by the sum of a classical intersection number plus a finite number of "quantum" corrections. This is the starting point in the definition of the quantum cohomology of X, see [23] for details.

2.2 Closed topological strings

In the above considerations on topological sigma models we have focused on $g = 0$. For $g = 1$ and a Calabi–Yau manifold, the only vacuum expectation value (vev) that may lead to a nontrivial answer is that of the unit operator, i.e. the partition function itself, while for $g > 1$ the virtual dimension of the moduli space is negative and the above theory is no longer useful to study the enumerative geometry of the target space X. This corresponds mathematically to the fact that, for a generic metric on the Riemann surface Σ_g, there are no holomorphic maps at genus $g > 1$. In order to circumvent this problem, we have to couple the theory to two-dimensional gravity, which means considering all possible metrics on the Riemann surface. The resulting model is called a *topological string* theory. We will start by giving a general idea from a more mathematical point of view (see [23] for a rigorous discussion), and then we will present the physical construction.

The moduli space of possible metrics (or equivalently, complex structures) on a Riemann surface with punctures is the famous Deligne–Mumford space $\overline{M}_{g,n}$ of stable curves with n marked points (the definition of what stable means can be found for example in [43]). The moduli space we have to consider in the theory of topological strings also involves maps. It consists on one hand of a point in $\overline{M}_{g,n}$, i.e. a Riemann surface with n punctures, $(\Sigma_g, p_1, \ldots, p_n)$, and this involves a choice of complex structure on Σ_g. On the other hand, we have a map $x : \Sigma_g \to X$ which is holomorphic with respect to the choice of complex structure on Σ_g.

Let us now fix the topological sector of the holomorphic map, i.e. the homology class $\beta = x_*[\Sigma_g]$. In general, there will be many maps in this sector. The set given by the possible data $(x, \Sigma_g, p_1, \ldots, p_n)$ associated to the class β can be promoted to a moduli space $\overline{M}_{g,n}(X, \beta)$, provided a certain number of conditions are satisfied. This is the basic moduli space we will need in the theory of topological strings. Its (complex) virtual dimension is given by:

$$(1 - g)(d - 3) + n + \int_{\Sigma_g} x^*(c_1(X)). \tag{2.8}$$

If we compare (2.8) to (2.6), we see that there is an extra $3(g-1)+n$ which comes from the Mumford–Deligne space $\overline{M}_{g,n}$. The moduli space $\overline{M}_{g,n}(X, \beta)$ comes equipped with the natural maps

$$\pi_1 : \overline{M}_{g,n}(X, \beta) \longrightarrow X^n,$$
$$\pi_2 : \overline{M}_{g,n}(X, \beta) \longrightarrow \overline{M}_{g,n}. \tag{2.9}$$

The first map is easy to define: given a point $(x, \Sigma_g, p_1, \ldots, p_n)$ in $\overline{M}_{g,n}(X, \beta)$, we just compute $(x(p_1), \ldots, x(p_n))$. The second map sends $(x, \Sigma_g, p_1, \ldots, p_n)$ to $(\Sigma_g, p_1, \ldots, p_n)$, i.e. forgets the information about the map and leaves the punctured curve (there are some subtleties with this map, associated to the stability conditions; see [23]). We can now formally define the Gromov–Witten invariant $I_{g,n,\beta}$ as follows. Let us consider cohomology classes $\phi_1, \ldots, \phi_n$ in $H^*(X)$. The map π_1 induces a map $\pi_1^* : H^*(X)^n \to H^*(\overline{M}_{g,n}(X, \beta))$, and we can pullback $\phi_1 \otimes \cdots \otimes \phi_n$ to get a differential form on the moduli space of holomorphic maps. This form can be integrated as long as there is a well-defined fundamental class for this space, and the result is the Gromov–Witten invariant $I_{g,n,\beta}(\phi_1, \ldots, \phi_n)$:

$$I_{g,n,\beta}(\phi_1, \ldots, \phi_n) = \int_{\overline{M}_{g,n}(X,\beta)} \pi_1^*(\phi_1 \otimes \cdots \otimes \phi_n). \tag{2.10}$$

By using the Gysin map $\pi_{2!}$, one can reduce this to an integral over the moduli space of curves $\overline{M}_{g,n}$. The Gromov–Witten invariant $I_{g,n,\beta}(\phi_1, \ldots, \phi_n)$ vanishes unless the degree of the form equals the dimension of the moduli space. Therefore, we have the following selection rule:

$$\frac{1}{2} \sum_{i=1}^n \deg(\phi_i) = (1 - g)(d - 3) + n + \int_{\Sigma_g} x^*(c_1(X)) \tag{2.11}$$

Notice that Calabi–Yau threefolds play a special role in the theory, since for those targets the virtual dimension only depends on the number of punctures, and therefore the above condition is always satisfied if the forms ϕ_i have degree 2. The invariants (2.10) generalize the invariants obtained from topological sigma models. In particular, $I_{0,3,\beta}$ are the invariants involved in the evaluation of correlation functions of the topological sigma model with a Calabi–Yau threefold as its target in (2.7). When $n = 0$, one gets an invariant $N_{g,\beta} = I_{g,0,\beta}$ which does not require any insertions. We will refer to this as the Gromov–Witten invariant of the Calabi–Yau threefold X at genus g and in the class β. These are the only (closed) Gromov–Witten invariants that we will deal with here. It can be also shown that, for genus 0 [23],

$$I_{0,3,\beta}(\phi_1, \phi_2, \phi_3) = N_{0,\beta} \int_\beta \phi_1 \int_\beta \phi_2 \int_\beta \phi_3, \tag{2.12}$$

so from these Gromov–Witten invariants one can recover as well the information about the three-point functions of the topological sigma model.

The physical point of view on the Gromov–Witten invariants $N_{g,\beta}$ comes about as follows. It is clear that we have to couple the topological sigma model to two dimensional gravity in order to get nontrivial invariants. To do that, one realizes [26, 15] that the structure of the twisted theory is tantalizingly close to that of the bosonic string. In the bosonic string, there is a nilpotent BRST operator, Q_{BRST}, and the energy-momentum tensor turns out to be a Q_{BRST}-commutator: $T(z) = \{Q_{BRST}, b(z)\}$. This is precisely the same structure that we found in (2.4), so the field $b_{\alpha\beta}$ plays the role of a ghost. Therefore, one can just follow the prescription of coupling to gravity for the bosonic string and define a genus g free energy as follows:

$$F_g = \int_{\overline{M}_g} \langle \prod_{k=1}^{6g-6} (b, \mu_k) \rangle, \tag{2.13}$$

where

$$(b, \mu_k) = \int d^2z (b_{zz}(\mu_k)_{\bar{z}}^z + b_{\bar{z}\bar{z}}(\overline{\mu}_k)_z^{\bar{z}}), \tag{2.14}$$

and μ_k are the usual Beltrami differentials. The vev in (2.13) refers to the path integral over the fields of the twisted sigma model. The result, which depends on the choice of complex structure of the Riemann surface, is then integrated over the moduli space $\overline{M}_g$. F_g can be evaluated again, like in the topological sigma model, as a sum over instanton sectors. It turns out [15] that F_g is a generating functional for the Gromov–Witten invariants $N_{g,\beta}$, or more precisely,

$$F_g(t) = \sum_\beta N_{g,\beta} q^\beta. \tag{2.15}$$

It is also useful to introduce a generating functional for the all-genus free energy:

$$F(g_s, t) = \sum_{g=0}^\infty F_g(t) g_s^{2g-2}. \tag{2.16}$$

The parameter g_s can be regarded as a formal variable, but in the context of type II strings it is nothing but the string coupling constant.

The first term in (2.15) corresponds to the contribution of constant maps, with $\beta = 0$. It was shown in [15] (see also [35]) that, for $g \geq 2$, this contribution can be expressed as an integral over $\overline{M}_g$. The result is as follows: on $\overline{M}_g$ there is a complex vector bundle $\mathbb{E}$ of rank g, called the Hodge bundle, whose fiber at a point

Σ is $H^0(\Sigma, K_\Sigma)$. The contribution of constant maps to F_g is then given by

$$N_{g,0} = (-1)^g \frac{\chi(X)}{2} \int_{\overline{M}_g} c_{g-1}^3(\mathbb{E}), \quad g \geq 2, \tag{2.17}$$

where c_{g-1} is the $(g-1)$-th Chern class of $\mathbb{E}$, and $\chi(X)$ is the Euler characteristic of the target space.

In general, Gromov–Witten invariants can be computed by using the localization techniques pioneered by Kontsevich [55]. These techniques are easier to implement in the case of non-compact Calabi–Yau manifolds (the so-called *local* case), where one can compute $N_{g,\beta}$ for arbitrary genus. For example, let us consider the non-compact Calabi–Yau manifold $\mathcal{O}(-3) \to \mathbb{P}^2$. This is the total space of $\mathbb{P}^2$ together with its anticanonical bundle, and it has $b_2 = 1$, corresponding to the hyperplane class of $\mathbb{P}^2$. Therefore, the class β is labelled by a single integer, the degree of the curve in $\mathbb{P}^2$. By using the localization techniques of Kontsevich, adapted to the noncompact case, one finds [20, 53]:

$$F_0(q) = -\frac{t^3}{18} + 3q - \frac{45\,q^2}{8} + \frac{244\,q^3}{9} - \frac{12333\,q^4}{64} \cdots$$

$$F_1(q) = -\frac{t}{12} + \frac{q}{4} - \frac{3\,q^2}{8} - \frac{23\,q^3}{3} + \frac{3437\,q^4}{16} \cdots \tag{2.18}$$

$$F_2(q) = \frac{\chi(X)}{5720} + \frac{q}{80} + \frac{3\,q^3}{20} - \frac{514\,q^4}{5} \cdots$$

and so on. In (2.18), t is the Kähler class of the manifold, $\chi(X) = 2$ is the Euler characteristic of the local $\mathbb{P}^2$, and $q = e^{-t}$. The first term in F_2 is the contribution of constant maps, and we will provide later on a universal expression for it.

It should be mentioned that there is of course a very powerful method to compute F_g, namely mirror symmetry (the B-model). In the B-model, the F_g amplitudes are deeply related to the variation of complex structures on the Calabi–Yau manifold (Kodaira–Spencer theory) and can be computed through the holomorphic anomaly equations of [15]. B-model computations of Gromov–Witten invariants and F_g amplitudes can be found for example in [15, 20, 44, 53, 50]. Finally, it should be mentioned that, when type II theory is compactified on a Calabi–Yau manifold, the F_g appear naturally as the couplings of some special set of F-terms of the low-energy supergravity action [15, 7]. This point of view has shown to be extremely important in understanding the properties of topological strings.

2.3 Open topological strings

Let us now consider open topological strings. The natural starting point is a topological sigma model in which the worldsheet is now a Riemann surface $\Sigma_{g,h}$ of genus g with h holes. Such models were analyzed in detail in [95]. The main issue is of course to specify boundary conditions for the maps $x : \Sigma_{g,h} \to X$. It turns out that, for the A-model, the relevant boundary conditions are Dirichlet, supported on Lagrangian submanifolds of the Calabi–Yau X. If we denote by C_i, $i = 1, \ldots, h$ the holes of $\Sigma_{g,h}$ (*i.e.* the disconnected components of the boundary $\partial \Sigma_{g,h}$), we have to pick Lagrangian submanifolds $\mathcal{L}_i$, and consider maps such that

$$x(C_i) \subset \mathcal{L}_i. \tag{2.19}$$

These boundary conditions are a consequence of requiring Q-invariance at the boundary. One also has boundary conditions on the fermionic fields of the theory, which require that χ and ψ at the boundary C_i take values on $x^*(T\mathcal{L}_i)$. We can also couple the theory to Chan–Paton degrees of freedom on the boundaries, giving rise to a $\bigotimes_i U(N_i)$ gauge symmetry. The model can then be interpreted as a topological open string theory in the presence of N_i topological D-branes wrapping the Lagrangian submanifolds $\mathcal{L}_i$. Notice that, in contrast to physical D-branes in Calabi–Yau manifolds, which wrap special Lagrangian submanifolds [13, 75], in the topological framework the conditions are relaxed to just Lagrangian.

Once boundary conditions have been specified, one can define the free energy of the topological string theory similarly to what we did in the closed case. Let us consider for simplicity the case in which one has a single Lagrangian submanifold $\mathcal{L}$, so that all the boundaries of $\Sigma_{g,h}$ are mapped to $\mathcal{L}$. Now, in order to specify the topological sector of the map, we have to give two different kinds of data: the boundary part and the bulk part. For the bulk part, the topological sector is labelled by relative homology classes, since we are requiring the boundaries of $x_*[\Sigma_{g,h}]$ to end on $\mathcal{L}$. Therefore, we will set

$$x_*[\Sigma_{g,h}] = \mathcal{Q}, \quad \mathcal{Q} \in H_2(X, \mathcal{L}) \tag{2.20}$$

To specify the topological sector of the boundary, we will assume that $b_1(\mathcal{L}) = 1$, so that $H_1(\mathcal{L})$ is generated by a nontrivial one cycle γ. We then have

$$x_*[C_i] = w_i \gamma, \quad w_i \in \mathbb{Z}, \ i = 1, \ldots, h, \tag{2.21}$$

in other words, w_i is the winding number associated to the map x restricted to C_i. We will collect these integers into a single vector h-uple denoted by $w = (w_1, \ldots, w_h)$.

There are various generating functionals that we can consider, depending on the topological data that we want to keep fixed. It is very useful to fix g, h and the winding numbers, and sum over all bulk classes. This produces the following generating functional of open Gromov–Witten invariants:

$$F_{w,g}(t) = \sum_{\mathcal{Q}} F_{w,g}^{\mathcal{Q}} e^{-\mathcal{Q} \cdot t}. \tag{2.22}$$

In this equation, we have labelled the relative cohomology classes $\mathcal{Q}$ of embedded Riemann surfaces by a vector $\mathcal{Q}$ of $b_2(X)$ integers defined as

$$\int_{\mathcal{Q}} \omega = \mathcal{Q} \cdot t, \tag{2.23}$$

where $t = (t_1, \ldots, t_{b_2(X)})$ are the complexified Kähler parameters of the Calabi–Yau manifold. In many examples relevant to knot theory, the entries Q are naturally chosen to be half-integers. Finally, the quantities $F_{w,g}^Q$ are the open string Gromov–Witten invariants, and they "count" in an appropriate sense the number of holomorphically embedded Riemann surfaces of genus g in X with Lagrangian boundary conditions specified by $\mathcal{L}$ and in the class represented by Q, w. These are in general rational numbers.

We can now consider the total free energy, which is the generating functional for all topological sectors:

$$F(V) = \sum_{g=0}^{\infty} \sum_{h=1}^{\infty} \sum_{w_1, \ldots, w_h} \frac{i^h}{h!} g_s^{2g-2+h} F_{g,w}(t) \mathrm{Tr}\, V^{w_1} \ldots \mathrm{Tr}\, V^{w_h}, \qquad (2.24)$$

where g_s is the string coupling constant, and V is a matrix source that keeps track of the topological sector at the boundary. The factor i^h is very convenient in order to compare to the Chern–Simons free energy, as we will see later. The factor $h!$ is a symmetry factor which takes into account that the holes are indistinguishable (or one could have absorbed them into the definition of $F_{g,w}$).

In order to compare open Gromov–Witten invariants to knot invariants, it is useful to introduce the following notation. When all w_i are positive, one can label w in terms of a vector $\vec{k}$. Given an h-uple $w = (w_1, \ldots, w_h)$, we define a vector $\vec{k}$ as follows: the i-th entry of $\vec{k}$ is the number of w_j's which take the value i. For example, if $w_1 = w_2 = 1$ and $w_3 = 2$, this corresponds to $\vec{k} = (2, 1, 0, \ldots)$. In terms of $\vec{k}$, the number of holes and the total winding number are

$$h = |\vec{k}| \equiv \sum_j k_j, \quad \ell = \sum_i w_i = \sum_j j k_j. \qquad (2.25)$$

Note that a given $\vec{k}$ will correspond to many w's which differ by permutation of entries. In fact there are $h!/\prod_j k_j!$ h-tuples w which give the same vector $\vec{k}$ (and the same amplitude). We can then write the total free energy for positive winding numbers as:

$$F(V) = \sum_{g=0}^{\infty} \sum_{\vec{k}} \frac{i^{|\vec{k}|}}{\prod_j k_j!} g_s^{2g-2+h} F_{g,\vec{k}}(t) \Upsilon_{\vec{k}}(V) \qquad (2.26)$$

where

$$\Upsilon_{\vec{k}}(V) = \prod_{j=1}^{\infty} (\mathrm{Tr}\, V^j)^{k_j}. \qquad (2.27)$$

Although a rigorous theory of open Gromov–Witten invariants is not available, localization techniques make possible to compute them in various situations [51, 67, 39, 73, 16, 52]. It is also possible to use mirror symmetry to compute disc invariants (i.e. when $g = 0$, $h = 1$), as it was first shown in [4] and subsequently explored in [2, 72, 65, 47, 38]. Finally, we also mention that the open string amplitudes $F_{g,w}$ also appear as low-energy couplings of type II superstrings compactified on Calabi–Yau manifolds in the presence of D-branes [15, 89].

2.4 Integer invariants from topological strings

The closed and open Gromov–Witten invariants that have been introduced are both rational, due to the orbifold structure of the moduli spaces. On the other hand, these invariants are deeply related to questions in enumerative geometry, but the relation between the invariants and the number of holomorphic curves of a given genus and in a given homology class is far from being simple. An obvious reason for this is *multicovering*. Suppose you have found a holomorphic map $x : \mathbb{S}^2 \to X$ in genus zero of degree d. Then, simply by composing this with a degree k cover $\mathbb{S}^2 \to \mathbb{S}^2$, you get another holomorphic map of degree kd. Therefore, at every degree, in order to count the actual number of "primitive" holomorphic curves, one should subtract the contributions coming from multicovering of curves with lower degree. On top of that, the contribution of a k-cover appears in $N_{0,kd}$ with weight k^{-3}. Therefore, although in genus zero the Gromov–Witten invariants are not integer, this is due to the effects of multicovering, and once this has been taken into account one extracts integer numbers that correspond in many cases to actual numbers of rational curves. The multicovering phenomenon at genus 0 was found experimentally in [19] and later on derived in [8].

Another geometric effect that has to be taken into account is bubbling [14, 15]. Imagine that you found a map $x : \Sigma_g \to X$ from a genus g surface to a Calabi–Yau threefold. By gluing to Σ_g a small Riemann surface of genus h, and making it very small, you get an approximate holomorphic map from a Riemann surface whose genus is topologically $g + h$. This means that "primitive" maps at genus g contribute to all genera $g' > g$, and in order to count curves properly we should take this effect into account.

These facts suggest that, although the Gromov–Witten invariants are not in general integer numbers, they have some hidden integrality structure, and that one can extract from them integer invariants that are related to a counting problem. But it turns out that, instead of deriving the various effects of multicovering and bubbling from a geometrical point of view, the underlying integral structure of the Gromov–Witten invariants is better revealed when the F_g is regarded as a low-energy coupling in a compactification of type IIA theory on a Calabi–Yau manifold. Using this approach, Gopakumar and Vafa showed [36] that one can write the generating functional $F(g_s, t)$ in terms of contributions associated to BPS states, and they used type IIA/M-theory duality to obtain a completely new point of view on topological strings. They showed in particular that Gromov–Witten invariants of closed strings can be written in terms of some new, *integer* invariants known as *Gopakumar–Vafa invariants*. These invariants count in a very precise way the number of BPS states that arise in the Calabi–Yau compactification of type IIA theory. We will now describe this result in some detail and provide some examples.

The result of Gopakumar and Vafa concerns the overall structure of $F(g_s, t)$. According to [36], the generating functional (2.16) can be written as

$$F(g_s, t) = \sum_{g=0}^{\infty} \sum_{\beta} \sum_{d=1}^{\infty} n_\beta^g \frac{1}{d} \left(2 \sin \frac{d g_s}{2} \right)^{2g-2} q^{d\beta}, \qquad (2.28)$$

where n_β^g, which are the Gopakumar–Vafa invariants, are integer numbers. In (2.28), t denotes the set of $b_2(X)$ Kähler parameters, and q^β is defined as in (2.7). It is very illuminating to expand (2.28) in powers of g_s and extract from it the structure of a given F_g. One easily obtains, for $g = 0$, the well-known structure of the prepotential [19, 8]:

$$F_0 = \frac{1}{3!} \int_X \omega^3 + \int_X c_2(X) \wedge \omega + \chi(X) \frac{\zeta(3)}{2} + \sum_{\beta} n_\beta^0 \text{Li}_3(q^\beta), \qquad (2.29)$$

up to the polynomial terms in t. Here $\chi(X)$, $c_2(X)$ denote respectively the Euler characteristic and the second Chern class of the Calabi–Yau target. We recall that Li_j denotes the polylogarithm of index j, which is defined by:

$$\text{Li}_j(x) = \sum_{n=1}^{\infty} \frac{x^n}{n^j}. \qquad (2.30)$$

Notice that $\text{Li}_1(x) = -\log(1-x)$, while for $j \leq 0$, $\text{Li}_j(x)$ is a rational function of x:

$$\text{Li}_j(x) = \left(x \frac{d}{dx} \right)^{|j|} \frac{1}{1-x} = |j|! \frac{x^{|j|}}{(1-x)^{|j|+1}} + \cdots. \qquad (2.31)$$

For $g = 1$, one obtains:

$$F_1 = \frac{1}{24} \int_X c_2(X) \wedge \omega + \sum_{\beta} \left(\frac{1}{12} n_\beta^0 + n_\beta^1 \right) \text{Li}_1(q^\beta). \qquad (2.32)$$

Finally, for $g > 1$, the Gopakumar–Vafa result gives:

$$F_g(t) = \frac{(-1)^g \chi(X) |B_{2g} B_{2g-2}|}{4g(2g-2)(2g-2)!} \qquad (2.33)$$

$$+ \sum_{\beta} \left(\frac{|B_{2g}| n_\beta^0}{2g(2g-2)!} + \frac{2(-1)^g n_\beta^2}{(2g-2)!} \pm \cdots - \frac{g-2}{12} n_\beta^{g-1} + n_\beta^g \right) \text{Li}_{3-2g}(q^\beta).$$

In this equation, B_n denote the Bernoulli numbers. The first term in (2.33) is the contribution to F_g associated to maps from Σ_g to a single point. Comparing it with

(2.17) we find that the Gopakumar–Vafa structure result predicts:

$$\int_{\overline{M}_g} c^3_{g-1}(\mathbb{E}) = \frac{|B_{2g}B_{2g-2}|}{2g(2g-2)(2g-2)!}.$$ (2.34)

This expression was conjectured by Faber [30], derived in [70] from heterotic/type IIA duality, and proved in [31].

The polylogarithm in (2.33) indicates that the degree k multicover of a curve of genus g contributes with a factor k^{2g-3} to F_g. This generalizes the results of [19] for genus 0 and results for genus 1 in [14]. The multicover contribution was also found in [70] by using heterotic/type II duality. But equation (2.33) also takes into account in a precise way the effect of bubbling on F_g: at every genus g, one has to take into account all the previous genera $g' < g$ in order to extract the Gopakumar–Vafa invariants n^g_β.

The Gopakumar–Vafa invariants contain all the information of the Gromov–Witten invariants, and vice versa: if one knows the Gopakumar–Vafa invariants n^g_β for all g and β, one can deduce the $N_{g,\beta}$, and the other way around. This follows just by comparing (2.28) with (2.16), and it is worked out in detail in [17], where explicit formulae for the relation between $N_{g,\beta}$ and n^g_β are given. But one remarkable aspect of the Gopakumar–Vafa picture is that, in many situations, the integer invariants n^g_β can be computed much more easily than their Gromov–Witten counterparts [36, 50]. In fact, their computation involves in many cases just classical algebraic geometry, so one gets rid of the complications of the moduli space of maps. The physical reason behind is that in the Gopakumar–Vafa picture one looks at worldsheet instantons using the physical gauge approach (in the terminology of [97]), *i.e.* one views the worldsheet instanton as a submanifold of the target, and not as a map embedding a Riemann surface Σ_g inside a Calabi–Yau. Related developments can be found in [45].

Let us consider some simple examples of the Gopakumar–Vafa invariants. The simplest one refers to the noncompact Calabi–Yau manifold $\mathcal{O}(-1) \oplus \mathcal{O}(-1) \to \mathbb{P}^1$, also known as the resolved conifold, which will play an important role later on. This manifold is toric, and can be described as the zero locus of

$$|x_1|^2 + |x_4|^2 - |x_2|^2 - |x_3|^2 = s$$ (2.35)

quotiented by a $U(1)$ that acts as

$$x_1, x_2, x_3, x_4 \to e^{i\alpha}x_1, e^{-i\alpha}x_2, e^{-i\alpha}x_3, e^{i\alpha}x_4$$ (2.36)

This is the description that appears naturally in the linear sigma model of [96]. Notice that, for $x_2 = x_3 = 0$, (2.35) describes a $\mathbb{P}^1$ whose area is proportional to s. Therefore, (x_1, x_4) can be taken as homogeneous coordinates of the $\mathbb{P}^1$ which is the basis of the fibration, while x_2, x_3 can be regarded as coordinates for the fibers. This manifold has $b_2(X) = 1$, corresponding to the $\mathbb{P}^1$ in the base, and its total free energy turns out to be

$$F(g_s, t) = \sum_{d=1}^{\infty} \frac{1}{d\left(2\sin\frac{dg_s}{2}\right)^2} q^d, \tag{2.37}$$

where $q = e^{-t}$ and t is the complexified area of the $\mathbb{P}^1$. We see that the only nonzero Gopakumar–Vafa invariant is $n_1^0 = 1$. On the other hand, this model already has an infinite number of nontrivial $N_{g,\beta}$ invariants, but these are all due to bubbling and multicovering: the model only has one true "primitive" curve, which is just $\mathbb{P}^1$, and this is what the Gopakumar–Vafa invariant is computing.

A more complicated example is the local $\mathbb{P}^2$ geometry considered before, which already has an infinite number of nontrivial Gopakumar–Vafa invariants. These have been computed in [53, 50, 3] using the A-model, the B-model, and the duality with Chern–Simons theory that we will explain in section 6. Some results are presented in Table 2.4. In this table, the integer d labels the class β, and corresponds to the degree of the curve in $\mathbb{P}^2$. Notice that the first Gromov–Witten invariants are $N_{0,1} = 3$, and $N_{0,2} = -45/8$, as listed in (2.18), therefore using the multicovering/bubbling formula one finds $n_1^0 = N_{0,1} = 3$, and $N_{0,2} = n_1^0/8 + n_2^0$, which gives $n_2^0 = -6$.

Table 1. Gopakumar–Vafa invariants n_d^g for $\mathcal{O}(-3) \to \mathbb{P}^2$.

d	$g = 0$	1	2	3	4
1	3	0	0	0	0
2	−6	0	0	0	0
3	27	−10	0	0	0
4	−192	231	−102	15	0
5	1695	−4452	5430	−3672	1386

For open topological strings one can derive a similar expression relating open Gromov–Witten invariants to a new set of integer invariants, that we will denote by $n_{w,g,Q}$. The corresponding multicovering/bubbling formula was derived in [76, 63], following arguments similar to those in [36], and states that the free energies of open topological string theory in the sector labelled by w can be written in terms of the integer invariants $n_{w,g,Q}$ as follows:

$$\sum_{g=0}^{\infty} g_s^{2g-2+h} F_{w,g}(t) = \tag{2.38}$$

$$\frac{1}{\prod_i w_i} \sum_{g=0}^{\infty} \sum_{d|w} \sum_Q (-1)^{h+g}\, n_{w/d,g,Q}\, d^{h-1} \left(2\sin\frac{dg_s}{2}\right)^{2g-2} \prod_i \left(2\sin\frac{w_i g_s}{2}\right) e^{-dQ\cdot t}.$$

Notice there is one such identity for each w. In this expression, the sum is over all integers d which satisfy that $d|w_i$ for all $i = 1, \ldots, h$. When this is the case, we define the h-uple w/d whose i-th component is w_i/d. The expression (2.38) can be expanded to give a set of multicovering/bubbling formulae for different genera. Up to genus 2 one finds,

$$F_{w,g=0}^Q = (-1)^h \sum_{d|w} d^{h-3} n_{w/d,0,Q/d},$$

$$F_{w,g=1}^Q = -(-1)^h \sum_{d|w} \left(d^{h-1} n_{w/d,1,Q/d} - \frac{d^{h-3}}{24} \left(2d^2 - \sum_i w_i^2 \right) n_{w/d,0,Q/d} \right),$$

$$F_{\vec{k},g=2}^Q = (-1)^h \sum_{d|w} \left(d^{h+1} n_{w/d,2,Q/d} + \frac{d^{h-1}}{24} n_{w/d,1,Q/d} \sum_i w_i^2 \right. \qquad (2.39)$$

$$\left. + \frac{d^{h-3}}{5760} \left(24 d^4 - 20 d^2 \sum_i w_i^2 - 2 \sum_i w_i^4 + 5 \sum_{i_1,i_2} w_{i_1}^2 w_{i_2}^2 \right) n_{w/d,0,Q/d} \right).$$

In these equations, the integer d has to divide the vector w (in the sense explained above) and it is understood that $n_{wd,g,Q/d}$ is zero if Q/d is not a relative homology class.

It is important to notice that the integer invariants $n_{w,g,Q}$ are not the most fundamental ones. When all the winding numbers are positive, we can represent w by a vector $\vec{k} = (k_1, k_2, \ldots)$, as we explained in 2.3. Such a vector can be interpreted as a label for a conjugacy class $C(\vec{k})$ of the symmetric group S_ℓ, where $\ell = \sum_j j k_j$ is the total winding number: $C(\vec{k})$ is the conjugacy class with k_1 one-cycles, k_2 two-cycles, and so on. The invariant $n_{w,g,Q}$ will be denoted as $n_{\vec{k},g,Q}$, and D-brane physics states that it can be written as

$$n_{\vec{k},g,Q} = \sum_R \chi_R(C(\vec{k})) N_{R,g,Q}, \qquad (2.40)$$

where $N_{R,g,Q}$ are integer numbers labelled by representations of the symmetric group, i.e. by Young tableaux, and χ_R is the character of S_ℓ in the representation R. The above relation is invertible, since by orthonormality of the characters one has

$$N_{R,g,Q} = \sum_{\vec{k}} \frac{\chi_R(C(\vec{k}))}{z_{\vec{k}}} n_{\vec{k},g,Q}, \qquad (2.41)$$

where

$$z_{\vec{k}} = \frac{\ell!}{|C(\vec{k})|} = \prod_j k_j! \prod_j j^{k_j}. \qquad (2.42)$$

Notice that integrality of $N_{R,g,Q}$ implies integrality of $n_{\vec{k},q,Q}$, but not the other way around. In that sense, the invariants $N_{R,g,Q}$ are more fundamental. We will further clarify this issue in section 4.

3 Chern–Simons theory and knot invariants

In this section we make a short review of Chern–Simons theory and its relations to knot invariants.

3.1 Chern–Simons theory: basic ingredients

Chern–Simons theory, introduced by Witten in [92], provides a quantum field theory description of a wide class of invariants of three-manifolds and of knots and links in three-manifolds. The Chern–Simons action with gauge group G on a generic three-manifold M is defined by

$$S = \frac{k}{4\pi} \int_M \text{Tr}\left(A \wedge dA + \frac{2}{3} A \wedge A \wedge A\right). \tag{3.1}$$

Here, k is the coupling constant, and A is a G-gauge connection on the trivial bundle over M. We will assume for simplicity that G is a simply-laced group, unless otherwise stated. As noticed by Witten, since this action does not involve the metric, the resulting quantum theory is topological, at least formally. In particular, the partition function

$$Z_k(M) = \int [\mathcal{D}A] e^{iS} \tag{3.2}$$

should define a topological invariant of the manifold M. A detailed analysis [92] shows that this is in fact the case, with an extra subtlety: the invariant depends on the three-manifold *and* of a choice of framing, *i.e.* a choice of trivialization of the bundle $TM \oplus TM$ (this should be called, strictly speaking, a 2-framing, but we will refer to it as framing, following standard practice). As explained in [9], for every three-manifold there is a canonical choice of framing, and the different choices are labelled by an integer $s \in \mathbb{Z}$ in such a way that $s = 0$ corresponds to the canonical framing. In the following all the results will be presented in the canonical framing.

The partition function of Chern–Simons theory can be computed in a variety of ways. One can for example use perturbation theory and produce an asymptotic series in k around a classical solution to the action. The classical solutions of Chern–Simons theory are just flat connections $F(A) = 0$ on M. Let us assume that these are a discrete set of points (this happens, for example, if M is a rational homology sphere). In that situation, one expresses $Z_k(M)$ as a sum of terms associated to stationary points:

$$Z_k(M) = \sum_c Z_k^{(c)}(M), \tag{3.3}$$

where c labels the different flat connections $A^{(c)}$ on M. The structure of the perturbative series was analyzed in various papers [92, 83, 11] and is given by the following expression:

$$Z_k^{(c)}(M) = Z_{1-\text{loop}}^{(c)}(M) \cdot \exp\left\{ \sum_{\ell=1}^{\infty} S_\ell^{(c)} x^\ell \right\}. \tag{3.4}$$

In this equation, x is the effective expansion parameter:

$$x = \frac{2\pi i}{k + y}, \tag{3.5}$$

where y is the dual Coxeter of the group, and we will set $l = k + y$. For $G = SU(N)$, $y = N$. The one-loop correction $Z_{1-\text{loop}}^{(c)}(M)$ was first analyzed in [92], and studied in great detail since then. It involves some important normalization factors of the path-integral, and determinants of differential operators. After some work it can be written in terms of topological invariants of the three-manifold and the flat connection $A^{(c)}$,

$$Z_{1-\text{loop}}^{(c)}(M) = \frac{(2\pi x)^{\frac{1}{2}(\dim H_{A^{(c)}}^0 - \dim H_{A^{(c)}}^1)}}{\text{vol}(H_c)} e^{-\frac{1}{x} S_{\text{CS}}(A^{(c)}) - \frac{i\pi}{4}\varphi} \sqrt{|\tau_R^{(c)}|}, \tag{3.6}$$

where $H_{A^{(c)}}^{0,1}$ are the de Rham cohomology groups with values in the Lie algebra of G and associated to the flat connection $A^{(c)}$, $\tau_R^{(c)}$ is the Reidemeister–Ray–Singer torsion of $A^{(c)}$, H_c is the isotropy group of $A^{(c)}$, and φ is a certain phase. More details about the structure of this term can be found in [92, 32, 48, 83]. The terms $S_\ell^{(c)}$ in (3.4) correspond to connected diagrams at $\ell + 1$ loops, and since they involve evaluation of group factors of Feynman diagrams, they depend explicitly on the gauge group G and the isotropy subgroup H_c. In the $SU(N)$ or $U(N)$ case, and for $A^{(c)} = 0$ (the trivial flat connection) they are polynomials in N. For the trivial flat connection, one also has that $\dim H_{A^{(c)}}^0 = \dim G$, $\dim H_{A^{(c)}}^1 = 0$, and $H_c = G$. The terms $S_\ell^{(c)}$ are also topological invariants associated to the three-manifold and the flat connection, and they emerge naturally from the perturbative analysis of Chern–Simons theory.

As Witten showed in [92], it is also possible to use nonperturbative methods to obtain a combinatorial formula for (3.2). This goes as follows. By canonical quantization, one associates a Hilbert space $\mathcal{H}(\Sigma)$ to any two-dimensional compact manifold that arises as the boundary of a three-manifold, so that the path-integral over a manifold with boundary gives a state in the corresponding Hilbert space. In order to compute the partition function of a three-manifold M, one can perform a Heegard splitting i.e. represent M as the connected sum of two three-manifolds M_1 and M_2 sharing a common boundary Σ, where Σ is a Riemann surface. If $f : \Sigma \to \Sigma$ is a homeomorphism, we will write $M = M_1 \cup_f M_2$, so that M is obtained by gluing M_1 to M_2 through their common boundary by using the homeomorphism f. This is represented in Fig. 1. We can then compute the full path integral (3.2) over M by computing first the path integral over M_1 and M_2. This produces two wavefunctions $|\Psi_{M_1}\rangle, |\Psi_{M_2}\rangle$ in $\mathcal{H}(\Sigma)$. On the other hand, the homeomorphism $f : \Sigma \to \Sigma$ will be

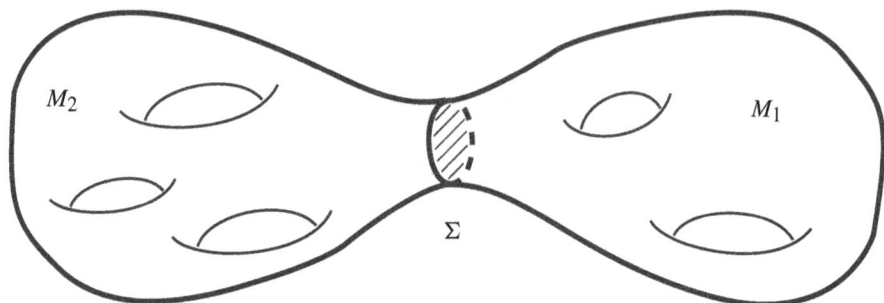

Figure 1. Heegard splitting of a three-manifold M into two three manifolds M_1 and M_2 with a common boundary Σ.

represented by an operator in the Hilbert space,

$$U_f : \mathcal{H}(\Sigma) \to \mathcal{H}(\Sigma) \tag{3.7}$$

and the partition function can then be evaluated as

$$Z_k(M) = \langle \Psi_{M_2} | U_f | \Psi_{M_1} \rangle. \tag{3.8}$$

In order to use this method, we have to find first the Hilbert space associated to a boundary. There is one special case in which this can be done quite systematically, namely when $\Sigma = \mathbb{T}^2$, a two-torus. As it was first shown in [92] (and worked out in detail in [28, 64, 59]), the states of the Hilbert space of Chern–Simons theory associated to the torus, $\mathcal{H}(\mathbb{T}^2)$, are in one to one correspondence with the integrable representations of the Wess–Zumino–Witten (WZW) model with gauge group G at level k [1]. A representation given by a highest weight Λ is integrable if the weight $\rho + \Lambda$ is in the Weyl alcove $\mathcal{F}_l$, where $l = k + y$ and ρ denotes as usual the Weyl vector, given by the sum of the fundamental weights. The Weyl alcove is given by $\Lambda_{\mathrm{w}}/l\Lambda_{\mathrm{r}}$ modded out by the action of the Weyl group. For example, in SU(N) a weight $p = \sum_{i=1}^{r} p_i \lambda_i$ is in $\mathcal{F}_l$ if

$$\sum_{i=1}^{r} p_i < l, \quad \text{and} \quad p_i > 0, \qquad i = 1, \ldots, r. \tag{3.9}$$

---

[1] We will use the following notations in the following: the fundamental weights of G will be denoted by λ_i, the simple roots by α_i, with $i = 1, \ldots, r$, and r denotes the rank of G. The weight and root lattices of G are denoted by Λ_{w} and Λ_{r}, respectively, and $|\Delta_+|$ denotes the number of positive roots.

In the following, the basis of integrable representations will be labelled by the weights in $\mathcal{F}_l$, and the states in the Hilbert state of the torus $\mathcal{H}(\mathbb{T}^2)$ will be denoted by $|p\rangle = |\rho+\Lambda\rangle$ where Λ, as we have stated, is an integrable representation of the WZW model at level l. The states $|p\rangle$ can be chosen to be orthonormal [28, 64, 59].

There is a special class of homeomorphisms of $\mathbb{T}^2$ that have a simple expression as operators in $\mathcal{H}(\mathbb{T}^2)$. These are $\mathrm{Sl}(2, \mathbb{Z})$ transformations, whose generators T and S have the following simple matrix elements in the above basis:

$$T_{\alpha\beta} = \delta_{\alpha\beta} e^{2\pi i (h_\alpha - c/24)},$$

$$S_{\alpha\beta} = \frac{i^{|\Delta_+|}}{(k+y)^{r/2}} \left(\frac{\mathrm{Vol}\, \Lambda^w}{\mathrm{Vol}\, \Lambda^r}\right)^{\frac{1}{2}} \sum_{w \in W} \epsilon(w) \exp\left(-\frac{2\pi i}{k+y} \alpha \cdot w(\beta)\right). \quad (3.10)$$

In the first equation, h_α is the conformal weight of the primary field associated to α:

$$h_\alpha = \frac{\alpha^2 - \rho^2}{2(k+y)}, \quad (3.11)$$

and c is the central charge of the WZW model. In the second equation, the sum over w is a sum over the elements of the Weyl group W, and $\epsilon(w)$ is the signature of w. These explicit formulae allow us to compute the partition function of any three-manifold that admits a Heegard splitting along a torus, like for example a lens space. The case of $\mathbb{S}^3$ is particularly simple. It is well-known that $\mathbb{S}^3$ can be obtained by gluing two solid tori along their boundaries through an S transformation. The wavefunction associated to the solid torus is simply the vacuum, which corresponds to $|\rho\rangle$, and we find

$$Z(\mathbb{S}^3) = \langle \rho | S | \rho \rangle = S_{\rho\rho}. \quad (3.12)$$

By using Weyl's denominator formula,

$$\sum_{w \in W} \epsilon(w) e^{w(\rho)} = \prod_{\alpha > 0} 2 \sinh \frac{\alpha}{2}, \quad (3.13)$$

one finds

$$Z(\mathbb{S}^3) = \frac{1}{(k+y)^{r/2}} \left(\frac{\mathrm{Vol}\, \Lambda^w}{\mathrm{Vol}\, \Lambda^r}\right)^{\frac{1}{2}} \prod_{\alpha > 0} 2 \sin\left(\frac{\pi(\alpha \cdot \rho)}{k+y}\right). \quad (3.14)$$

Besides providing invariants of three-manifolds, Chern–Simons theory also provides invariants of knots and links inside three-manifolds (for a survey of modern knot theory, see [68, 80]). Some examples of knots and links are depicted in Fig. 2. When dealing with knots, we will always consider that the Chern–Simons gauge group is $G = SU(N)$ or $U(N)$. Given a knot $\mathcal{K}$ in $\mathbb{S}^3$, we can consider the trace of the holonomy of the gauge connection around $\mathcal{K}$ in a given irreducible representation R of $SU(N)$, which gives the Wilson loop operator:

$$W_R^{\mathcal{K}}(A) = \mathrm{Tr}_R\left(\mathrm{P} \exp \oint_\gamma A\right), \quad (3.15)$$

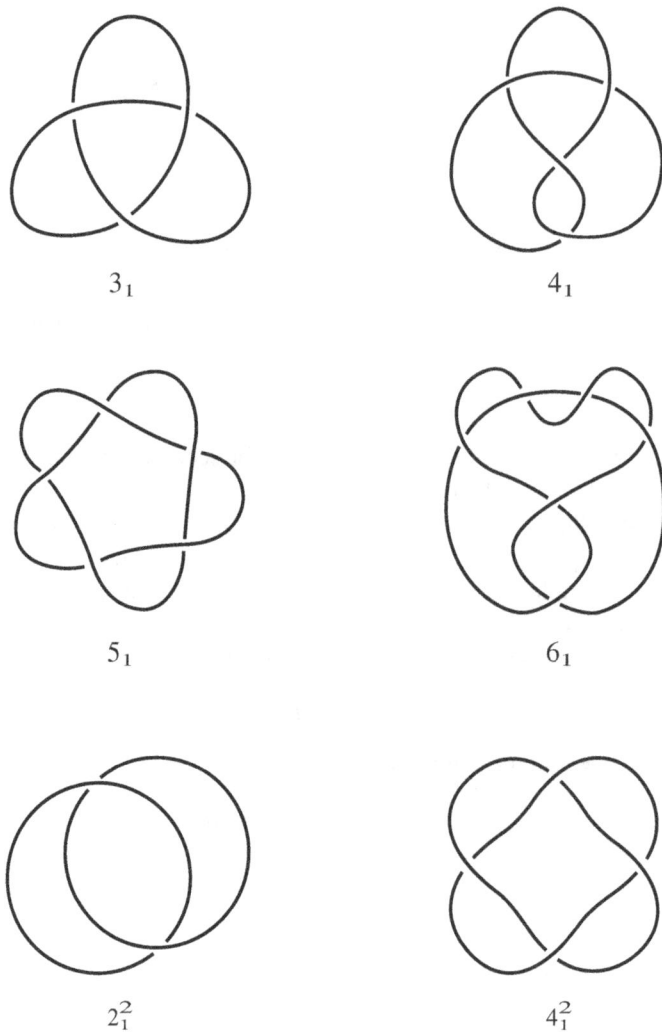

Figure 2. Some knots and links. In the notation x_n^L, x indicates the number of crossings, L the number of components (in case it is a link with $L > 1$) and n is a number used to enumerate knots and links in a given set characterized by x and L. The knot 3_1 is also known as the trefoil knot, while 4_1 is known as the figure-eight knot. The link 2_1^2 is called the Hopf link.

where P denotes path-ordered exponential. This is a gauge invariant operator whose definition does not involve the metric on the three-manifold. The irreducible representations of $SU(N)$ can be labelled by highest weights or equivalently by the lengths of rows in a Young tableau, l_i, where $l_1 \geq l_2 \geq \cdots$. If we now consider a link $\mathcal{L}$ with components $\mathcal{K}_i$, $i = 1, \ldots, L$, we can in principle compute the correlation function,

$$W_{(R_1, \ldots, R_L)}(\mathcal{L}) = \langle W_{R_1}^{\mathcal{K}_1} \ldots W_{R_L}^{\mathcal{K}_L} \rangle = \frac{1}{Z(M)} \int [\mathcal{D}A] \Big(\prod_{i=1}^{L} W_{R_i}^{\mathcal{K}_i} \Big) e^{iS}. \tag{3.16}$$

The topological character of the action, and the fact that the Wilson loop operators can be defined without using any metric on the three-manifold, indicate that (3.16) is a topological invariant of the link $\mathcal{L}$. These correlation functions can be studied in a variety of ways. The nonperturbative approach pioneered by Witten in [92], by exploiting the relation with WZW model, shows that these correlation functions are rational functions of $q^{\pm \frac{1}{2}}, \lambda^{\pm \frac{1}{2}}$, where

$$q = e^x = \exp\left(\frac{2\pi i}{k + N}\right), \quad \lambda = q^N. \tag{3.17}$$

It turns out that the correlation function (3.16) is the quantum group invariant of the link $\mathcal{L}$ associated to the irreducible representations $R_1, \ldots, R_L$ of $U_q(su(N))$ (see for example [82] for a general definition of the quantum group invariant).

The invariants of knots and links obtained as correlation functions in Chern–Simons theory include and generalize the HOMFLY polynomial [33] (which is a generalization itself of the Jones polynomial). The HOMFLY polynomial of a link $\mathcal{L}$, $P_{\mathcal{L}}(q, \lambda)$, can be defined through the so-called *skein relation*. This goes as follows. Let $\mathcal{L}$ be a link in $\mathbb{S}^3$, and let us focus on one of the crossings in its plane projection. The crossing can be an overcrossing, like the one depicted in L_+ in Fig. 3, or an undercrossing,

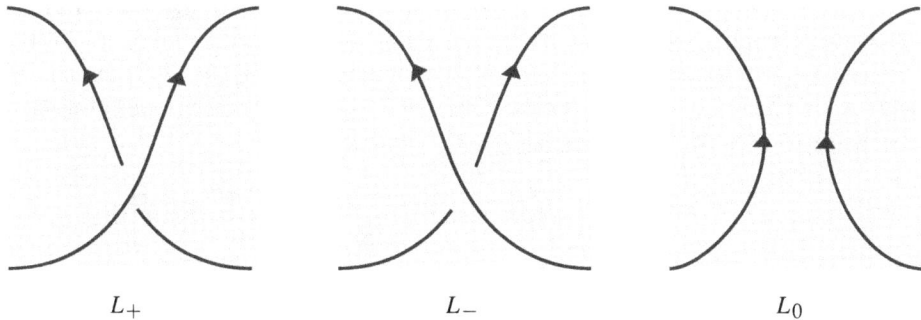

$$L_+ \qquad\qquad L_- \qquad\qquad L_0$$

Figure 3. Skein relations for the HOMFLY polynomial.

like the one depicted in L_-. If the crossing is L_+, we can form two other links either by undoing the crossing (and producing L_0 of Fig. 3) or by changing L_+ into L_-. In both cases the rest of the link is left unchanged. Similarly, if the crossing is L_-, we form two links by changing L_- into L_+ or into L_0. The links produced in this way will be in general topologically inequivalent to the original one (they can even have a different number of components). The skein relation

$$\lambda^{\frac{1}{2}} P_{L_+} - \lambda^{-\frac{1}{2}} P_{L_-} = (q^{\frac{1}{2}} - q^{-\frac{1}{2}}) P_{L_0} \tag{3.18}$$

expresses the HOMFLY polynomial of the original link in terms of the links that are obtained by changing the crossing. By using recursively this relation, one can undo all the crossings and express the polynomial in terms of its value on the unknot, or trivial knot. This value is usually taken to be $P = 1$. The HOMFLY polynomial corresponds to a Chern–Simons $SU(N)$ link invariant with all the components in the fundamental representation $R_\alpha = \square$:

$$W_{(\square,\ldots,\square)}(\mathcal{L}) = \lambda^{\mathrm{lk}(\mathcal{L})} \left(\frac{\lambda^{\frac{1}{2}} - \lambda^{-\frac{1}{2}}}{q^{\frac{1}{2}} - q^{-\frac{1}{2}}} \right) P_{\mathcal{L}}(q, \lambda) \tag{3.19}$$

where $\mathrm{lk}(\mathcal{L})$ is the linking number of $\mathcal{L}$. This can be shown, as in [92], by proving that the vev in the fundamental representation satisfies the skein relation.

The link invariants defined in (3.16) can be computed in many different ways. A particularly useful framework is the formalism of knot operators [59]. In this formalism, one constructs operators that "create" knots wrapped around a Riemann surface in the representation R of the gauge group associated to the highest weight Λ:

$$W_\Lambda^{\mathcal{K}} : \mathcal{H}(\Sigma) \to \mathcal{H}(\Sigma). \tag{3.20}$$

Notice that the topology of Σ restricts the type of knots that one can consider. So far these operators have been constructed in the case when $\Sigma = \mathbb{T}^2$. The knots that can be put on a torus are called *torus knots*, and they are labelled by two integers (n, m) that specify the number of times that they wrap the two cycles of the torus. Here, n refers to the winding number around the noncontractible cycle of the solid torus, while m refers to the contractible one. The trefoil knot 3_1 in Fig. 2 is the $(2, 3)$ torus knot, and the knot 5_1 is the $(2, 5)$ torus knot. The operator that creates the (n, m) torus knot will be denoted by $W_\Lambda^{(n,m)}$, and it has a fairly explicit expression:

$$W_\Lambda^{(n,m)} |p\rangle = e^{2\pi i n m h_{\rho+\Lambda}} \sum_{\mu \in M_\Lambda} \exp\left[-i\pi \mu^2 \frac{nm}{k+N} - 2\pi i \frac{m}{k+N} p \cdot \mu \right] |p + n\mu\rangle. \tag{3.21}$$

In this equation, $h_{\rho+\Lambda}$ is the conformal weight, and M_Λ is the set of weights corresponding to the irreducible representation with highest weight Λ. This equation allows us to compute the vev of the Wilson loop around a torus knot in $\mathbb{S}^3$ as follows: first of all, one makes a Heegard splitting of $\mathbb{S}^3$ into two solid tori, as we explained before. Then, one puts the torus knot on the surface of one of the solid tori by acting with the knot operator (3.21) on the vacuum $|\rho\rangle$. Finally, one glues together the tori by performing an S-transformation. The normalized vev of the Wilson loop is then given by:

$$\langle W_\Lambda^{(n,m)} \rangle = \frac{\langle \rho | SW_\Lambda^{(n,m)} | \rho \rangle}{\langle \rho | S | \rho \rangle}. \tag{3.22}$$

One can show that [59]

$$W_\Lambda^{(1,0)} | \rho \rangle = | \rho + \Lambda \rangle. \tag{3.23}$$

On the other hand, the operator $W_\Lambda^{(1,0)}$ clearly creates a trivial knot, or *unknot*, on the torus, therefore the states $|\rho + \Lambda\rangle$ are obtained by doing the path integral over the solid torus with an insertion of a Wilson loop around the noncontractible loop in the representation Λ, as shown in [92]. We can now evaluate easily the corresponding Chern–Simons invariant. Using the explicit expression in (3.10), we find:

$$W_{R_\Lambda}(\text{unknot}) = \frac{\langle \rho | SW_\Lambda^{(1,0)} | \rho \rangle}{\langle \rho | S | \rho \rangle} = \frac{\sum_{w \in W} \epsilon(w) e^{-\frac{2\pi i}{k+N} \rho \cdot w(\Lambda + \rho)}}{\sum_{w \in W} \epsilon(w) e^{-\frac{2\pi i}{k+N} \rho \cdot w(\rho)}}. \tag{3.24}$$

Using Weyl's denominator formula, the vacuum expectation value can be written as a character

$$W_{R_\Lambda}(\text{unknot}) = \text{ch}_\Lambda \left[-\frac{2\pi i}{k+N} \rho \right]. \tag{3.25}$$

Moreover, using (3.13), we can finally write

$$W_{R_\Lambda}(\text{unknot}) = \prod_{\alpha > 0} \frac{\sin\left(\frac{\pi}{k+N} \alpha \cdot (\Lambda + \rho)\right)}{\sin\left(\frac{\pi}{k+N} \alpha \cdot \rho\right)}. \tag{3.26}$$

Notice that, in the limit $k + N \to \infty$ (*i.e.* in the semiclassical limit), this becomes the dimension of the representation R. For this reason, the above quantity is called the *quantum dimension* of R, denoted by $\dim_q R$. It can be explicitly written as follows. Define the q-numbers:

$$[x] = q^{\frac{x}{2}} - q^{-\frac{x}{2}}, \quad [x]_\lambda = \lambda^{\frac{1}{2}} q^{\frac{x}{2}} - \lambda^{-\frac{1}{2}} q^{-\frac{x}{2}}. \tag{3.27}$$

If R has a Young tableau with c_R rows of lengths l_i, $i = 1, \dots, c_R$, then the quantum dimension can be explicitly written as:

$$\dim_q R = \prod_{1 \le i < j \le c_R} \frac{[l_i - l_j + j - i]}{[j - i]} \prod_{i=1}^{c_R} \frac{\prod_{v=-i+1}^{l_i - i} [v]_\lambda}{\prod_{v=1}^{l_i} [v - i + c_R]}. \tag{3.28}$$

This gives the Chern–Simons invariant of the unknot in the representation R.

What about other torus knots? When acting with the knot operator (3.21) on the vacuum, we get the set of weights $\rho + n\mu$, where $\mu \in M_\Lambda$. These weights will have representatives in the Weyl alcove $\mathcal{F}_l$, which can be obtained by a series of Weyl reflections. The set of representatives in $\mathcal{F}_l$ will be denoted by $\mathcal{M}(n, \Lambda)$, and it depends on the irreducible representation with highest weight Λ, and on the integer number n. Using the fact that $\rho + n\mu = w(\rho + \xi)$ for some $w \in \mathcal{W}$, we conclude that the Chern–Simons invariant of a torus knot (n, m) can be written as:

$$e^{2\pi i nmh_{\rho+\Lambda}} \sum_{\rho+\xi \in \mathcal{M}(n, \Lambda)} \exp\left[-\frac{i\pi m}{n(k+N)}\xi \cdot (\xi + 2\rho)\right] \mathrm{ch}_\xi\left[-\frac{2\pi i}{k+N}\rho\right]. \quad (3.29)$$

Notice that, since the representatives $\rho + \xi$ live in $\mathcal{F}_l$, the weights ξ can be considered as highest weights for a representation, hence (3.29) makes sense. As an example of this procedure, one can compute the invariant in the fundamental representation. By performing Weyl reflections, one can show that $\mathcal{M}(n, \lambda_1)$ is given by the following weights [60]:

$$\rho + (n - i)\lambda_1 + \lambda_i, \quad i = 1, \ldots, N. \quad (3.30)$$

The computation of the characters is now straightforward (they are just the quantum dimensions of the weights (3.30)), and one finally obtains:

$$W_\square^{(n,m)} = t^{\frac{1}{2}}\lambda^{-\frac{1}{2}} \frac{(\lambda t^{-1})^{\frac{(m-1)(n-1)}{2}}}{t^n - 1} \sum_{\substack{p+i+1=n \\ p,i\geq 0}} (-1)^i t^{-mi+\frac{1}{2}p(p+1)} \frac{\prod_{j=-p}^{i}(\lambda - t^j)}{(i)!(p)!}$$

$$(3.31)$$

This is in fact the unnormalized HOMFLY polynomial of an (n, m) torus knot. If we divide by the vev of the unknot, we find the expression for the HOMFLY polynomial first obtained in [49]. For the trefoil one has for example:

$$W_\square = \frac{1}{q^{\frac{1}{2}} - q^{-\frac{1}{2}}}\left(- 2\lambda^{\frac{1}{2}} + 3\lambda^{\frac{3}{2}} - \lambda^{\frac{5}{2}}\right) + (q^{\frac{1}{2}} - q^{-\frac{1}{2}})(-\lambda^{\frac{1}{2}} + \lambda^{\frac{3}{2}}). \quad (3.32)$$

With more effort one can obtain invariants of torus knots and links in arbitrary representations [60, 61, 63]. For the trefoil in representations with two boxes one finds:

$$W_{\square\square} = \frac{(\lambda - 1)(\lambda q - 1)}{\lambda(q^{\frac{1}{2}} - q^{-\frac{1}{2}})^2 (1 + q)}((\lambda q^{-1})^2(1 - \lambda q^2 + q^3$$
$$- \lambda q^3 + q^4 - \lambda q^5 + \lambda^2 q^5 + q^6 - \lambda q^6))$$
$$W_{\square\hspace{-2pt}\square} = \frac{(\lambda - 1)(\lambda - q)}{\lambda(q^{\frac{1}{2}} - q^{-\frac{1}{2}})^2 (1 + q)}((\lambda q^{-2})^2(1 - \lambda - \lambda q$$
$$+ \lambda^2 q + q^2 + q^3 - \lambda q^3 - \lambda q^4 + q^6)) \quad (3.33)$$

For the Hopf link, one finds:

$$W_{(\square,\square)} = \left(\frac{\lambda^{\frac{1}{2}} - \lambda^{-\frac{1}{2}}}{q^{\frac{1}{2}} - q^{-\frac{1}{2}}}\right)^2 - \lambda^{-1}(\lambda - 1), \quad (3.34)$$

which can be also easily obtained using the skein relations of the HOMFLY polynomial (3.18) together with (3.19).

3.2 Framing dependence

In the above discussion on the correlation functions of Wilson loops we have missed an important ingredient. We mentioned that, in order to define the partition function of Chern–Simons theory at the quantum level, one has to specify a framing of the three-manifold. It turns out that the evaluation of correlation functions like (3.16) also involves a choice of framing of the knots, as Witten discovered in [92]. Since this is important in the duality with topological strings, we will explain it in some detail.

A good starting point to understand the framing is to take Chern–Simons theory with gauge group $U(1)$. This is also useful to understand $U(N)$ versus $SU(N)$ Chern–Simons theory, and to get a concrete feeling of how to deal with correlation functions like (3.16). The Abelian Chern–Simons theory turns out to be extremely simple, since the cubic term in (3.1) drops out, and we are left with a Gaussian theory [79]. The different representations are labelled by integers, and in particular the vevs of Wilson loop operators can be computed exactly. In order to compute them, however, one has to choose a framing for each of the knots $\mathcal{K}_i$. This arises as follows: in evaluating the vev, contractions of the holonomies corresponding to different $\mathcal{K}_i$ produce the following integral:

$$\mathrm{lk}(\mathcal{K}_i, \mathcal{K}_j) = \frac{1}{4\pi} \oint_{\mathcal{K}_i} dx^\mu \oint_{\mathcal{K}_j} dy^\nu \epsilon_{\mu\nu\rho} \frac{(x-y)^\rho}{|x-y|^3}. \tag{3.35}$$

This is in fact a topological invariant, *i.e.* it is invariant under deformations of the knots $\mathcal{K}_i$, $\mathcal{K}_j$, and it is in fact their linking number $\mathrm{lk}(\mathcal{K}_i, \mathcal{K}_j)$. On the other hand, contractions of the holonomies corresponding to the same knot $\mathcal{K}$ involve the integral

$$\phi(\mathcal{K}) = \frac{1}{4\pi} \oint_{\mathcal{K}} dx^\mu \oint_{\mathcal{K}} dy^\nu \epsilon_{\mu\nu\rho} \frac{(x-y)^\rho}{|x-y|^3}. \tag{3.36}$$

This integral is well-defined and finite (see, for example, [42]), and it is called the cotorsion of $\mathcal{K}$. The problem is that the cotorsion is not invariant under deformations of the knot. In order to preserve topological invariance one has to choose another definition of the composite operator $(\int_{\mathcal{K}} A)^2$ by means of a framing. A framing of the knot consists of choosing another knot $\mathcal{K}^f$ around $\mathcal{K}$, specified by a normal vector field n. The cotorsion $\phi(\mathcal{K})$ becomes then

$$\phi_f(\mathcal{K}) = \frac{1}{4\pi} \oint_{\mathcal{K}} dx^\mu \oint_{\mathcal{K}^f} dy^\nu \epsilon_{\mu\nu\rho} \frac{(x-y)^\rho}{|x-y|^3} = \mathrm{lk}(\mathcal{K}, \mathcal{K}^f). \tag{3.37}$$

The correlation function that we obtain in this way is a topological invariant (a linking number) but the price that we have to pay is that our regularization depends on a set of integers $p_i = \mathrm{lk}(\mathcal{K}_i, \mathcal{K}_i^f)$ (one for each knot). The vev (3.16) in the Abelian case can now be computed, after choosing the framings, as follows:

$$\left\langle \prod_i \exp\left(n_i \int_{\gamma_i} A\right) \right\rangle = \exp\left(\frac{\pi i}{k} \sum_i n_i^2 p_i + \frac{\pi i}{k} \sum_{i \neq j} n_i n_j \mathrm{lk}(\mathcal{K}_i, \mathcal{K}_j)\right). \tag{3.38}$$

This regularization is nothing but the 'point-splitting' method familiar in the context of QFT's.

Let us now consider Chern–Simons theory with gauge group $SU(N)$, and suppose that you want to compute a correlation function like (3.16). If you try to do it in perturbation theory, for example, you will find very soon that self-contractions of the holonomies lead to the same kind of ambiguities that we found in the Abelian case, *i.e.* you will have to make a choice of framing for each knot $\mathcal{K}_i$. The only difference is that the self contraction comes with a group factor $\mathrm{Tr}_{R_i}(T_a T_a)$ for each knot $\mathcal{K}_i$, where T_a is a basis of the Lie algebra [42]. The precise result can be better stated as the effect on the correlation function (3.16) under a change of framing, and it says that, under a change of framing of $\mathcal{K}_i$ by p_i units, the vev of the product of Wilson loops changes as follows [92]:

$$W_{(R_1,\dots,R_L)} \to \exp\left[2\pi i \sum_i p_i h_{R_i}\right] W_{(R_1,\dots,R_L)}, \tag{3.39}$$

In this equation, h_R is the conformal weight of the WZW primary field corresponding to the representation R. In (3.11) we labelled R through $\alpha = \rho + \Lambda$, where Λ is the highest weight of R. In fact, one can write (3.11) as

$$h_R = \frac{C_R}{2(k+N)}, \tag{3.40}$$

where $C_R = \mathrm{Tr}_R(T_a T_a)$ is the quadratic Casimir in the representation R. For $SU(N)$, one has

$$C_R^{SU(N)} = N\ell + \kappa_R - \frac{\ell^2}{N}, \tag{3.41}$$

where ℓ is the total number of boxes in the tableau, and

$$\kappa_R = \ell + \sum_i (l_i^2 - 2i l_i). \tag{3.42}$$

We then see that the evaluation of vacuum expectation values of Wilson loop operators in Chern–Simons theory depends on a choice of framing for knots. It turns out that for knots and links in $\mathbb{S}^3$, there is a *standard* or canonical framing, defined by requiring that the self-linking number is zero. The expressions listed in (3.33) and (3.34) are all in the standard framing, and the skein relations for the HOMFLY polynomial produce invariants in the standard framing as well. Once the value of the invariant is known in the standard framing, the value in any other framing specified by nonzero integers p_i can be easily obtained from (3.39).

Let us now consider a $U(N)$ Chern–Simons theory. The $U(1)$ factor decouples from the $SU(N)$ theory, and all the vevs factorize into an $U(1)$ and an $SU(N)$ piece. Representations of $U(N)$ are also labelled by Young tableaux, and they decompose into a representation of $SU(N)$ corresponding to that tableau, and a representation of

U(1) with charge:

$$n = \frac{\ell}{\sqrt{N}},\tag{3.43}$$

where ℓ is the number of boxes in the Young tableau. In order to compute the vevs associated to the U(1) of U(N), one has to take also into account that the coupling constant k is shifted as $k \to k + N$. We then find that the vev of a product of U(N) Wilson loops in representations R_i is given by:

$$W^{U(N)}_{(R_1,\dots,R_L)} = \exp\left(\frac{\pi i}{N(k+N)}\sum_i \ell_i^2 p_i + \frac{\pi i}{N(k+N)}\sum_{i\neq j} \ell_i \ell_j \mathrm{lk}(\mathcal{K}_i, \mathcal{K}_j)\right) W^{SU(N)}_{(R_1,\dots,R_L)},\tag{3.44}$$

where the SU(N) vev is computed in the framing specified by p_i. Notice that, in the case of knots, the SU(N) and U(N) computations differ in a factor which only depends on the choice of framing, while for links the answers also differ in a topological piece involving the linking numbers. The change of framing for vacuum expectation values in the U(N) theory is again governed by (3.39) and (3.40), but now the quadratic Casimir is given by

$$C^{U(N)}_R = N\ell + \kappa_R,\tag{3.45}$$

Notice that the difference between the change of SU(N) and U(N) vevs under the change of framing is consistent with (3.44). In terms of the variables (3.17) we see that U(N) vevs change, under the change of framing, as

$$W_{(R_1,\dots,R_L)} \to q^{\frac{1}{2}\sum_i \kappa_{R_i} p_i} \lambda^{\frac{1}{2}\sum_i \ell_i p_i} W_{(R_1,\dots,R_L)}.\tag{3.46}$$

3.3 Generating functionals for Wilson loops

As we will see, the relation between Chern–Simons theory and string theory involves the vacuum expectation values for arbitrary irreducible representations of U(N), so it is convenient to have a generating functional that encodes all the information about them. We will for simplicity consider the case in which one has just a single knot. We then have to find a suitable basis for the Wilson loop operators. There are two natural basis for the problem: the basis labelled by representations R, and the basis labelled by conjugacy classes $C(\vec{k})$ of the symmetric group. Let U be the holonomy of the gauge connection around the knot $\mathcal{K}$, and consider the operator $\Upsilon_{\vec{k}}(U)$ defined as in (2.27). The vevs of these operators give the "$\vec{k}$-basis" for the vacuum expectation values of the Wilson loops:

$$W_{\vec{k}} = \langle \Upsilon_{\vec{k}}(U)\rangle = \sum_R \chi_R(C(\vec{k})) W_R\tag{3.47}$$

where χ_R are characters of the permutation group S_ℓ in the representation R, and we have used Frobenius formula

$$\mathrm{Tr}_R(U) = \sum_{\vec{k}} \frac{\chi_R(C(\vec{k}))}{z_{\vec{k}}} \Upsilon_{\vec{k}}(U), \tag{3.48}$$

and $z_{\vec{k}}$ has been defined in (2.42). If V is a $U(M)$ matrix (a "source" term), one can define the following operator, which was introduced in [76] and is known sometimes as the Ooguri–Vafa operator:

$$Z(U, V) = \exp\left[\sum_{n=1}^{\infty} \frac{1}{n} \mathrm{Tr}\, U^n \, \mathrm{Tr}\, V^n\right]. \tag{3.49}$$

When expanded, this operator can be written in the k-basis as follows,

$$Z(U, V) = 1 + \sum_{\vec{k}} \frac{1}{z_{\vec{k}}} \Upsilon_{\vec{k}}(U) \Upsilon_{\vec{k}}(V). \tag{3.50}$$

We see that $Z(U, V)$ includes all possible Wilson loop operators $\Upsilon_{\vec{k}}(U)$ associated to a knot $\mathcal{K}$. One can also use Frobenius formula to show that

$$Z(U, V) = \sum_R \mathrm{Tr}_R(U) \mathrm{Tr}_R(V), \tag{3.51}$$

where the sum over representations starts with the trivial one. In $Z(U, V)$ we assume that U is the holonomy of a dynamical gauge field and that V is a source. The vacuum expectation value $Z(V) = \langle Z(U, V)\rangle$ has then information about the vevs of the Wilson loop operators, and by taking its logarithm one can define the connected vacuum expectation values $W_{\vec{k}}^{(c)}$:

$$F_{\mathrm{CS}}(V) = \log Z(V) = \sum_{\vec{k}} \frac{1}{z_{\vec{k}}!} W_{\vec{k}}^{(c)} \Upsilon_{\vec{k}}(V) \tag{3.52}$$

One has, for example:

$$W_{(2,0,\dots)}^{(c)} = \langle(\mathrm{Tr}U)^2\rangle - \langle\mathrm{Tr}U\rangle^2 = W_{\square\square} + W_{\square\!\square} - W_{\square}^2.$$

The free energy $F_{\mathrm{CS}}(V)$, which is a generating functional for connected vevs $W_{\vec{k}}^{(c)}$, will be the relevant object for the duality with topological strings.

4 Chern–Simons theory and large N transitions

4.1 The $1/N$ expansion

As 't Hooft pointed out in [86] (see [21] for a nice review), given a theory with U(N) or SU(N) gauge symmetry one can always perform a $1/N$ expansion of the free energy and the correlation functions. To do that, one writes the Feynman diagrams of the theory as "fatgraphs" or ribbon graphs. The amplitude associated to these ribbon graphs depends on the coupling constant x and on the rank of the gauge group (through its group factor). Let us consider for example the expansion of the free energy. This will involve connected vacuum bubbles with V vertices, E propagators and h loops of internal indices, and therefore will have a factor

$$x^{E-V} N^h = x^{2g-2+h} N^h = x^{2g-2} t^h, \tag{4.1}$$

where $t = Nx$ is the so called '*t Hooft parameter*. In writing this equation we regard the fatgraph as a Riemann surface with holes, *i.e.* each internal loop represents the boundary of a hole, and we used Euler's relation $E - V + h = 2g - 2$. In Fig. 4 we show a fatgraph with $g = 1$ and $h = 9$, and in Fig. 5 the Riemann surface that can be associated to it. We can then write,

$$F^{\mathrm{p}} = \sum_{g=0}^{\infty} \sum_{h=1}^{\infty} F^{\mathrm{p}}_{g,h} x^{2g-2} t^h. \tag{4.2}$$

The superscript p means that this is the perturbative contribution to the free energy. The full free energy may also have a nonperturbative contribution. This is easily

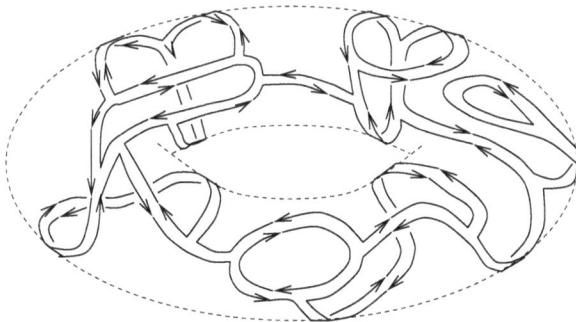

Figure 4. This figure, taken from [77], shows a fatgraph with $h = 9$ and $g = 1$.

seen, in the case of Chern–Simons theory, in (3.4): the free energy has a perturbative contribution coming from the S_ℓ, but there is a nonperturbative contribution due to the one-loop prefactor (which also depends on N, x) and involves one-loop determinants as well as the precise normalization of the path integral. In (4.2) we have written the

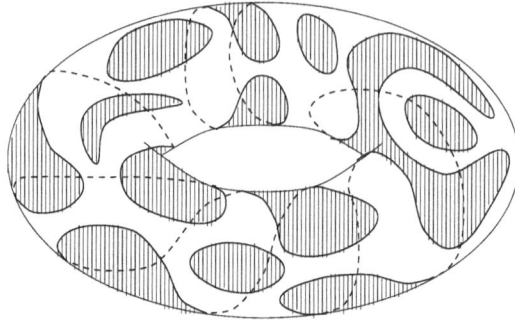

Figure 5. The Riemann surface associated to the fatgraph of the previous figure.

diagrammatic series as an expansion in x around $x = 0$, keeping t fixed. Equivalently, we can regard it as an expansion in $1/N$ for fixed t, and then the N dependence appears as N^{2g-2}. The above expansion can be interpreted as the perturbative expansion of an *open* string theory, where $F^{\mathrm{p}}_{g,h}$ corresponds to some amplitude on a Riemann surface of genus g with h holes like the one depicted in Fig. 5. If we now introduce the function

$$F^{\mathrm{p}}_g(t) = \sum_{h=1}^{\infty} F^{\mathrm{p}}_{g,h} t^h, \qquad (4.3)$$

the total perturbative free energy can be written as

$$F^{\mathrm{p}}(x,t) = \sum_{g=0}^{\infty} x^{2g-2} F^{\mathrm{p}}_g(t), \qquad (4.4)$$

which looks like a *closed* string expansion where t is some modulus of the theory. Notice that in writing (4.3) we have grouped together all open Riemann surfaces with the same bulk topology but with different number of holes, so by "summing over all holes" we "fill up the holes" to produce a closed Riemann surface. This leads to 't Hooft's idea [86] that, given a gauge theory, one should be able to find a string theory interpretation in the way we have described, namely, the fatgraph expansion of the free energy is resummed to give a function of the 't Hooft parameter $F_g(t)$ at every genus that is then interpreted as a closed string amplitude.

We can now ask what is the interpretation of the vacuum expectation values of Wilson loop operators in this context. Using standard large N techniques (as reviewed for example in [21]), it is easy to see that the vevs that have a well-defined behavior in the $1/N$ expansion are the connected vevs $W^{(c)}_{\vec{k}}$ introduced in (3.52). One finds that these vevs admit an expansion of the form,

$$W_{\vec{k}}^{(c)} = \sum_{g=0}^{\infty} W_{g,\vec{k}}(t) x^{2g-2+|\vec{k}|}. \tag{4.5}$$

This can be regarded as an open string expansion, where $W_{g,\vec{k}}(t)$ are interpreted as amplitudes in an open string theory at genus g and with $h = |\vec{k}|$ holes. The vector $\vec{k}$ specifies the winding numbers of the holes around a one-cycle in the target space of the theory, according to the rule we gave in subsection 2.3. We could say that the Wilson loop "creates" a one-cycle in the target space where the boundaries of Riemann surfaces can end, and the generating functional for connected vevs (3.52) is interpreted as the total free energy of an open string, as in (2.24). These open strings shouldn't be confused with the ones that we associated to the expansion (4.2). The open strings underlying (4.5) should be regarded as an open string sector in the closed string theory associated to the resummed expansion (4.4).

This is then the program to interpret gauge theories with U(N) or SU(N) symmetry in terms of a string theory. So far this program has been led to completion in just a few examples. A first example is a class of gauge theories in zero dimensions, the matrix models of Kontsevich, which are equivalent to topological minimal matter in two dimensions coupled to gravity [54], *i.e.* to topological strings in $d < 1$ dimensions. Another example is Yang–Mills theory in two dimensions, which also has a string theory description [41, 22]. Finally, $\mathcal{N} = 4$ supersymmetric Yang–Mills theory is equivalent to type IIB string theory on $\mathbb{S}^5 \times \mathrm{AdS}_5$ [6]. The last example shows very clearly that the target of the string theory is not necessarily the spacetime where the gauge theory lives, and that the string description may need "extra" dimensions. The question we want to address now is the following: is there a string description of Chern–Simons theory? As we will see, at least for Chern–Simons on the three-sphere, the answer is yes. The resulting description provides a very nice realization of 't Hooft ideas, and as we will show, leads to new insights on knot and link invariants[2].

4.2 Chern–Simons theory as an open string theory

In order to give a string theory interpretation of Chern–Simons theory on $\mathbb{S}^3$, a good starting point is to give an open string interpretation to the $1/N$ expansion of the free energy (4.2). This was done by Witten in [95], and we will summarize here the main points of the argument.

First of all, we have to recall that open bosonic strings have a spacetime description in terms of the cubic open string field theory introduced in [90]. The action of this theory is given by

$$S = \frac{1}{g_s} \int \left(\frac{1}{2} \Psi \star Q_{\mathrm{BRST}} \Psi + \frac{1}{3} \Psi \star \Psi \star \Psi \right). \tag{4.6}$$

---

[2]Other attempts to find a string theory interpretation of Chern–Simons theory can be found in [78, 27].

In this equation, Ψ is the string field, $\star$ is the associative, noncommutative product obtained by gluing strings, and the integration is a map $\int : \Psi \to \mathbb{R}$ that involves the gluing of the two halves of the string field (more details can be found in [90]). If we add Chan–Paton factors, the string field is promoted to a $U(N)$ matrix of string fields, and the integration includes Tr. This action has all the information about the spacetime dynamics of open bosonic strings, with or without D-branes. In particular, one can derive the Born-Infeld action describing the dynamics of D-branes from this cubic string field theory (see for example [85]).

Consider now a three-manifold M. The total space of its cotangent bundle T^*M is a noncompact Calabi–Yau manifold. Moreover, it is easy to see that M is a Lagrangian submanifold in T^*M. We can then consider a system of N topological D-branes wrapping M, thus providing Dirichlet boundary conditions for the open strings. We want to obtain a spacetime action describing the dynamics of these topological D-branes. To do this, we can exploit again the analogy between open topological strings and the open bosonic string that we used to define the coupling of topological sigma models to gravity (i.e., that both have a nilpotent BRST operator and an energy-momentum tensor that is Q_{BRST}-exact). Using the fact that both theories have a similar structure, one can argue [95] that the dynamics of topological D-branes in T^*M is governed as well by (4.6). However, one has to work out what is exactly the string field, the $\star$ algebra and so on in the context of topological open strings. It turns out that the string field is simply a $U(N)$ gauge connection A on M, the integration of string functionals becomes ordinary integration of forms on M, and the star product becomes the usual wedge product of forms. We then have the following dictionary:

$$\Psi \to A, \quad Q_{\text{BRST}} \to d$$
$$\star \to \wedge, \quad \int \to \int_M . \tag{4.7}$$

The resulting action (4.6) is then the usual Chern–Simons action, and we have the following relation between the string coupling constant and the Chern–Simons coupling

$$g_s = \frac{2\pi}{k + N}, \tag{4.8}$$

after accounting for the usual shift $k \to k + N$. Notice that, in the open bosonic string, the string field involves an infinite tower of string excitations. For the open topological string, the topological character of the model implies that all excitations are Q-exact (and therefore decouple), except for the lowest lying one, which is a $U(N)$ gauge connection. In other words, the usual reduction to a finite number of degrees of freedom that occurs in topological theories downsizes the string field to a single excitation.

The topological open string theory that we are obtaining has some important differences with the one that we described in section 2. As Witten pointed out in [95], there are no honest worldsheet instantons in this geometry! To be precise, worldsheet instantons whose boundaries lie in M must have zero area, and one would then conclude that the only contributions come from constant maps. A detailed analysis shows however that there are nontrivial worldsheet instantons contributing to the path integral, but they are degenerate and belong to the boundary of the moduli space of holomorphic maps. These degenerate instantons look just like fatgraphs, and in fact they correspond to the Feynman diagrams of the $1/N$ expansion of Chern–Simons theory! In particular, to characterize topologically these degenerate instantons we just need their genus g and number of holes h, which are of course the same ones of the associated fatgraph. There are no winding numbers to specify.

The outcome of this discussion is that, for topological open strings on noncompact Calabi–Yau manifolds of the form T^*M, the dynamics is governed by the usual Chern–Simons action on M. In particular, the coefficient $F^{\mathrm{p}}_{g,h}$ in (4.2) can be interpreted as the free energy of an open string of genus g and h holes propagating on T^*M and with Lagrangian boundary conditions specified by M.

This result can be extended [95], and the more general picture will be extremely useful later on. Consider a Calabi–Yau manifold X together with some Lagrangian submanifolds $M_i \subset X$, with N_i D-branes wrapped over M_i. In this case the topological open strings will have contributions from degenerate holomorphic curves, which are captured by Chern–Simons theories in the way we explained for T^*M, as well as some honest holomorphic curves. As shown in [95], these honest holomorphic curves are open Riemann surfaces whose boundaries are embedded knots inside the three-manifolds M_i and give rise to Wilson loops. Each holomorphic curve with area A ending on M_i over the knot $\mathcal{K}_i$ will contribute $e^{-A} \prod_i \mathrm{Tr} U_{\mathcal{K}_i}$ to the free energy, where $U_{\mathcal{K}_i}$ denotes the holonomy of the Chern–Simons $U(N_i)$ gauge connection A_i around the knot $\mathcal{K}_i$. We can then take into account the contributions of all curves by including the corresponding Chern–Simons theories $S_{\mathrm{CS}}(A_i)$, which account for the degenerate curves, coupled in an appropriate way to the honest holomorphic curves. The spacetime action will then have the form

$$S(A_i) = S_{\mathrm{CS}}(A_i) + F_{\mathrm{ndg}}(U_{\mathcal{K}_i}) \tag{4.9}$$

where

$$F_{\mathrm{ndg}} = \sum_{\mathrm{instantons}} e^{-A} \prod_i \mathrm{Tr} U_{\mathcal{K}_i} \tag{4.10}$$

denotes the contribution of the non-degenerate holomorphic curves, and it is a sum over honest open worldsheet instantons. Notice that all the Chern–Simons theories $S_{\mathrm{CS}}(A_i)$ have the same coupling constant, equal to the string coupling constant. More precisely,

$$\frac{2\pi}{k_i + N_i} = g_s. \tag{4.11}$$

In the action (4.9), the honest holomorphic curves are put "by hand" in F_{ndg}, and in principle one has to solve a nontrivial enumerative problem to find them. Once they are included in the action, the path integral over the Chern–Simons connections will join degenerate instantons to these honest worldsheet instantons: if we have a nondegenerate worldsheet instanton ending on a knot $\mathcal{K}$, it will give rise to a Wilson loop operator in (4.10), and the evaluation of the vacuum expectation value will generate, in the $1/N$ expansion, all possible fatgraphs Γ joined to the knot $\mathcal{K}$, as it is well-known in Chern–Simons perturbation theory in the presence of Wilson loops (see for example [56]). These fatgraphs are interpreted as degenerate instantons. Therefore, the path integral with the action (4.9) will be a sum of contributions coming from partial degenerations of Riemann surfaces, in which a surface $\Sigma_{g,h}$ degenerates to another surface $\Sigma_{g',h'}$ whose boundary ends on a knot $\mathcal{K}$, together with a fatgraph whose external legs end in $\mathcal{K}$ as well. An example of this situation is depicted in Fig. 6, where a disc ends on an unknot, and the fatgraph generated by Chern–Simons perturbation theory gives

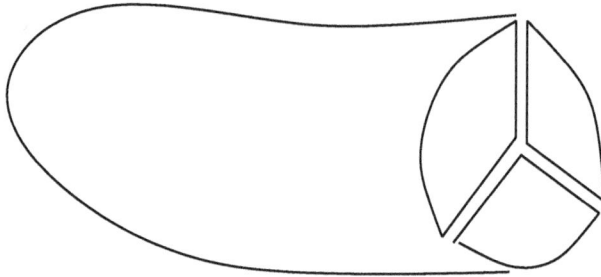

Figure 6. This figure shows a partially degenerated worldsheet instanton of genus $g = 0$ and with $h = 3$ ending on an unknot. The instanton is made out of a honest holomorphic disk and the degenerate piece, which is a fatgraph.

in the end a Riemann surface of $g = 0$ and $h = 3$. This more complicated scenario was explored in [5, 24, 25, 3], and we will provide examples of (4.9) in section 6.

4.3 The conifold transition

We have learned that Chern–Simons theory on $\mathbb{S}^3$ is a topological open string theory on $T^*\mathbb{S}^3$. Notice that the target of the string theory is different from (and has higher dimensionality than) the spacetime of the gauge theory, as in the string description of $\mathcal{N} = 4$ Yang–Mills theory. The next step is to see if there is a *closed* string theory leading to the resummation (4.4). As shown by Gopakumar and Vafa in an important paper [37], the answer is yes.

One way to motivate their result is as follows: since the holes of the Riemann surfaces are due to the presence of D-branes, "filling the holes" to get the closed strings means getting rid of the D-branes. But this is precisely what happens in the AdS/CFT correspondence [6], where type IIB theory in flat space in the presence of D-branes is conjectured to be equivalent to type IIB theory in AdS$_5 \times \mathbb{S}^5$ with no

D-branes. The reason for that is that, at large N, the presence of the D-branes can be traded by a deformation of the background geometry, and the radius of the $\mathbb{S}^5$ is related to the number of D-branes. In other words, we can make the branes disappear if we change the background geometry at the same time. As Gopakumar and Vafa have pointed out, large N dualities relating open and closed strings should involve transitions in the geometry. This reasoning suggests to look for a transition involving the background $T^*\mathbb{S}^3$. It turns out that such a transition is well-known in the physical and the mathematical literature, and it is called the conifold transition (see for example [18]). Let us explain this in detail.

Although we have regarded $T^*\mathbb{S}^3$ as the total space of the cotangent space bundle of the three-sphere, this background can be also regarded as the deformed conifold geometry, which is usually described by the algebraic equation

$$\sum_{\mu=1}^{4} \eta_\mu^2 = a. \tag{4.12}$$

To see this equivalence, let us write $\eta_\mu = x_\mu + i p_\mu$, where x_μ, p_μ are real coordinates. We find the two equations

$$\sum_{\mu=1}^{4} (x_\mu^2 - p_\mu^2) = a,$$

$$\sum_{\mu=1}^{4} x_\mu p_\mu = 0. \tag{4.13}$$

The first equation indicates that the locus $p_\mu = 0$, $\mu = 1, \dots, 4$, describes a sphere $\mathbb{S}^3$ of radius $R^2 = a$, and the second equation shows that the p_μ are coordinates for the cotangent space. Therefore, (4.12) is nothing but $T^*\mathbb{S}^3$.

It is useful to rewrite the deformed conifold in yet another way. Introduce the following complex coordinates:

$$x = \eta_1 + i\eta_2, \qquad v = i(\eta_3 - i\eta_4),$$
$$u = i(\eta_3 + i\eta_4), \qquad y = \eta_1 - i\eta_2. \tag{4.14}$$

The deformed conifold can be now written as

$$xy = uv + a. \tag{4.15}$$

Notice that in this parameterization the geometry has a $\mathbb{T}^2$ fibration

$$x, y, u, v \to xe^{i\theta_a}, ye^{-i\theta_a}, ue^{i\theta_b}, ve^{-i\theta_b} \tag{4.16}$$

where the θ_a and θ_b actions above can be taken to generate the $(1, 0)$ and $(0, 1)$ cycles of the $\mathbb{T}^2$. The $\mathbb{T}^2$ fiber can degenerate to $\mathbb{S}^1$ by collapsing one of its one-cycles. In the equation above, for example, the $U(1)_a$ action fixes $x = 0 = y$ and therefore fails to generate a circle there. In the total space, the locus where this happens, i.e. the

$x = 0 = y$ subspace of X, is a cylinder $uv = -a$. Similarly, the locus where the other circle collapses, $u = 0 = v$, gives another cylinder $xy = a$. Therefore, we can regard the whole geometry as a $\mathbb{T}^2 \times \mathbb{R}$ fibration over $\mathbb{R}^3$: if we define $z = uv$, the $\mathbb{R}^3$ of the base is given by $\mathrm{Re}(z)$ and the axes of the two cylinders. The fiber is given by the circles of the two cylinders, and by $\mathrm{Im}(z)$. It is very useful to represent the above geometry by depicting the singular loci of the torus action in the base $\mathbb{R}^3$. The loci where the cycles of the torus collapse, which are cylinders, project to lines in the base space. Notice that $\mathbb{S}^3$ can be regarded as a torus fibration over an interval, with singular loci at the endpoints. In Fig. 7, the three-sphere of the deformed conifold

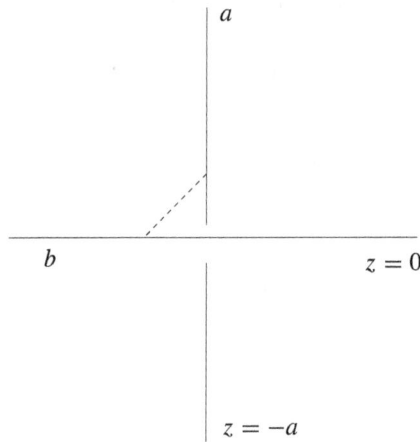

Figure 7. This figure represents $T^*\mathbb{S}^3$, regarded as a $\mathbb{T}^2 \times \mathbb{R}$ fibration of $\mathbb{R}^3$. Two of the directions represent the axes of the two cylinders, and the third direction represents the real axis of the z-plane.

geometry is represented by a dashed line in the z-plane between $z = 0$ and $z = -a$, together with the θ_a and the θ_b circles that degenerate over the endpoints.

The conifold singularity appears when $a = 0$ and the three-sphere collapses. This is described by the equation:

$$xy = uv. \tag{4.17}$$

In algebraic geometry, singularities can be avoided in two ways, in general. The first way is to deform the complex geometry. This leads in our case to the deformed conifold (4.12). The other way is to resolve the singularity, for example by performing a blow up, and this leads to the resolved conifold geometry (see for example [18]). The resolution of the geometry can be explained as follows. When $a = 0$, (4.15) says that $xy = uv$. We can solve (4.17) by setting

$$x = \lambda v, \quad u = \lambda y \tag{4.18}$$

where λ is regarded as an inhomogeneous coordinate in $\mathbb{P}^1$. The space described by the complex coordinates x, y, λ, u, v together with the relations (4.18) is the resolved conifold, and it turns out to be the bundle $\mathcal{O}(-1) \oplus \mathcal{O}(-1) \rightarrow \mathbb{P}^1$, as one can see from (4.18) [18]. To make contact with the toric description given in (2.35), we put $x = x_1 x_2$, $y = x_3 x_4$, $u = x_1 x_3$ and $v = x_2 x_4$. We then see that $\lambda = x_1/x_4$ is the inhomogeneous coordinate for the $\mathbb{P}^1$ described in (2.35) by $|x_1|^2 + |x_4|^2 = s$. It is instructive to represent the resolved conifold by solving the constraint (2.35) in the first octant of $\mathbb{R}^3$, and depicting the fixed point locus of the isometries above. In terms of the coordinates $x_1, \ldots, x_4$, the $\mathbb{T}^2$ action (4.16) is given by

$$x_1, x_2, x_3, x_4 \rightarrow e^{i(\theta_a + \theta_b)} x_1, \, e^{-i\theta_a} x_2, \, e^{-i\theta_b} x_3, x_4, \tag{4.19}$$

and the fixed loci are depicted in Fig. 8. In the conifold transition, the three-sphere of

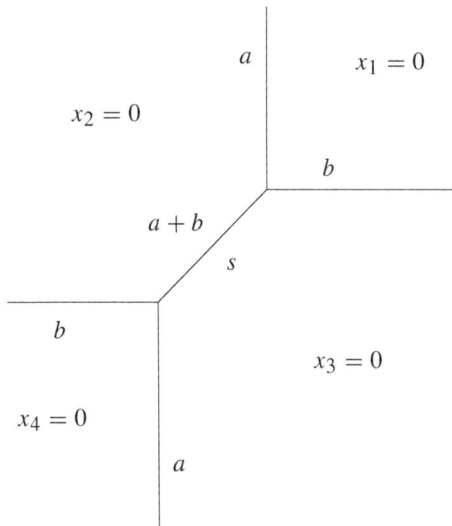

Figure 8. This figure represents the resolved conifold $\mathcal{O}(-1) \oplus \mathcal{O}(-1) \rightarrow \mathbb{P}^1$ and the fixed point loci of the $\mathbb{T}^2$ action.

the deformed conifold shrinks to zero size as a goes to zero, and then a two-sphere of size s blows up giving the resolved conifold.

We know that Chern–Simons theory is an open topological string on the deformed conifold geometry with N topological D-branes wrapping the three-sphere. The conjecture of Gopakumar and Vafa is that at large N the D-branes induce a conifold transition in the background geometry, so that we end up with the resolved conifold and no D-branes. But in the absence of D-branes that enforce boundary conditions we just have a theory of closed topological strings. Therefore, *Chern–Simons theory on* $\mathbb{S}^3$ *is equivalent to closed topological string theory on the resolved conifold.*

This conjecture has been proved by embedding the duality in type II superstring theory [89] and lifting it to M-theory [1, 10], and more recently a worldsheet derivation has been presented in [77]. In the remaining of this section, we will give evidence for the conjecture at the level of the free energy.

4.4 First test of the duality: the free energy

A nontrivial test of the duality advocated by Gopakumar and Vafa is to verify that the free energy of U(N) Chern–Simons theory on the sphere agrees with the free energy of closed topological strings on the resolved conifold. The partition function of CS with gauge group U(N) on the sphere is a slight modification of (3.14):

$$
Z = \frac{1}{(k+N)^{N/2}} \prod_{\alpha>0} 2 \sin\left(\frac{\pi(\alpha \cdot \rho)}{k+N}\right). \tag{4.20}
$$

and differs from it in an overall factor $N^{1/2}/(k+N)^{1/2}$ which is the partition function for the U(1) factor (recall that $U(N) = U(1) \otimes SU(N)/\mathbb{Z}_N$). Using the explicit description of the positive roots of $SU(N)$, one gets

$$
F = \log Z = -\frac{N}{2} \log(k+N) + \sum_{j=1}^{N-1}(N-j) \log\left[2\sin\frac{\pi j}{k+N}\right]. \tag{4.21}
$$

We can now write the sin as

$$
\sin \pi z = \pi z \prod_{n=1}^{\infty}\left(1 - \frac{z^2}{n^2}\right), \tag{4.22}
$$

and we find that the free energy is the sum of two pieces. One of them is the nonperturbative piece:

$$
F^{\mathrm{np}} = -\frac{N^2}{2} \log(k+N) + \frac{1}{2}N(N-1)\log 2\pi + \sum_{j=1}^{N-1}(N-j)\log j, \tag{4.23}
$$

and the other piece is the perturbative one:

$$
F^{\mathrm{p}} = \sum_{j=1}^{N-1}(N-j) \sum_{n=1}^{\infty} \log\left[1 - \frac{j^2 g_s^2}{4\pi^2 n^2}\right], \tag{4.24}
$$

where g_s corresponds to the open string coupling constant and it is given by (4.8). To see that (4.23) corresponds to the nonperturbative piece of the free energy, we notice that the volume of $U(N)$ can be written as (see for example [77]):

$$\text{vol}(U(N)) = \frac{(2\pi)^{\frac{1}{2}N(N+1)}}{G_2(N+1)} \tag{4.25}$$

where $G_2(z)$ is the Barnes function, defined by

$$G_2(z+1) = \Gamma(z)G_2(z), \quad G_2(1) = 1. \tag{4.26}$$

It is now easy to see that

$$F^{\text{np}} = \log \frac{(2\pi g_s)^{\frac{1}{2}N^2}}{\text{vol}(U(N))} \tag{4.27}$$

so it is given by the log of (3.6), where $A^{(c)}$ is in this case the trivial flat connection. Therefore, F^{np} is the log of the prefactor associated to the normalization of the path integral, which is not captured by Feynman diagrams.

Let us work out the perturbative piece (4.24). By expanding the log, using that $\sum_{n=1}^{\infty} n^{-2k} = \zeta(2k)$, and the formula

$$\sum_{j=1}^{N-1} j^{2k} = -\frac{N^{2k}}{2} + \sum_{l=0}^{k} \binom{2k+1}{2l} \frac{B_{2l}}{2k+1} N^{2k+1-2l} \tag{4.28}$$

we find that (4.24) can be written as

$$F^{\text{p}} = \sum_{g=0}^{\infty} \sum_{h=2}^{\infty} F_{g,h}^{\text{p}} g_s^{2g-2+h} N^h, \tag{4.29}$$

where $F_{g,h}^{\text{p}}$ is given by:

$$\begin{aligned}
F_{0,h}^{\text{p}} &= -\frac{2\zeta(h-2)}{(2\pi)^{h-2}(h-2)h(h-1)}, \\
F_{1,h}^{\text{p}} &= \frac{1}{6}\frac{\zeta(h)}{(2\pi)^h h}, \\
F_{g,h}^{\text{p}} &= 2\frac{\zeta(2g-2+h)}{(2\pi)^{2g-2+h}}\binom{2g-3+h}{h}\frac{B_{2g}}{2g(2g-2)}, \quad g \geq 2.
\end{aligned} \tag{4.30}$$

This gives the contribution of connected diagrams with two loops and beyond to the free energy of Chern–Simons on the sphere, so we can write

$$\sum_{l=1}^{\infty} S_l(N)x^l = \sum_{g=0}^{\infty} \sum_{h=2}^{\infty} (-1)^{g-1+h/2} F_{g,h} x^{2g-2+h} N^h, \tag{4.31}$$

where x is given by (3.5), and we have explicitly indicated the dependence of S_l on N. Notice that the only nonzero $F_{g,h}$ have h even. One can check that the $F_{g,h}$ that we have obtained in (4.30) are in agreement with known results of perturbative Chern–Simons theory on the sphere (see for example [12, 69]). The nonperturbative piece also admits an expansion that can be easily worked out from the asymptotics of the Barnes function [78, 77]. One finds:

$$F^{\mathrm{np}} = \frac{N^2}{2}\left(\log(Ng_s) - \frac{3}{2}\right) - \frac{1}{12}\log N + \zeta'(-1) + \sum_{g=2}^{\infty} \frac{B_{2g}}{2g(2g-2)} N^{2-2g}. \quad (4.32)$$

So far, what we have uncovered is the open string expansion of Chern–Simons theory, which is (order by order in x) determined by the perturbative expansion. In order to find a closed string interpretation, we have to sum over the holes, as in (4.3). Define the 't Hooft parameter t as:

$$t = ig_s N = xN, \quad (4.33)$$

then

$$F_g^{\mathrm{p}}(t) = \sum_{h=1}^{\infty} F_{g,h}^{\mathrm{p}}(-it)^h. \quad (4.34)$$

We will now focus on $g \geq 2$. To perform the sum explicitly, we write again the ζ function as $\zeta(2g - 2 + 2p) = \sum_{n=1}^{\infty} n^{2-2g-2p}$, and use the binomial series,

$$\frac{1}{(1-z)^q} = \sum_{n=0}^{\infty} \binom{q+n-1}{n} z^n, \quad (4.35)$$

to obtain

$$F_g^{\mathrm{p}}(t) = \frac{(-1)^g |B_{2g} B_{2g-2}|}{2g(2g-2)(2g-2)!} + \frac{B_{2g}}{2g(2g-2)} \sum_{n\in\mathbb{Z}}' \frac{1}{(-it+2\pi n)^{2g-2}}, \quad (4.36)$$

where $'$ means that we omit $n = 0$. Now we notice that, if we write

$$F^{\mathrm{np}} = \sum_{g=0}^{\infty} F_g^{\mathrm{np}}(t) g_s^{2g-2} \quad (4.37)$$

then for, $g \geq 2$,

$$F_g^{\mathrm{np}}(t) = \frac{B_{2g}}{2g(2g-2)}(-it)^{2-2g},$$

which is precisely the $n = 0$ term missing in (4.36). We then define:

$$F_g(t) = F_g^{\mathrm{p}}(t) + F_g^{\mathrm{np}}(t). \quad (4.38)$$

Finally, since

$$\sum_{n\in\mathbb{Z}} \frac{1}{n+z} = \frac{2\pi i}{1-e^{-2\pi i z}}, \quad (4.39)$$

by taking derivatives w.r.t. z we can write

$$F_g(t) = \frac{(-1)^g |B_{2g} B_{2g-2}|}{2g(2g-2)(2g-2)!} + \frac{|B_{2g}|}{2g(2g-2)!} \mathrm{Li}_{3-2g}(\mathrm{e}^{-t}), \qquad (4.40)$$

again for $g \geq 2$. If we now compare (4.40) to (2.33), we see that it has *precisely* the structure of the free energy of a closed topological string, with $n_1^0 = 1$, and the rest of the Gopakumar–Vafa invariants being zero. Also, from the first term, which gives the contribution of the constant maps, we find that $\chi(X) = 2$. In fact, (4.40) is the F_g amplitude of the resolved conifold. One can also work out the expressions for $F_0(t)$ and $F_1(t)$ and find agreement with the corresponding results for the resolved conifold [37]. This is a remarkable check of the conjecture.

5 Wilson loops and large N transitions

5.1 Incorporating Wilson loops

How do we incorporate Wilson loops in the large N duality for Chern–Simons theory? As we discussed in the previous section, once one has a closed string description of the $1/N$ expansion, Wilson loops are related to the open string sector in the closed string geometry. Since the string description involves topological strings, it is natural to assume that Wilson loops are going to be described by open topological strings in the resolved conifold, and this means that we need a Lagrangian submanifold specifying boundary conditions.

These issues were addressed in an important paper by Ooguri and Vafa [76]. In order to give boundary conditions for the open strings in the resolved conifold, Ooguri and Vafa constructed a Lagrangian submanifold $\widehat{\mathcal{C}}_{\mathcal{K}}$ in $T^*\mathbb{S}^3$ for any knot $\mathcal{K}$ in $\mathbb{S}^3$. This Lagrangian is rather canonical, and it is called the conormal bundle of $\mathcal{K}$. The details are as follows: suppose that the knot is parameterized by a curve $q(s)$, where $s \in [0, 2\pi)$, for example. The conormal bundle of $\mathcal{K}$ is then the space

$$\widehat{\mathcal{C}}_{\mathcal{K}} = \left\{ (q(s), p) \in T^*\mathbb{S}^3 \ \Big|\ \sum_i p_i \dot{q}_i = 0, \ 0 \leq s < 2\pi \right\} \qquad (5.1)$$

where p_i are coordinates for the cotangent bundle, and $\dot{q}_i$ denote the derivatives w.r.t. s. This space is an $\mathbb{R}^2$-fibration of the knot itself, where the fiber on the point $q(s)$ is given by the two-dimensional subspace of $T_q^*\mathbb{S}^3$ of planes orthogonal to $\dot{q}(s)$. $\widehat{\mathcal{C}}_{\mathcal{K}}$ has in fact the topology of $\mathbb{S}^1 \times \mathbb{R}^2$, and intersects $\mathbb{S}^3$ along the knot $\mathcal{K}$.

One can now consider, together with the N branes wrapping $\mathbb{S}^3$, a set of M probe branes wrapping $\widehat{\mathcal{C}}_{\mathcal{K}}$, and study the effective theory that one obtains in this way. On the N branes wrapping $\mathbb{S}^3$ we have $\mathrm{U}(N)$ Chern–Simons theory. But the strings stretched between the N branes and the M branes give an extra state in topological string field theory, which turns out to be a massless complex scalar field ϕ in the bifundamental representation $(N, \overline{M})$, and living in the intersection of the two branes, $\mathcal{K}$. If A denotes the U(N) gauge connection on $\mathbb{S}^3$, and $\widetilde{A}$ denotes the U(M) gauge connection on $\widehat{\mathcal{C}}_{\mathcal{K}}$, the action for the scalar is given by

$$\oint_{\mathcal{K}} \operatorname{Tr} \bar{\phi} \, D\phi, \tag{5.2}$$

where $D = d + A - \widetilde{A}$. Here we regard $\widetilde{A}$ as a source. We can now proceed to integrate out ϕ [76]. This is just a one loop computation giving

$$\exp\left[-\log \det D\right] = \exp\left[-\operatorname{Tr} \log\left(U^{-\frac{1}{2}} \otimes V^{\frac{1}{2}} - U^{\frac{1}{2}} \otimes V^{-\frac{1}{2}}\right)\right]. \tag{5.3}$$

In this equation, U, V are the holonomies of A, $\widetilde{A}$ around the knot $\mathcal{K}$. To obtain this equation, we have diagonalized A, $\widetilde{A}$ and taken into account that

$$\log \det\left[\frac{d}{ds} + i\theta\right] = \sum_{n=-\infty}^{\infty} \log(n + \theta) = \log \sin(\pi\theta) + \text{const.}, \tag{5.4}$$

where use has been made of (4.22). In this way we obtain the effective action for the A field

$$S_{\text{CS}}(A) + \sum_{n=1}^{\infty} \frac{1}{n} \operatorname{Tr} U^n \operatorname{Tr} V^{-n} \tag{5.5}$$

where $S_{\text{CS}}(A)$ is the Chern–Simons action for A associated to the N branes on the three-sphere [3]. Therefore, in the presence of the probe branes, the action gets deformed by the Ooguri–Vafa operator that we introduced in (3.49). Since we are regarding the M branes as a probe, the holonomy V is an arbitrary source, and we will put $V^{-1} \to V$.

Let us now follow this system through the geometric transition. The N branes disappear, and the background geometry becomes the resolved conifold. However, the M probe branes are still there. The first conjecture of Ooguri and Vafa is that these branes are wrapping a Lagrangian submanifold $\mathcal{C}_{\mathcal{K}}$ of $\mathcal{O}(-1) \oplus \mathcal{O}(-1) \to \mathbb{P}^1$ that can be obtained from $\widehat{\mathcal{C}}_{\mathcal{K}}$ through the geometric transition. The final outcome is therefore the existence of a map

$$\{\text{knots in } \mathbb{S}^3\} \to \{\text{Lagrangian submanifolds in } \mathcal{O}(-1) \oplus \mathcal{O}(-1) \to \mathbb{P}^1\} \tag{5.6}$$

which sends

$$\mathcal{K} \to \mathcal{C}_{\mathcal{K}}. \tag{5.7}$$

Moreover, one has $b_1(\mathcal{C}_{\mathcal{K}}) = 1$. This conjecture is clearly well-motivated in the physics of the problem, and some aspects of the map (5.6) are already well understood: in [76] Ooguri and Vafa constructed $\mathcal{C}_{\mathcal{K}}$ explicitly when $\mathcal{K}$ is the unknot, and [63] proposed Lagrangian submanifolds for certain algebraic knots and links (including torus knots). Taubes has generalized this proposal [84] and constructed in detail a

---

[3] In the above equation we have factored out a contribution involving the U(1) pieces of U(N), U(M). These can be reabsorbed in a change of framing.

map from a wide class of knots to Lagrangian submanifolds in the resolved conifold. Later on we will discuss the case of the unknot.

The resulting Lagrangian submanifold $\mathcal{C}_{\mathcal{K}}$ in the resolved geometry provides boundary conditions for open strings, and therefore it gives the open string sector that is needed in order to extend the large N duality to Wilson loops. The second conjecture of [76] states that the free energy of open topological strings (2.24) with boundary conditions specified by $\mathcal{C}_{\mathcal{K}}$ is identical to the free energy of the deformed Chern–Simons theory with action (5.5), which is nothing but (3.52):

$$F_{\text{string}}(V) = F_{\text{CS}}(V). \tag{5.8}$$

Notice that, since $b_1(\mathcal{C}_{\mathcal{K}}) = 1$, the topological sectors of maps with positive winding numbers correspond to vectors $\vec{k}$ labelling the connected vevs, and one finds

$$i^{|\vec{k}|} \sum_{g=0}^{\infty} F_{g,\vec{k}}(t) g_s^{2g-2+|\vec{k}|} = \frac{1}{\prod_j j^{k_j}} W_{\vec{k}}^{(c)}. \tag{5.9}$$

Of course, $F_{g,\vec{k}}(t)$ are (up to constants) the functions of the 't Hooft parameter that appeared in (4.5). The variable λ defined in (3.17) that appears in the Chern–Simons invariants of knots and links is related to the 't Hooft parameter through

$$\lambda = e^t.$$

Notice that the Chern–Simons invariants are labelled by vectors $\vec{k}$, therefore they only give rise to positive winding numbers in the string side. At the same time, they involve both positive and negative powers of λ, while in the string side we only have negative powers. Therefore, in order to make (5.8) precise, we further need some sort of analytic continuation that gives an appropriate matching of the variables. In the cases where both sides of the equality are known, there is such an analytic continuation, and it is expected that this will be the case in more general situations. Up to these subtleties, (5.9) tells us that the Chern–Simons invariant in the left-hand side is a generating function for open Gromov–Witten invariants, for all degrees and genera, but with fixed boundary data (*i.e.* the number of holes and the winding numbers). To extract a particular open Gromov–Witten invariant from the Chern–Simons invariant, we consider the connected vev labelled by the vector $\vec{k}$ associated to the boundary data, we write it in terms of $\lambda = e^t$ and $q = e^x$, and then we expand the result in powers of $x = ig_s$. The coefficients of this series, which are polynomials in λ, are then equated to the generating function of open Gromov–Witten invariants at fixed genus g.

We should mention that, although we have focused on knots for simplicity, all these results can be extended to links, as shown in [63].

5.2 BPS invariants for open strings from knot invariants

In section 2 we have learned that Gromov–Witten invariants can be written in terms of integer, or BPS invariants. We will now find what is the precise relation between Chern–Simons invariants and these integer invariants. This will lead to some surprising structure results for the Chern–Simons invariants of knots.

The first step is to introduce the so-called f-polynomials, through the relation:

$$F_{CS}(V) = \sum_{n=1}^{\infty} \sum_{R} \frac{1}{n} f_R(q^n, \lambda^n) \mathrm{Tr}_R V^n. \tag{5.10}$$

As shown in [61, 62], the f_R polynomials are completely determined by this equation, and can be expressed in terms of the usual vevs of Wilson loops W_R by:

$$f_R(q, \lambda) = \sum_{d,m=1}^{\infty} (-1)^{m-1} \frac{\mu(d)}{dm} \sum_{\vec{k}_1,\dots,\vec{k}_m} \sum_{R_1,\dots,R_m} \chi_R \left(C\left(\left(\sum_{j=1}^{l} \vec{k}_j \right)_d \right) \right)$$

$$\times \prod_{j=1}^{m} \frac{\chi_{R_j}(C(\vec{k}_j))}{z_{\vec{k}_j}} W_{R_j}(q^d, \lambda^d), \tag{5.11}$$

where $\vec{k}_d$ is defined as follows: $(\vec{k}_d)_{di} = k_i$ and has zero entries for the other components. Therefore, if $\vec{k} = (k_1, k_2, \dots)$, then

$$\vec{k}_d = (0, \dots, 0, k_1, 0, \dots, 0, k_2, 0, \dots)$$

where k_1 is in the d-th entry, k_2 is in the $2d$-th entry, and so on. The sum over $\vec{k}_1, \dots, \vec{k}_m$ is over all vectors with $|\vec{k}_j| > 0$. In (5.11), $\mu(d)$ denotes the Moebius function. Recall that the Moebius function is defined as follows: if d has the prime decomposition $d = \prod_{i=1}^{a} p_i^{m_i}$, then $\mu(d) = 0$ if any of the m_i is greater than one. If all $m_i = 1$ (i.e. d is square-free) then $\mu(d) = (-1)^a$. Some examples of (5.11) are

$$f_\square(q, \lambda) = W_\square(q, \lambda),$$

$$f_{\square\square}(q, \lambda) = W_{\square\square}(q, \lambda) - \frac{1}{2} \left(W_\square(q, \lambda)^2 + W_\square(q^2, \lambda^2) \right),$$

$$f_{\Box\Box}(q, \lambda) = W_{\Box\Box}(q, \lambda) - \frac{1}{2} \left(W_\square(q, \lambda)^2 - W_\square(q^2, \lambda^2) \right). \tag{5.12}$$

Therefore, given a representation R with ℓ boxes, the polynomial f_R is given by W_R, plus some "lower order corrections" that involve $W_{R'}$ where R' has $\ell' < \ell$ boxes. One can then easily compute these polynomials starting from the results for vevs of Wilson loops in Chern–Simons theory. Although we are calling f_R polynomials, they are not, strictly speaking. In fact, it follows from the multicovering/bubbling formula that the f_R have the structure

$$f_R(q, \lambda) = \frac{P_R(q, \lambda)}{q^{\frac{1}{2}} - q^{-\frac{1}{2}}}. \tag{5.13}$$

But we can be more precise about the structure of f_R. As shown in [63], one can write the f_R in terms of even more basic objects, that were denoted by $\widehat{f}_R$. The precise relation between them is

$$f_R = \sum_{R'} M_{RR'} \widehat{f}_{R'} \tag{5.14}$$

where the sum in R' runs over all representations with the same number of boxes than R, and the matrix $M_{RR'}$ is given by:

$$M_{RR'} = \sum_{R''} C_{RR'R''} S_{R''}(q). \tag{5.15}$$

In this equation, $C_{RR'R''}$ are the Clebsch–Gordon coefficients of the symmetric group. They can be explicitly written in terms of characters [34]:

$$C_{RR'R''} = \sum_{\vec{k}} \frac{|C(\vec{k})|}{\ell!} \chi_R(C(\vec{k})) \chi_{R'}(C(\vec{k})) \chi_{R''}(C(\vec{k})). \tag{5.16}$$

The $S_R(q)$ are monomials defined as follows. If R is a hook or L-shaped representation of the form

$$\text{(5.17)}$$

with ℓ boxes in total, and $\ell - d$ boxes in the first row, then

$$S_R(q) = (-1)^d q^{-\frac{\ell-1}{2}+d}, \tag{5.18}$$

and $S_R(q) = 0$ for the rest of the representations. For example, for the case of two boxes one has that $S_{\square\square}(q) = q^{-1/2}$ and $S_{\square\atop\square}(q) = -q^{1/2}$, while for $\ell = 3$ one has

$$S_{\square\square\square}(q) = q^{-1}, \quad S_{\square\atop\square}(q) = -1, \quad S_{\square\atop\square\atop\square}(q) = q. \tag{5.19}$$

The square matrix $M_{RR'}$ that relates f_R to $\widehat{f}_R$ is invertible. This can be easily seen: define the polynomials $P_{\vec{k}}(q)$, labelled by conjugacy classes, as the character transforms of the monomials $S_R(q)$:

$$P_{\vec{k}}(q) = \sum_R \chi_R(C(\vec{k})) S_R(q). \tag{5.20}$$

It can be seen that

$$P_{\vec{k}}(q) = \frac{\prod_j (q^{-\frac{j}{2}} - q^{\frac{j}{2}})^{k_j}}{q^{-\frac{1}{2}} - q^{\frac{1}{2}}}. \tag{5.21}$$

In terms of these polynomials, the matrix $M_{RR'}$ is written as

$$M_{RR'} = \sum_{\vec{k}} \frac{1}{z_{\vec{k}}} \chi_R(C(\vec{k})) \chi_{R'}(C(\vec{k})) P_{\vec{k}}(q), \qquad (5.22)$$

and using the orthogonality of the characters one can see that

$$M_{RR'}^{-1} = \sum_{\vec{k}} \frac{1}{z_{\vec{k}}} \chi_R(C(\vec{k})) \chi_{R'}(C(\vec{k}))(1/P_{\vec{k}}(q)). \qquad (5.23)$$

Therefore, one can obtain the polynomials $\widehat{f}_R$ from the f_R, i.e. one can obtain the polynomials $\widehat{f}_R$ from the knot invariants of Chern–Simons theory. The claim is now that the $\widehat{f}_R$ are generating functions for the BPS invariants $N_{R,g,Q}$ that were introduced in (2.40). More precisely, one has

$$\widehat{f}_R(q, \lambda) = \sum_{g \geq 0} \sum_Q N_{R,g,Q} \left(q^{-\frac{1}{2}} - q^{\frac{1}{2}}\right)^{2g-1} \lambda^Q \qquad (5.24)$$

Therefore, this gives a very precise way to compute the BPS invariants $N_{R,g,Q}$ from Chern–Simons theory: compute the usual vevs W_R, extract f_R through the relation (5.11), compute $\widehat{f}_R$, and expand them as in (5.24).

We would like to point out two important things. First, the fact that one can extract the integer invariants $N_{R,g,Q}$ from Chern–Simons theory in the way we have just described is by no means obvious and constitutes a strong check of the large N duality between Chern–Simons theory and topological strings. We will see examples of this in the next subsection. Another important comment is that the statement that $\widehat{f}_R$ have the structure predicted in (5.24) is equivalent to the multicovering/bubbling formula for open string invariants (2.38) (more precisely, it is equivalent to the strong version of this formula, which says that in addition to (2.38) one can write the $n_{\vec{k},g,Q}$ in terms of integer $N_{R,g,Q}$ through (2.40)). This is easily seen by noticing that, according to (5.14) and (5.15), f_R is given by

$$f_R(q, \lambda) = \sum_{g \geq 0} \sum_Q \sum_{R',R''} C_{R\,R'\,R''} S_{R'}(q) N_{R'',g,Q} \left(q^{-\frac{1}{2}} - q^{\frac{1}{2}}\right)^{2g-1} \lambda^Q. \qquad (5.25)$$

If we now write the exponent in the r.h.s. of (5.10) in the $\vec{k}$ basis, it is easy to see that one obtains precisely (2.38), after making use of (5.21).

The physical origin of the structure of f_R (and therefore of the multicovering/bubbling formula for open Gromov–Witten invariants) can be easily understood in physical terms. We will give a short account, referring the reader to [63] for more details. In the D-brane approach to open string instantons, one regards the open Riemann surfaces ending on a Lagrangian submanifold as D2-branes ending on M D4-branes wrapping the Lagrangian submanifold. Following the approach of [36], we have to study the moduli space of D2-branes ending on D4-branes. This moduli space is the product of three factors: the moduli of Abelian gauge fields on the worldvolume of the D2 brane, the moduli of geometric deformations of the D2's in the ambient space, and finally the Chan–Paton factors associated to the boundaries of the D2 which appear in the D4 as magnetic charges [76]. If the D2's are genus g surfaces with ℓ holes in the relative cohomology class labelled by Q, the moduli space of Abelian gauge fields gives rise to the Jacobian $J_{g,\ell} = \mathbb{T}^{2g+\ell-1}$, and the moduli of geometric deformations will be a manifold $\mathcal{M}_{g,\ell,Q}$. Finally, for the Chan–Paton degrees of freedom we get a factor of F (the fundamental representation of $SU(M)$) from each hole. The Hilbert space is obtained by computing the cohomology of these moduli, and we obtain

$$F^{\otimes \ell} \otimes H^*(J_{g,\ell}) \otimes H^*(\mathcal{M}_{g,\ell,Q}). \tag{5.26}$$

An important point is that this Hilbert space is associated with the moduli space of ℓ *distinguished* holes, which is not physical, and we have to mod out by the action of the permutation group S_ℓ. We can factor out the cohomology of the Jacobian $\mathbb{T}^{2g}$ of the "bulk" Riemann surface, $H^*(\mathbb{T}^{2g})$, since the permutation group does not act on it. The projection onto the symmetric piece can be easily done using the Clebsch–Gordon coefficients $C_{R\,R'\,R''}$ of the permutation group S_ℓ [34]:

$$\mathrm{Sym}\big(F^{\otimes \ell} \otimes H^*((\mathbb{S}^1)^{\ell-1}) \otimes H^*(\mathcal{M}_{g,\ell,Q})\big)$$
$$= \sum_{R\,R'\,R''} C_{R\,R'\,R''} \mathbb{S}_R(F^{\otimes \ell}) \otimes \mathbb{S}_{R'}(H^*((\mathbb{S}^1)^{\ell-1})) \otimes \mathbb{S}_{R''}(H^*(\mathcal{M}_{g,\ell,Q})) \tag{5.27}$$

where $\mathbb{S}_R$ is the Schur functor that projects onto the corresponding subspace. The space $\mathbb{S}_R(F^{\otimes \ell})$ is nothing but the vector space underlying the irreducible representation R of $SU(M)$. $\mathbb{S}_{R'}(H^*((\mathbb{S}^1)^{\ell-1}))$ gives the hook Young tableau, and the Euler characteristic of $\mathbb{S}_{R''}(H^*(\mathcal{M}_{g,\ell,Q}))$ is the integer invariant $N_{R'',g,Q}$. Therefore, the above decomposition corresponds very precisely to (5.25).

All the results above have been stated for knot invariants in the canonical framing. The situation for arbitrary framing was analyzed in detail in [71]. Suppose that we consider a knot in $\mathbb{S}^3$ in the framing labelled by an integer p (the canonical framing corresponds to $p = 0$). Then, the integer invariants $N_{R,g,Q}(p)$ are obtained from (5.11) but with the vevs

$$W_R^{(p)}(q, \lambda) = (-1)^{\ell p} q^{\frac{1}{2}p\kappa_R} W_R(q, \lambda), \tag{5.28}$$

where κ_R is defined in (3.42). One has, for example,

$$f_\square^{(p)}(q, \lambda) = (-1)^p W_\square(q, \lambda),$$
$$f_{\square\square}^{(p)}(q, \lambda) = q^p W_{\square\square}(q, \lambda) - \frac{1}{2}\big(W_\square(q, \lambda)^2 + (-1)^p W_\square(q^2, \lambda^2)\big), \tag{5.29}$$
$$f_{\substack{\square\\\square}}^{(p)}(q, \lambda) = q^{-p} W_{\substack{\square\\\square}}(q, \lambda) - \frac{1}{2}\big(W_\square(q, \lambda)^2 - (-1)^p W_\square(q^2, \lambda^2)\big),$$

and so on. Notice that the right framing factor in order to match the topological string theory prediction is (3.45), and not (3.41). This is yet another indication that the duality of [36] involves the $U(N)$ gauge group, not the $SU(N)$ group. The rationale for introducing the extra sign $(-1)^p$ is not completely clear in the context of Chern–

Simons theory, and it was introduced by consistency with the results for the B-model in [2]. This sign is crucial for integrality of $N_{R,g,Q}(p)$.

All the above results on f-polynomials, integer invariant structure, etc., can be extended to links, see [63, 62].

5.3 Tests involving Wilson loops

There are two types of tests of the large N duality involving Wilson loops: a test in the strong sense, in which one verifies that the open Gromov–Witten invariants agree with the Chern–Simons amplitude, and a test in the weak sense, in which one verifies that the Chern–Simons knot invariants satisfy the integrality properties that follow from the conjectured dual description.

The only test so far of the duality in the strong sense is for the framed unknot. In this case, we know both sides of the duality in detail and we can compare the results. Let us start with the string description. The first thing we need is a construction of the Lagrangian submanifold $\mathcal{C}_{\mathcal{K}}$ that corresponds to the unknot in $\mathbb{S}^3$. This was done by Ooguri and Vafa in [76]. The construction goes as follows. Let us start with $T^*\mathbb{S}^3$ expressed as (4.12), and consider the following anti-holomorphic involution on it.

$$\eta_{1,2} = \bar{\eta}_{1,2}, \quad \eta_{3,4} = -\bar{\eta}_{3,4}. \tag{5.30}$$

The symplectic form ω changes its sign under the involution, therefore its fixed point set is a Lagrangian submanifold of $T^*\mathbb{S}^3$. If we write $\eta_\mu = x_\mu + ip_\mu$, the invariant locus of the action (5.30) is

$$p_{1,2} = 0, \quad x_{3,4} = 0 \tag{5.31}$$

and intersects the deformed conifold at

$$x_1^2 + x_2^2 = a + p_3^2 + p_4^2. \tag{5.32}$$

Therefore, the fixed point locus intersects $\mathbb{S}^3$ along the equator, which is an unknot described by the equations

$$x_1^2 + x_2^2 = a, \quad x_3 = x_4 = 0.$$

We conclude that, if we denote by $\mathcal{U}$ the unknot in $\mathbb{S}^3$, the above fixed point locus defined by (5.30) is the Lagrangian submanifold $\widehat{\mathcal{C}}_{\mathcal{U}}$. Now we want to construct the Lagrangian submanifold $\mathcal{C}_{\mathcal{U}}$, obtained from $\widehat{\mathcal{C}}_{\mathcal{U}}$ after the conifold transition. To do that, we continue to identify it with the invariant locus of the anti-holomorphic involution. We can describe this explicitly by using the coordinates (x, u, z) or (y, v, z^{-1}) defined in (4.14) and (4.18). In these coordinates, $\widehat{\mathcal{C}}_{\mathcal{U}}$ is characterized by

$$x = \bar{y}, \quad u = \bar{v}, \tag{5.33}$$

and the conifold equation (4.17) restricted to $\widehat{\mathcal{C}}_{\mathcal{U}}$ becomes

$$x\bar{x} = u\bar{u}. \tag{5.34}$$

The complex coordinate on the base $\mathbb{P}^1$ defined by (4.18) is

$$z = \frac{x}{u}, \qquad (5.35)$$

, but since $|x| = |u|$, z is a phase. We then find that $\mathcal{C}_u$ is a line bundle over the equator $|z| = 1$ of $\mathbb{P}^1$, and the fiber over z is the subspace of $\mathcal{O}(-1) + \mathcal{O}(-1)$ given by $x = z\bar{u}$ (remember that x, u are complex coordinates for the fibers). In particular, $\mathcal{C}_u$ intersects with the $\mathbb{P}^1$ at the base along $|z| = 1$, see Fig. 9.

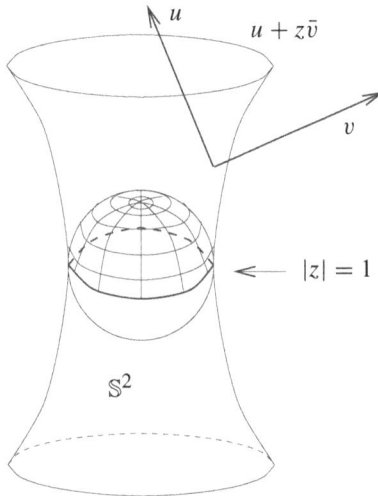

Figure 9. This figure [76] represents the Lagrangian submanifold in $\mathcal{O}(-1) \oplus \mathcal{O}(-1) \to \mathbb{P}^1$ that corresponds to the unknot in $\mathbb{S}^3$. The notation is as in [76], and is related to ours by $u \to x$ and $v \to -u$.

The open Gromov–Witten invariants associated to open strings in $\mathcal{O}(-1)\oplus\mathcal{O}(-1) \to \mathbb{P}^1$ whose boundaries end in the above Lagrangian submanifold have been computed in [51, 67, 73] (see also [39]). The procedure relies on localization formulae, as in the closed string case. However, in the open string case, it has been realized that the open invariants depend on an extra choice of an integer (the calculation depends on the weights on the localizing torus action). This is precisely the dependence we expect on Chern–Simons theory, since there is a choice of framing also labelled by an integer. This framing ambiguity in the context of open strings was first discovered in the B-model [2], and subsequently confirmed in the A-model computation of [51] as well as in other examples [39, 73]. Let us now make a detailed comparison of the answers. Katz and Liu [51] compute the open Gromov–Witten invariants $F^Q_{w,g}$ for

$Q = \ell/2$, where $\ell = \sum_i w_i$, and obtain:

$$F_{w,g}^{\ell/2} = (-1)^{p\ell+1}(p(p+1))^{h-1}\left(\prod_{i=1}^{h}\frac{\prod_{j=1}^{w_i-1}(j+w_ip)}{(w_i-1)!}\right)$$

$$\cdot \operatorname{Res}_{u=0}\int_{\overline{M}_{g,h}}\frac{c_g(\mathbb{E}^\vee(u))c_g(\mathbb{E}^\vee((-p-1)u))c_g(\mathbb{E}^\vee(pu))u^{2h-4}}{\prod_{i=1}^{h}(u-w_i\psi_i)}.$$

(5.36)

In this formula, $\overline{M}_{g,h}$ is the Deligne–Mumford moduli space of genus g stable curves with h marked points, $\mathbb{E}$ is the Hodge bundle over $\overline{M}_{g,h}$, and its dual is denoted by $\mathbb{E}^\vee$. The Chern classes of the Hodge bundle will be denoted by:

$$\lambda_j = c_j(\mathbb{E}). \tag{5.37}$$

In (5.36), we have written

$$c_g(\mathbb{E}^\vee(u)) = \sum_{i=0}^{g} c_{g-i}(\mathbb{E}^\vee)u^i, \tag{5.38}$$

and similarly for the other two factors. The integral in (5.36) also involves the ψ_i classes of two-dimensional topological gravity, which are constructed as follows. We first define the line bundle $\mathcal{L}_i$ over $\overline{M}_{g,h}$ to be the line bundle whose fiber over each stable curve Σ is the cotangent space of Σ at x_i (where x_i is the i-th marked point). We then have,

$$\psi_i = c_1(\mathcal{L}_i), \quad i = 1, \dots, h. \tag{5.39}$$

The integrals of the ψ classes can be obtained by the results of Witten and Kontsevich on 2d topological gravity [94, 54], while the integrals involving ψ and λ classes (the so-called Hodge integrals) can be in principle computed by reducing them to pure ψ integrals [29]. Explicit formulae for some Hodge integrals can be found in [35].

In the above formula (5.36), p is an integer that parameterizes the ambiguity in the open string calculation. A particularly simple case of the above expression is when $p = 0$, i.e. the standard framing. The only contribution comes from $h = 1$, and the above integral boils down to

$$\operatorname{Res}_{u=0}\int_{\overline{M}_{g,1}}\frac{\lambda_g c_g(\mathbb{E}^\vee(u))c_g(\mathbb{E}^\vee(-u))u^{2h-4}}{(u-w\psi_1)}, \tag{5.40}$$

where w is the winding number. The Mumford relations [74] give $c(\mathbb{E})c(\mathbb{E}^\vee) = 1$, which implies

$$c_g(\mathbb{E}^\vee(u))c_g(\mathbb{E}^\vee(-u)) = (-1)^g u^{2g} \tag{5.41}$$

After taking the residue, we end up with the following expression for the open Gromov–Witten invariant:

$$F_{w,g}^{w/2} = -w^{2g-2} \int_{\overline{M}_{g,1}} \psi_1^{2g-2} \lambda_g.$$ (5.42)

The above Hodge integral has been computed in [31], and it is given by b_g, where b_g is defined by the generating functional

$$\sum_{g=0}^{\infty} b_g x^g = \frac{x/2}{\sin(x/2)}.$$ (5.43)

We can now sum over all genera and all positive winding numbers to obtain [51]

$$F(V) = -\sum_{d=1}^{\infty} \frac{e^{dt/2}}{2d \sin\left(\frac{dg_s}{2}\right)} \mathrm{Tr} V^d.$$ (5.44)

Notice that the above open Gromov–Witten invariants correspond to a disk instanton wrapping the northern hemisphere of $\mathbb{P}^1$, with its boundary on the equator, together with all the multicoverings and bubblings at genus g [4]. Let us now compare to the Chern–Simons computation. In the case of the unknot in the canonical framing, Ooguri and Vafa showed [76] that the generating function (3.52) can be explicitly computed to all orders. The reason is that the quantum dimension in the representation R can be regarded as the trace in the representation R of an $N \times N$ diagonal matrix U_0 whose i-th diagonal entry is

$$\exp\left[-\frac{\pi i}{k+N}(N-2i-1)\right].$$ (5.45)

This is easily seen by remembering that ρ lives in the dual of the Cartan subalgebra H, and by using the natural isomorphism between H and H^* induced by the Killing form we obtain the above result from (3.25). Notice that U_0 is like a "master field" that gives the right answer by evaluating a "classical" trace. Therefore, one can compute $F_{CS}(V)$ by substituting $\mathrm{Tr} U_0^n$ in (3.49), to obtain

$$F(V) = -i \sum_{d=1}^{\infty} \frac{e^{dt/2} - e^{-dt/2}}{2d \sin\left(\frac{dg_s}{2}\right)} \mathrm{Tr} V^d.$$ (5.46)

The answer from Chern–Simons theory contains the contribution given in (5.44), together with a similar contribution (with $e^{t/2}$) that corresponds to holomorphic maps wrapping the southern hemisphere of the $\mathbb{P}^1$.

What happens for $p \neq 0$? In that case, it is no longer possible to sum up all the correlation functions, but we can still compute the connected vevs $W_{\vec{k}}^{(c)}$ at arbitrary framing [71]. To do that, remember that the W_R for the unknot in the canonical

---

[4]In this equation we have chosen the sign for the instantons wrapping the northern hemisphere in such a way that one has e^{dt} in the generating function, in order to compare to the results in [71].

framing are just the quantum dimensions of the representation R given in (3.28). We have to correct them with the framing factor as prescribed in (5.28), compute the $W_{\vec{k}}$ with Frobenius formula, and then extract the connected piece by using:

$$\frac{1}{z_{\vec{k}}} W_{\vec{k}}^{(c)} = \sum_{n \geq 1} \frac{(-1)^{n-1}}{n} \sum_{\vec{k}_1, \dots, \vec{k}_n} \delta_{\sum_{i=1}^{n} \vec{k}_i, \vec{k}} \prod_{i=1}^{n} \frac{W_{\vec{k}_i}}{z_{\vec{k}_i}}. \tag{5.47}$$

In this equation, the second sum is over n vectors $\vec{k}_1, \dots, \vec{k}_n$ such that $\sum_{i=1}^{n} \vec{k}_i = \vec{k}$ (as indicated by the Kronecker delta), and therefore the right hand side of (5.47) involves a finite number of terms. The generating functional for the open Gromov–Witten invariants is then explicitly given by

$$\sum_{Q} \sum_{g=0}^{\infty} F_{\vec{k},g}^{Q} g_s^{2g-2+|\vec{k}|} e^{Qt} =$$

$$(-1)^{p\ell} i^{-|\vec{k}|-\ell} \prod_{j} k_j! \sum_{n \geq 1} \frac{(-1)^n}{n} \sum_{\vec{k}_1, \dots, \vec{k}_n} \delta_{\sum_{\sigma=1}^{n} \vec{k}_\sigma, \vec{k}} \sum_{R_\sigma} \prod_{\sigma=1}^{n} \frac{\chi_{R_\sigma}(C(\vec{k}_\sigma))}{z_{\vec{k}_\sigma}} \tag{5.48}$$

$$\cdot \, e^{ip\kappa_{R_\sigma} g_s/2} \prod_{1 \leq i < j \leq c_{R_\sigma}} \frac{\sin[(l_i^\sigma - l_j^\sigma + j - i)g_s/2]}{\sin[(j - i)g_s/2]}$$

$$\prod_{i=1}^{c_{R_\sigma}} \frac{\prod_{v=-i+1}^{l_i^\sigma - i} \left(e^{\frac{t}{2} + \frac{i v g_s}{2}} - e^{-\frac{t}{2} - \frac{i v g_s}{2}} \right)}{\prod_{v=1}^{l_i^\sigma} 2 \sin[(v - i + c_{R_\sigma})g_s/2]}.$$

Let us compare this expression with the result of Katz and Liu in some simple examples with $h = 1$. Notice that the Chern–Simons result is slightly more general, since it gives the answer for any Q, while (5.36) only computes $Q = \ell/2$. For Riemann surfaces with one hole the homotopy class of the map is given by a single winding number w. For $g = 1$, one finds from (5.36):

$$F_{w,1}^{w/2} = \frac{(-1)^{pw}}{(w-1)!} \prod_{l=1}^{w-1} (l + wp) \left(\left(\int_{\overline{M}_{1,1}} \lambda_1 - w\psi_1 \right) p(p+1) + \int_{\overline{M}_{1,1}} \lambda_1 \right), \tag{5.49}$$

and for $g = 2$,

$$F_{w,2}^{w/2} = \frac{(-1)^{pw}}{(w-1)!} \prod_{l=1}^{w-1} (l + wp) \left(\left(\int_{\overline{M}_{2,1}} w^2 \psi_1^4 - w\psi_1^3 \lambda_1 + \psi_1^2 \lambda_2 \right) w^2 p^3 (p+2) \right.$$

$$+ \left(\int_{\overline{M}_{2,1}} w^3 \psi_1^4 - 2w^2 \psi_1^3 \lambda_1 - \psi_1 \lambda_1 \lambda_2 + 3w\psi_1^2 \lambda_2 \right) wp^2 \tag{5.50}$$

$$+ \left(\int_{\overline{M}_{2,1}} -w^2 \psi_1^3 \lambda_1 - \psi_1 \lambda_1 \lambda_2 + 2w\psi_1^2 \lambda_2 \right) wp + w^2 \int_{\overline{M}_{2,1}} \psi_1^2 \lambda_2 \right).$$

To obtain this expression, we have used the Mumford relation, which implies in particular $\lambda_2^2 = 0$ and $\lambda_1^2 = 2\lambda_2$. On the other hand, the Chern–Simons answer for the connected vevs when $w = 1$ and $w = 2$ is:

$$
i W_1^{(c)}(g_s) = \frac{(-1)^p}{g_s}\left(1 + \frac{1}{24}g_s^2 + \frac{7}{5760}g_s^4 + \mathcal{O}(g_s^6)\right),
$$

$$
\frac{i}{2} W_2^{(c)}(g_s) = \frac{1 + 2p}{g_s}\left(\frac{1}{4} - \frac{1}{24}(p^2 + p - 1)g_s^2\right. \tag{5.51}
$$

$$
\left. + \frac{1}{1440}(7 - 11p - 8p^2 + 6p^3 + 3p^4)g_s^4 + \mathcal{O}(g_s^6)\right),
$$

and so on. By using now the following values of the Hodge integrals for $g = 1$

$$
\int_{\overline{M}_{1,1}} \psi_1 = \int_{\overline{M}_{1,1}} \lambda_1 = \frac{1}{24} \tag{5.52}
$$

and for $g = 2$

$$
\int_{\overline{M}_{2,1}} \psi_1^4 = \frac{1}{1152}, \qquad \int_{\overline{M}_{2,1}} \psi_1^3 \lambda_1 = \frac{1}{480},
$$

$$
\int_{\overline{M}_{2,1}} \psi_1^2 \lambda_2 = \frac{7}{5760}, \qquad \int_{\overline{M}_{2,1}} \psi_1 \lambda_1 \lambda_2 = \frac{1}{2880},
$$

we find perfect agreement between (5.49) and (5.50) for $w = 1, 2$, and the Chern–Simons answer. Moreover, it is in principle possible to compute all the integrals over $\overline{M}_{g,h}$ that appear in (5.36) from the explicit expression (5.48). These Hodge integrals include an arbitrary number of ψ classes and up to three λ classes. Therefore, all correlation functions of two-dimensional topological gravity can in principle be extracted from (5.48). It should be noted, however, that some of the simple structural properties of (5.36) are not at all obvious from (5.48). For example, for $g = 0, h = 1$, (5.36) gives a fairly compact expression for the open Gromov–Witten invariant, and the fact that this equals the Chern–Simons answer amounts to a rather nontrivial combinatorial identity. It is also possible to check that the open Gromov–Witten invariants obtained in this way can be expressed in terms of BPS invariants, see [71] for more details.

Unfortunately, although there are proposals for the Lagrangian submanifolds that should correspond to other knots [63, 84], the associated open Gromov–Witten invariants have not been computed yet, so one is forced to test the conjecture in the "weak" sense of showing that one can extract integer invariants from the Chern–Simons invariants in the way described before. This was done in [61, 81, 63] for various knots and links and it was shown in all cases that indeed such invariants can be extracted in a highly nontrivial way. We will give a simple example of this, involving the trefoil knot. By using the known values for the Chern–Simons invariants (3.33), and the

Table 2. BPS invariants for the trefoil knot in the symmetric representation.

g	$Q = 1$	2	3	4	5
0	-2	8	-12	8	-2
1	-1	6	-10	6	-1
2	0	1	-2	1	0

Table 3. BPS invariants for the trefoil knot in the antisymmetric representation.

g	$Q = 1$	2	3	4	5
0	-44	16	-24	16	-4
1	-4	20	-32	20	-4
2	-1	8	-14	8	-1
3	0	1	-2	1	0

defining relations for the f-polynomials (5.12), one can easily obtain:

$$f_{\square}(q, \lambda) = \frac{q^{-\frac{1}{2}} \lambda (\lambda - 1)^2 \, (1 + q^2) \, (q + \lambda^2 q - \lambda (1 + q^2))}{q^{\frac{1}{2}} - q^{-\frac{1}{2}}},$$

$$f_{\boxminus}(q, \lambda) = -\frac{1}{q^3} f_{\square}(q, \lambda). \tag{5.53}$$

Notice that, although the Chern–Simons invariants have complicated denominators, the f-polynomials have indeed the structure (5.13). One can go further and extract the BPS invariants $N_{\square, g, Q}$, $N_{\boxminus, g, Q}$ from (5.53), by using (5.14). The results are presented in Table 5.3 and Table 5.3, respectively. The above results have been obtained in the canonical framing. Some integer invariants for the trefoil knot in arbitrary framing are listed in [71]. Results for the BPS invariants of other knots and links can be found in [63].

6 Large N transitions and toric geometry

The duality between Chern–Simons on $\mathbb{S}^3$ and closed topological strings on the resolved conifold gives a surprising point of view on Chern–Simons invariants of knots and links. However, from the "gravity" point of view we do not learn much about the closed string geometry, since the resolved conifold is quite simple (remember that it only has one nontrivial Gopakumar–Vafa invariant). It would be very interesting to find a topological gauge theory dual to more complicated geometries, in such a way that we could use our knowledge of the gauge theory side to learn about enumerative invariants of closed strings, and about closed strings in general.

Such a program was started by Aganagic and Vafa in [4]. Their basic idea is to construct geometries that locally contain T^*S^3's, and then follow geometric transitions to dual geometries where the "local" deformed conifolds are replaced by resolved conifolds. Remarkably, a large class of non-compact toric manifolds can be realized in this way, as it was made clear in [3]. Let us consider in detail an example that allows one to recover the local $\mathbb{P}^2$ geometry.

Recall from our discussion in section 4 that the deformed conifold can be represented by a graph where one indicates the degeneration loci of the cycles of the torus fiber. Following this idea, one can construct more general $\mathbb{T}^2 \times \mathbb{R}$ fibrations of $\mathbb{R}^3$ by specifying degeneration loci in the basis. An example of this is shown in Fig. 10. Notice that this geometry contains three $\mathbb{S}^3$'s, represented as dashed lines in

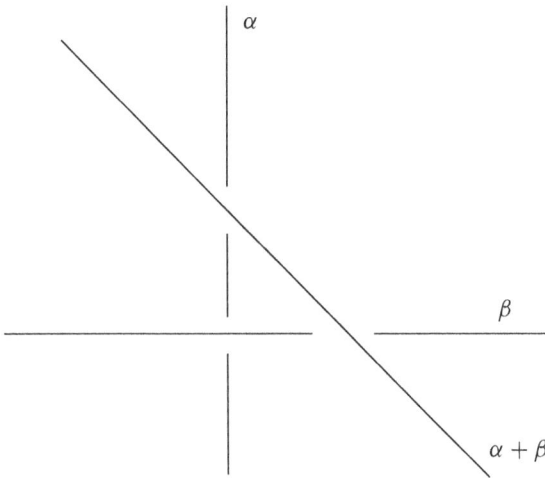

Figure 10. This shows a Calabi–Yau which is a $\mathbb{T}^2 \times \mathbb{R}$ fibration of $\mathbb{R}^3$, where the α, β, and $\alpha + \beta$ cycles of the torus degenerate at three lines.

Fig. 11. One can then think about a geometric transition where the three-spheres go to zero size, and then the corresponding singularities are blown-up to give a resolved geometry, as shown in Fig. 11.

The resolved geometry turns out to be toric, and in fact it can be obtained by three blowups of the Calabi–Yau manifold $\mathcal{O}(-3) \to \mathbb{P}^2$. Up to flops of the three $\mathbb{P}^1$'s, the resulting geometry is the noncompact Calabi–Yau manifold given by the del Pezzo surface $\mathbb{B}_3$ together with its canonical bundle. To recover the local $\mathbb{P}^2$ geometry, one just sends the sizes of the three $\mathbb{P}^1$'s to infinity. The remaining "triangle" is the toric diagram for the local $\mathbb{P}^2$ geometry, see [66, 2].

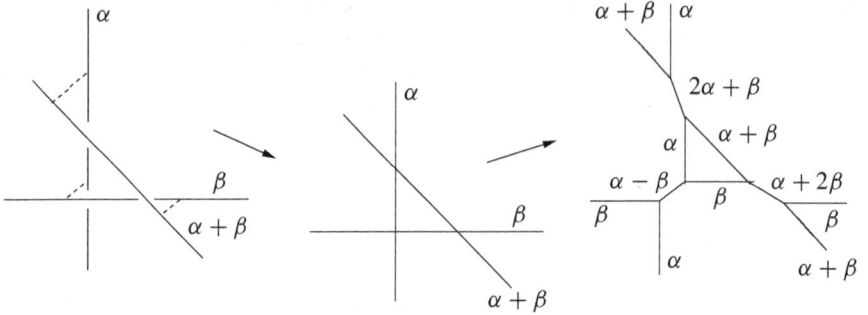

Figure 11. This shows the geometric transition of the Calabi–Yau in the previous figure. In the leftmost geometry there are three minimal 3-cycles. The lengths of the dashed lines are proportional to their sizes. The intermediate geometry is singular, and the figure on the right is the base of the smooth toric Calabi–Yau after the transition. This Calabi–Yau is related to $\mathbb{B}_3$ by flopping three $\mathbb{P}^1$'s.

Let us now wrap N_i branes, $i = 1, 2, 3$, around the three $\mathbb{S}^3$'s of the deformed geometry depicted in Fig. 10. What is the effective topological action describing the resulting open strings? For open strings with both ends on the same $\mathbb{S}^3$, the dynamics is described by Chern–Simons theory with gauge group $U(N_i)$, therefore we will have three Chern–Simons theories with groups $U(N_1)$, $U(N_2)$ and $U(N_3)$. However, there are new sectors of open strings stretched between two spheres, giving the nondegenerate instantons that we described in 4.2, following [95]. Instead of describing these open strings in geometric terms, it is better to use the spacetime physics associated to these strings. In fact, a similar situation was considered when we analyzed the incorporation of Wilson loops in the large N duality. There we had two sets of intersecting D-branes, giving a massless complex scalar field living in the intersection and in the bifundamental representation of the gauge groups. Now, if we focus, say, on the N_1, N_2 branes, we will get again a complex scalar ϕ in $(N_1, \overline{N}_2)$. This complex scalar is generically massive, and its mass is proportional to the "distance" between the two three-spheres, and it is given by a complexified Kähler parameter that will be denote by r. We can now integrate out this complex scalar field to obtain the correction to the Chern–Simons actions on the three-spheres due to the presence of the new sector of open strings, which is given by:

$$\mathcal{O}(U_1, U_2; r) = \exp\left[-\mathrm{Tr}\, \log\left(e^{r/2} U_1^{-1/2} \otimes U_2^{1/2} - e^{-r/2} U_1^{1/2} \otimes U_2^{-1/2}\right)\right]$$

$$= \exp\left\{\sum_{n=1}^{\infty} \frac{e^{-nr}}{n}\, \mathrm{Tr} U_1^n\, \mathrm{Tr} U_2^{-n}\right\}, \tag{6.1}$$

where $U_{1,2}$ are the holonomies of the corresponding gauge fields around a loop. Note that the operator $\mathcal{O}$ is the amplitude for a primitive annulus of size r together with its multicovers, as one can see from the first equation of (2.39) for $h = 2$. This annulus "connects" the two $\mathbb{S}^3$'s, *i.e.* one of its boundaries is in one three-sphere, and the other boundary is in the other sphere. The exponent in (6.1) is the contribution to F_{ndg} in (4.10) due to these configurations of open strings, and r is the complexified area of the annulus.

The problem now is to determine how many configurations like this one contribute to the full amplitude. It turns out that the only contributions come from open strings stretching along the degeneracy locus. This was found by Diaconescu, Florea and Grassi [25] using localization arguments, and derived in [3] by exploiting invariance under deformation of complex structures. This result simplifies the problem enormously, and gives a precise description of all the nondegenerate instantons contributing in this geometry: they are annuli stretching along the fixed lines of the $\mathbb{T}^2$ action, together with their multicoverings. This is illustrated in Fig. 12. The action

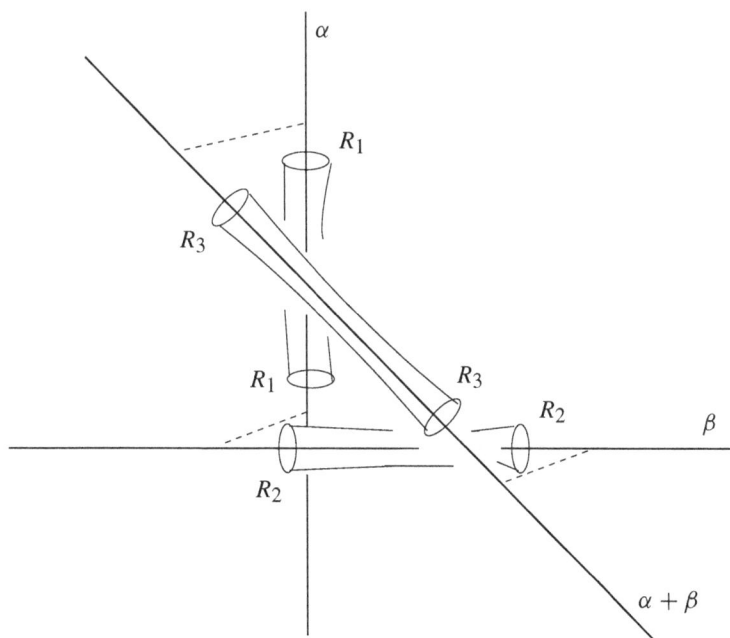

Figure 12. The only nondegenerate instantons contributing to the geometry depicted here are annuli stretching along the degeneracy locus.

describing the dynamics of topological D-branes is then:

$$S = \sum_{i=1}^{3} S_{CS}(A_i)$$

$$+ \sum_{n=1}^{\infty} \frac{1}{n} \left(e^{-nr_1} \, \mathrm{Tr} U_1^n \, \mathrm{Tr} U_2^{-n} + e^{-nr_2} \, \mathrm{Tr} U_2^n \, \mathrm{Tr} U_3^{-n} + e^{-nr_3} \, \mathrm{Tr} U_3^n \, \mathrm{Tr} U_1^{-n} \right),$$

$$(6.2)$$

where the A_i are $U(N_i)$ gauge connections on each of the $\mathbb{S}^3$'s, $i = 1, 2, 3$, and U_i are the corresponding holonomies around loops. There is a very convenient way to write the free energy of the theory with the above action. First notice that, by following the same steps that led to (3.51), one can write the operator (6.1) as

$$\mathcal{O}(U_1, U_2; r) = \sum_{R} \mathrm{Tr}_R U_1 e^{-\ell r} \mathrm{Tr}_R U_2^{-1}, \qquad (6.3)$$

where ℓ denotes the number of boxes of the representation R. In the situation depicted in Fig. 12, we see that there are two annuli ending on each three-sphere. The boundaries of these annuli give knots, so we have a two-component link in each $\mathbb{S}^3$. The holonomies around the components of these links will be in different representations of $U(N)$, as indicated in Fig. 12. Therefore, the free energy will be given by:

$$F = \sum_{i=1}^{3} F_{CS}(N_i, g_s) + \log \left\{ \sum_{R_1, R_2, R_3} e^{-\sum_{i=1}^{3} \ell_i r_i} W_{R_1, R_2}(\mathcal{L}_1) W_{R_2, R_3}(\mathcal{L}_2) W_{R_3, R_1}(\mathcal{L}_3) \right\},$$

$$(6.4)$$

where ℓ_i is the number of boxes in the representation R_i, and $F_{CS}(N_i, g_s)$ denotes the free energy of Chern–Simons theory with gauge group $U(N_i)$. These correspond to the degenerate instantons that come from each of the three-spheres.

Of course, in order to compute (6.4) we need some extra information: we have to know what are, topologically, the links $\mathcal{L}_i$, and also if there is some framing induced by the geometry. It turns out that these questions can be easily answered by looking at the geometry of the degeneracy locus. The key point is to note that in this geometry the three-spheres represented by dashed lines between two degeneracy loci have natural Heegard splittings into two tori, and the gluing instructions are determined by the $Sl(2, \mathbb{Z})$ transformation that maps the degenerating cycle at the end of the corresponding three-sphere, to the degenerating cycle at the other end [3]. For example, the three-sphere between the α and the β degenerating loci in Fig. 12 comes from gluing two tori with an S^{-1} transformation, which maps the α cycle into the β cycle. Following this procedure (see [3] for details) one finds that the $\mathcal{L}_i$ are all Hopf links (see Fig. 2), and that some of the components do actually have nontrivial framing. If we denote the components of $\mathcal{L}_i$ by $\mathcal{K}_i$ and $\mathcal{K}_i'$, $i = 1, 2, 3$, the framings turn out to be the following: $\mathcal{K}_1$, $\mathcal{K}_1'$ and $\mathcal{K}_3$ have framing zero, while the remaining knots have framing $p = 1$. This means that $\mathcal{L}_1$ is in the canonical framing, in $\mathcal{L}_2$

both components are framed, while in $\mathcal{L}_3$ only one of the components, $\mathcal{K}'_3$, is framed. This is depicted in Fig. 13.

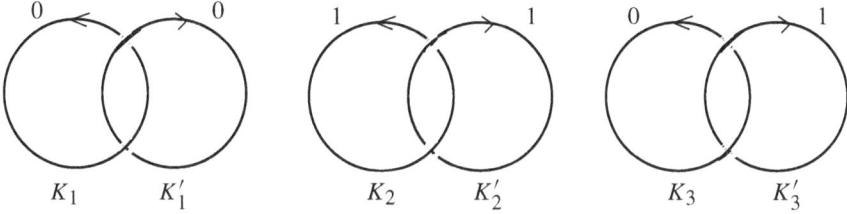

Figure 13. The figure shows the Hopf links $\mathcal{L}_i$, $i = 1, 2, 3$. The numbers indicate the framing of each knot.

What happens now if we go through the geometric transition of Fig. 11? As in the case originally studied by Gopakumar and Vafa, the string coupling constant gives the Chern–Simons "effective" coupling constant $g_s = 2\pi/(k_i + N_i)$ (which is the same for the three theories, see (4.11)), while the 't Hooft parameters $t_i = g_s N_i$ correspond to the sizes of the three outward legs of the toric diagram on the right side of Fig. 11. The free energy (6.4) is, due to the large N transition, the free energy of topological *closed* strings propagating in that toric geometry. In order to recover just local $\mathbb{P}^2$, we have to take the 't Hooft parameters to infinity, and "tune" the sizes of the annuli at the same time. It turns out that one has to perform a double scaling limit, taking both t_i and r_i to infinity in such a way that

$$r = r_1 - \frac{t_1 + t_3}{2} = r_2 - \frac{t_1 + t_2}{2} = r_3 - \frac{t_2 + t_3}{2} \tag{6.5}$$

remains finite. Then, r can be identified with the complexified Kähler parameter of local $\mathbb{P}^2$. We refer again to [3] for details. The free energy has in this limit the structure:

$$F = \log\left\{1 + \sum_{\ell=1}^{\infty} a_\ell(q)e^{-\ell r}\right\} = \sum_{\ell=1}^{\infty} a_\ell^{(c)}(q)e^{-\ell r} \tag{6.6}$$

where $q = e^{ig_s}$. The coefficients $a_\ell(q)$, $a_\ell^{(c)}(q)$ can be easily obtained in terms of the invariants of the Hopf link in arbitrary representations. One finds, for example [3],

$$a_1(q) = -\frac{3}{(q^{-\frac{1}{2}} - q^{\frac{1}{2}})^2},$$

$$a_2^{(c)}(q) = \frac{6}{(q^{-\frac{1}{2}} - q^{\frac{1}{2}})^2} + \frac{1}{2}a_1(q^2). \tag{6.7}$$

If we compare to (2.28) and take into account the effects of multicovering, we find the following values for the Gopakumar–Vafa invariants of $\mathcal{O}(-3) \to \mathbb{P}^2$:

$$
\begin{aligned}
n_1^0 &= 3, & n_1^g &= 0 & \text{for } g > 0, \\
n_2^0 &= -6, & n_2^g &= 0 & \text{for } g > 0,
\end{aligned}
\tag{6.8}
$$

in agreement with the results listed in Table 2.4. In fact, one can go much further with this method and compute the Gopakumar–Vafa invariants to high degree. The advantage of this procedure is that, in contrast to both the A and the B model computations, one gets the answer for *all genera*, see [3] for a complete listing of the invariants up to degree 12.

Although we have focused here on local $\mathbb{P}^2$, one can analyze in a similar way other toric geometries, including local $\mathbb{P}^1 \times \mathbb{P}^1$ and other local del Pezzo surfaces (see also [25, 46]). In fact, one can in principle recover all local toric geometries in this way. We then see that large N transitions produce gauge theory duals of topological strings propagating on various toric backgrounds. The gauge theory dual is given in general by a product of Chern–Simons theories together with complex scalars in bifundamental representations, and moreover the gauge theory data are nicely encoded in the toric diagram. Other aspects of these dualities for toric manifolds can be found in [3, 25, 46].

7 Conclusions

The remarkable connections between enumerative geometry and knot invariants that have been reviewed in this paper certainly deserve further investigation. Some directions for further research are the following:

1) The correspondence between knot invariants and open Gromov–Witten invariants has been tested only for the unknot. It would be very interesting to test nontrivial knots and improve our understanding of the map relating knots and links in $\mathbb{S}^3$ to Lagrangian submanifolds in the resolved conifold. This will certainly open new perspectives in the study of Chern–Simons knot invariants.

2) Another direction to explore is the correspondence between coupled Chern–Simons systems and closed string invariants that we explained in section 6. Extensions to more general local toric geometries, and even to compact geometries, would give a fascinating new point of view on the enumerative geometry of Calabi–Yau threefolds.

3) The "unreasonable effectiveness of physics in solving mathematical problems" [87] has given again surprising results connecting two seemingly unrelated areas of geometry, and we need a deeper mathematical understanding of these connections. For example, the results of section 6 may be understood in terms of the localization techniques introduced in [55], as suggested in [3].

Acknowledgments. I would like to thank Mina Aganagic, Jose Labastida and Cumrun Vafa for enjoyable collaborations on the topics discussed in this review, and for sharing their insights with me. I would also like to thank Andrew Neitzke for a careful reading of the manuscript. This work has been supported by NSF-PHY/98-02709.

References

[1]   B. S. Acharya, On realising $\mathcal{N} = 1$ super Yang-Mills in M theory, hep-th/0011089.

[2]   M. Aganagic, A. Klemm and C. Vafa, Disk instantons, mirror symmetry and the duality web, hep-th/0105045, *Z. Naturforsch.* A 57 (2002), 1.

[3]   M. Aganagic, M. Mariño and C. Vafa, All loop topological string amplitudes from Chern-Simons theory, hep-th/0206164.

[4]   M. Aganagic and C. Vafa, Mirror symmetry, D-branes and counting holomorphic discs, hep-th/0012041.

[5]   M. Aganagic and C. Vafa, G_2 manifolds, mirror symmetry and geometric engineering, hep-th/0110171.

[6]   O. Aharony, S. S. Gubser, J. M. Maldacena, H. Ooguri and Y. Oz, Large N field theories, string theory and gravity, hep-th/9905111, *Phys. Rep.* 323 (2000), 183.

[7]   I. Antoniadis, E. Gava, K. S. Narain and T. R. Taylor, Topological amplitudes in string theory, hep-th/9307158, *Nucl. Phys.* B 413 (1994), 162.

[8]   P. S. Aspinwall and D. R. Morrison, Topological field theory and rational curves, hep-th/9110048, *Commun. Math. Phys.* 151 (1993), 245.

[9]   M. Atiyah, On framings of three-manifolds, *Topology* 29 (1990), 1.

[10]   M. Atiyah, J. Maldacena and C. Vafa, An M-theory flop as a large N duality, hep-th/0011256, *J. Math. Phys.* 42 (2001), 3209.

[11]   S. Axelrod and I. Singer, Chern-Simons perturbation theory, in *Differential geometric methods in theoretical physics,* hep-th/9110056, World Scientific, 1991, 3.

[12]   D. Bar-Natan, S. Garoufalidis, L. Rozansky and D. Thurston, Wheels, wheeling, and the Kontsevich integral of the unknot, q-alg/9703025, *Israel J. Math.* 119 (2000), 217.

[13]   K. Becker, M. Becker and A. Strominger, Fivebranes, membranes, and nonperturbative string theory, hep-th/9507158, *Nucl. Phys.* B 456 (1995), 130.

[14]   M. Bershadsky, S. Cecotti, H. Ooguri and C. Vafa, Holomorphic anomalies in topological field theories, hep-th/9302103, *Nucl. Phys.* B 405 (1993), 279.

[15]   M. Bershadsky, S. Cecotti. H. Ooguri and C. Vafa, Kodaira-Spencer theory of gravity and exact results for quantum string amplitudes, hep-th/9309140, *Commun. Math. Phys.* 165 (1994), 311.

[16]   J. Blum, Calculation of nonperturbative terms in open string models, hep-th/0112139, *Nucl. Phys.* B 634 (2002), 3.

[17]   J. Bryan and R. Pandharipande, BPS states of curves in Calabi-Yau 3-folds, math.AG/0009025, *Geom. Topol.* 5 (2001), 287.

[18]   P. Candelas and X. C. De La Ossa, Comments on conifolds, *Nucl. Phys.* B 342 (1990), 246.

[19]   P. Candelas, X. C. De La Ossa, P. S. Green and L. Parkes, A pair of Calabi-Yau manifolds as an exactly soluble superconformal theory, *Nucl. Phys.* B 359 (1991), 21.

[20]   T. M. Chiang, A. Klemm, S. T. Yau and E. Zaslow, Local mirror symmetry: Calculations and interpretations, hep-th/9903053, *Adv. Theor. Math. Phys.* 3 (1999), 495.

[21]   S. Coleman, *Aspects of symmetry,* Cambridge University Press, 1985.

[22]   S. Cordes, G. Moore and S. Rangoolam, Large N 2-D Yang-Mills theory and topological string theory, hep-th/9402107, *Commun. Math. Phys.* 185 (1997), 543. Lectures on 2-d Yang-Mills theory, equivariant cohomology and topological field theories, hep-th/9411210, *Nucl. Phys. Proc. Suppl.* 41 (1995), 184.

[23]   D. Cox and S. Katz, *Mirror symmetry and algebraic geometry,* Math. Surveys Monogr. 68, Amer. Math. Soc., Providence, R.I. 1999.

[24]   D. E. Diaconescu, B. Florea and A. Grassi, Geometric transitions and open string instantons, hep-th/0205234.

[25]   D. E. Diaconescu, B. Florea and A. Grassi, Geometric transitions, del Pezzo surfaces and open string instantons, hep-th/0206163.

[26]   R. Dijkgraaf, E. Verlinde and H. Verlinde, Notes on topological string theory and two-dimensional topological gravity, in *String theory and quantum gravity* (Trieste, 1990), World Scientific Publishing, River Edge, N.J., 1991, 91.

[27]   M.R. Douglas, Chern-Simons-Witten theory as a topological Fermi liquid, hep-th/9403119.

[28]   S. Elitzur, G. Moore, A. Schwimmer and N. Seiberg, Remarks on the canonical quantization of the Chern-Simons-Witten theory, *Nucl. Phys.* B 326 (1989), 108.

[29]   C. Faber, Algorithms for computing intersection numbers of curves, with an application to the class of the locus of Jacobians, alg-geom/9706006.

[30]   C. Faber, A conjectural description of the tautological ring of the moduli space of curves, math.AG/9711218.

[31]   C. Faber and R. Pandharipande, Hodge integrals and Gromov-Witten theory, math.AG/9810173, *Invent. Math.* 139 (2000), 173.

[32]   D. S. Freed and R. E. Gompf, Computer calculation of Witten's three manifold invariant, *Commun. Math. Phys.* 141 (1991), 79.

[33]   P. Freyd, D. Yetter, J. Hoste, W. B. R. Lickorish, K. Millett and A.A. Ocneanu, A new polynomial invariant of knots and links, *Bull. Amer. Math. Soc.* (N.S.) 12 (1985), 239.

[34]   W. Fulton and J. Harris, *Representation theory: a first course,* Grad. Texts in Math. 129, Springer-Verlag, New York 1991.

[35]   E. Getzler and R. Pandharipande, Virasoro constraints and the Chern classes of the Hodge bundle, math.AG/9805114, *Nucl. Phys.* B 530 (1998), 701.

[36]   R. Gopakumar and C. Vafa, M-theory and topological strings. I, II, hep-th/9809187, hep-th/9812127.

[37] R. Gopakumar and C. Vafa, On the gauge theory/geometry correspondence, hep-th/9811131, *Adv. Theor. Math. Phys.* 3 (1999), 1415.

[38] S. Govindarajan, T. Jayaraman and T. Sarkar, Disc instantons in linear sigma models, hep-th/0108234.

[39] T. Graber and E. Zaslow, Open string Gromov-Witten invariants: Calculations and a mirror 'theorem', hep-th/0109075.

[40] A. Grassi and M. Rossi, Large N dualities and transitions in geometry, math.ag/0209044.

[41] D. Gross and W. Taylor, Two-dimensional QCD is a string theory, hep-th/9301068, *Nucl. Phys.* B 400 (1993), 181.

[42] E. Guadagnini, M. Martellini and M. Mintchev, Wilson lines in Chern-Simons theory and link invariants, *Nucl. Phys.* B 330 (1990), 575. E. Guadagnini, The universal link polynomial, *Internat. J. Modern Phys.* A 7 (1992) 877; *The link invariants of the Chern-Simons field theory,* Walter de Gruyter, Berlin 1993.

[43] J. Harris and I. Morrison, *Moduli of curves*, Grad. Texts in Math. 187, Springer-Verlag, New York 1998.

[44] S. Hosono, M. H. Saito and A. Takahashi, Holomorphic anomaly equation and BPS state counting of rational elliptic surface, hep-th/9901151, *Adv. Theor. Math. Phys.* 3 (1999), 177. S. Hosono, Counting BPS states via holomorphic anomaly equations, hep-th/0206206, in *Calabi-Yau varieties and mirror symmetry* (Toronto, ON, 2001), Fields Inst. Commun. 38, Amer. Math. Soc., Providence, RI, 2003, 57.

[45] S. Hosono, M. H. Saito and A. Takahashi, Relative Lefschetz action and BPS state counting, math.AG/0105148.

[46] A. Iqbal, All genus topological string amplitudes and 5-brane webs as Feynman diagrams, hep-th/0207114.

[47] A. Iqbal and A. K. Kashani-Poor, Discrete symmetries of the superpotential and calculation of disk invariants, hep-th/0109214.

[48] L. C. Jeffrey, Chern-Simons-Witten invariants of lens spaces and torus bundles, and the semiclassical approximation, *Commun. Math. Phys.* 147 (1992), 563.

[49] V. F. R. Jones, Hecke algebra representations of braid groups and link polynomials, *Ann. of Math.* 126 (1987), 335.

[50] S. Katz, A. Klemm and C. Vafa, M-theory, topological strings and spinning black holes, hep-th/9910181, *Adv. Theor. Math. Phys.* 3 (1999), 1445.

[51] S. Katz and C. C. Liu, Enumerative geometry of stable maps with Lagrangian boundary conditions and multiple covers of the disc, math.AG/0103074, *Adv. Theor. Math. Phys.* 5 (2002), 1.

[52] A. Klemm, unpublished.

[53] A. Klemm and E. Zaslow, Local mirror symmetry at higher genus, hep-th/9906046, in *Winter School on Mirror Symmetry, Vector bundles and Lagrangian Submanifolds* (Cambridge, Mass., 1999), AMS/IP Stud. Adv. Math. 23, Amer. Math. Soc., Providence, R.I., 2001, 183.

[54] M. Kontsevich, Intersection theory on the moduli space of curves and the matrix Airy function, *Commun. Math. Phys.* 147 (1992), 1.

[55] M. Kontsevich, Enumeration of rational curves via torus actions, hep-th/9405035, in *The moduli space of curves* (Texel Island, 1994), Progr. Math. 129, Birkhäuser, Boston 1995, 335.

[56] J. M. F. Labastida, Chern-Simons gauge theory: ten years after, hep-th/9905057.

[57] J. M. F. Labastida, Math and physics, hep-th/0107079; Knot theory from the perspective of field and string theory, hep-th/0201038.

[58] J. M. F. Labastida and P. M. Llatas, Topological matter in two dimensions, hep-th/9112051, *Nucl. Phys.* B 379 (1992), 220.

[59] J. M. F. Labastida, P. M. Llatas and A. V. Ramallo, Knot operators in Chern-Simons gauge theory, *Nucl. Phys.* B 348 (1991), 651.

[60] J. M. F. Labastida and M. Mariño, The HOMFLY polynomial for torus links from Chern-Simons gauge theory, hep-th/9402093, *Internat. J. Modern Phys.* A 10 (1995), 1045.

[61] J. M. F. Labastida and M. Mariño, Polynomial invariants for torus knots and topological strings, hep-th/0004196, *Commun. Math. Phys.* 217 (2001), 423.

[62] J. M. F. Labastida and M. Mariño, A new point of view in the theory of knot and link invariants," math.QA/0104180, *J. Knot Theory Ramifications* 11 (2002), 173.

[63] J. M. F. Labastida, M. Mariño and C. Vafa, Knots, links and branes at large N, hep-th/0010102, *J. High Energy Phys.* 0011 (2000), 007.

[64] J. M. F. Labastida and A. V. Ramallo, Operator formalism for Chern-Simons theories, *Phys. Lett.* B 227 (1989), 92.

[65] W. Lerche and P. Mayr, On $\mathcal{N} = 1$ mirror symmetry for open type II strings, hep-th/0111113. W. Lerche, P. Mayr and N. Warner, $\mathcal{N} = 1$ special geometry, mixed Hodge variations and toric geometry, hep-th/0208039.

[66] N. C. Leung and C. Vafa, Branes and toric geometry, hep-th/9711013, *Adv. Theor. Math. Phys.* 2 (1998), 91.

[67] J. Li and Y. S. Song, Open string instantons and relative stable morphisms, hep-th/0103100, *Adv. Theor. Math. Phys.* 5 (2002), 67.

[68] W. B. R. Lickorish, *An introduction to knot theory,* Grad. Texts in Math. 175, Springer-Verlag, New York 1998.

[69] M. Mariño, Chern-Simons theory, matrix integrals, and perturbative three-manifold invariants, hep-th/0207096.

[70] M. Mariño and G. Moore, Counting higher genus curves in a Calabi-Yau manifold, hep-th/9808131, *Nucl. Phys.* B 543 (1999), 592.

[71] M. Mariño and C. Vafa, Framed knots at large N, hep-th/0108064.

[72] P. Mayr, $\mathcal{N} = 1$ mirror symmetry and open/closed string duality, hep-th/0108229.

[73] P. Mayr, Summing up open string instantons and $\mathcal{N} = 1$ string amplitudes, hep-th/0203237.

[74] D. Mumford, Towards an enumerative geometry of the moduli space of curves, in *Arithmetic and geometry*, vol. 2, Progr. Math. 36, Birkhäuser, Boston, MA, 1983, 271.

[75] H. Ooguri, Y. Oz and Z. Yin, D-branes on Calabi-Yau spaces and their mirrors, hep-th/9606112, Nucl. Phys. B 477 (1996), 407.

[76] H. Ooguri and C. Vafa, Knot invariants and topological strings, hep-th/9912123, *Nucl. Phys.* B 577 (2000), 419.

[77] H. Ooguri and C. Vafa, Worldsheet derivation of a large *N* duality, hep-th/0205297, *Nucl. Phys.* B 641 (2002), 3.

[78] V. Periwal, Topological closed string interpretation of Chern-Simons theory, hep-th/9305115, *Phys. Rev. Lett.* 71 (1993), 1295.

[79] A. M. Polyakov, Fermi-Bose transmutations induced by gauge fields, *Mod. Phys. Lett.* A 3 (1988), 325.

[80] V. V. Prasolov and A. B. Sossinsky, *Knots, links, braids and 3-manifolds*, Transl. Math. Monogr. 154, Amer. Math. Soc., Providence, RI, 1997.

[81] P. Ramadevi and T. Sarkar, On link invariants and topological string amplitudes, hep-th/0009188, *Nucl. Phys.* B 600 (2001), 487.

[82] M. Rosso, Groupes quantiques et modèles à vertex de V. Jones en théorie des noeuds, *C. R. Acad. Sci. Paris Sér. I Math.* 307 (1988), 207.

[83] L. Rozansky, A large *k* asymptotics of Witten's invariant of Seifert manifolds, hep-th/9303099, *Commun. Math. Phys.* 171 (1995), 279. A contribution of the trivial connection to Jones polynomial and Witten's invariant of 3-D manifolds. 1, 2, hep-th/9401061, 9403021, *Commun. Math. Phys.* 175 (1996), 275, 297.

[84] C. H. Taubes, Lagrangians for the Gopakumar-Vafa conjecture, math.DG/0201219, *Adv. Theor. Math. Phys.* 5 (2001), 139.

[85] W. Taylor, D-brane effective field theory from string field theory, hep-th/0001201, *Nucl. Phys.* B 585 (2000), 171.

[86] G. 't Hooft, A planar diagram theory for strong interactions, *Nucl. Phys.* B 72 (1974), 461.

[87] C. Vafa, Geometric physics, hep-th/9810149.

[88] C. Vafa, Unifying themes in topological field theories, hep-th/0005180.

[89] C. Vafa, Superstrings and topological strings at large *N*, hep-th/0008142, *J. Math. Phys.* 42 (2001), 2798.

[90] E. Witten, Noncommutative geometry and string field theory, *Nucl. Phys.* B 268 (1986), 253.

[91] E. Witten, Topological sigma models, *Commun. Math. Phys.* 118 (1988), 411.

[92] E. Witten, Quantum field theory and the Jones polynomial, *Commun. Math. Phys.* 121 (1989), 351.

[93] E. Witten, Mirror manifolds and topological field theory, hep-th/9112056.

[94] E. Witten, Two-dimensional gravity and intersection theory on the moduli space, *Surveys in Differential Geometry* 1 (1991), 243.

[95] E. Witten, Chern-Simons gauge theory as a string theory, hep-th/9207094, in *The Floer memorial volume*, Prog. Math. 133, Birkhäuser, Basel 1995, 637.

[96] E. Witten, Phases of $\mathcal{N} = 2$ theories in two dimensions, hep-th/9301042, *Nucl. Phys.* B 403 (1993), 159.

[97] E. Witten, World-sheet corrections via D-instantons, *J. High Energy Phys.* 0002 (2000), 030, hep-th/9907041.

Gerbes, (twisted) K-theory, and the supersymmetric WZW model

Jouko Mickelsson

Mathematical Physics, Royal Institute of Technology, KTH,
SE-106 91, Stockholm, Sweden
email: `jouko@theophys.kth.se`

Abstract. The aim of this talk is to explain how symmetry breaking in a quantum field theory problem leads to a study of projective bundles, Dixmier–Douady classes, and associated gerbes. A gerbe manifests itself in different equivalent ways. Besides the cohomological description as a DD class, it can be defined in terms of a family of local line bundles or as a prolongation problem for an (infinite-dimensional) principal bundle, with the fiber consisting of (a subgroup of) projective unitaries in a Hilbert space. The prolongation aspect is directly related to the appearance of central extensions of (broken) symmetry groups. We also discuss the construction of twisted K-theory classes by families of supercharges for the supersymmetric Wess–Zumino–Witten model.

2000 Mathematics Subject Classification: 81T50, 58B25, 19K56.

Keywords: Gerbes, hamiltonian quantization, twisted K theory.

Introduction

In quantum field theory gerbes arise when one asks the question whether a given bundle of quantum mechanical projective spaces is a projectivization of a Hilbert space bundle. Nontrivial obstructions to the existence of the Hilbert bundle are generated by QFT anomalies.

An anomaly in field theory is a breakdown of a group of (gauge) symmetries in the quantization of a classical field theory model. A classical symmetry can be broken in the quantization of massless fermions in a classical background field. The background consists typically of a curved space-time metric or a Yang–Mills field. Because of the breakdown of the symmetry, in the quantized theory one cannot identify gauge or diffeomorphism equivalent Hilbert spaces. The quantized symmetry group acts only projectively, through a central (or an abelian) extension; for this reason modding out by the symmetry group leads to a bundle of projective spaces parametrized by the external field configurations.

The obstruction to replacing a projective bundle over X by a true vector bundle is given by a cohomology class in $H^3(X, \mathbb{Z})$, the *Dixmier–Douady class*. At the same time, the Dixmier–Douady class describes a (stable) equivalence class of a gerbe over X. For the general theory of gerbes, the reader is recommended to consult [Br]. Here our discussion is closely related to a specialized form, the *bundle gerbe,* introduced in [Mu], which is an abstraction of a quantum field theory problem involving massless fermions, [Mi].

A gerbe over X can be viewed as a collection of local line bundles L_{ij} over intersections $U_{ij} = U_i \cap U_j$ of open subsets of X. In addition, the gerbe data involves a family of isomorphisms

$$L_{ij} \otimes L_{jk} = L_{ik}$$

on the triple overlaps U_{ijk}. Given a family of curvature forms ω_{ij} for the local line bundles, satisfying the cocycle property $\omega_{ij} + \omega_{jk} = \omega_{ik}$, one can produce a closed integral 3-form ω defined on X using a standard construction in a Cech–de Rham double complex. The class $[\omega] \in H^3(X, \mathbb{Z})$ is de Rham form of the Dixmier–Douady class of the gerbe. (This is not the whole story, because there are cases when the DD class is pure torsion.)

The local line bundles L_{ij} arise in a natural way when trying to deprojectivisize a projective bundle over X. The transition functions on the overlaps U_{ij} of a projective bundle are given as functions g_{ij} taking values in a projective unitary group. The unitary group $U(H)$ is a central extension by the circle S^1 of the projective unitary group $PU(H)$. Thus the possible unitaries $\hat{g}_{ij}(x)$ representing $g_{ij}(x) \in PU(H)$ form a circle over the point $x \in U_{ij}$; replacing the circle by $\mathbb{C}$ we obtain a complex line $L_{ij}(x)$. The group product in $U(H)$ gives a natural identification of $L_{ij} \otimes L_{jk}$ as L_{ik} in the common domain.

Sections 2 and 3 contain an introduction to the (twisted) K-theory aspects of gerbes, from the quantum field theory point of view. K-theory arises naturally in (hamiltonian) quantization. We have a family of Hamilton operators parametrized by points in X. When the Hamilton operators are self-adjoint and have both positive and negative essential spectrum, they give (by definition) an element of $K^1(X)$. It is known that $K^1(X)$ is parametrized by the odd cohomology groups $H^{2k+1}(X, \mathbb{Z})$. The Dixmier–Douady class is then the projection to the 3-cohomology part. If we disregard torsion, we can describe this by a de Rham form of degree 3. This is also the starting point for constructing the twisted K-theory classes, [Ro]. In section 4 we give an explicit example using the supersymmetric WZW model.

Of course, in this short presentation I have left out many interesting topics; for recent discussions on the applications of gerbes and K-theory to strings and conformal field theory see e.g. [GR], [Se].

I want to thank Jens Hoppe for pointing out the reference [Ko] and Edwin Langmann for critical reading of the manuscript and telling me about [B-R]. This work was partially supported by the Erwin Schrödinger International Institute for Mathematical Physics and by the Royal Swedish Academy of Sciences, which is gratefully acknowledged.

1 Gerbes from canonical quantization

Let us first recall some basic facts about canonical anticommutation relations (CAR) in infinite dimensions, for an extensive review see [Ar]. Let H be a complex Hilbert space. To each vector $u \in H$ one associates a pair of elements $a(u)$ and $a^*(u)$ which are generators of a complex unital algebra, the CAR algebra based on H. The basic relations are

$$a(u)a^*(v) + a^*(v)a(u) = \langle u, v \rangle \mathbf{1}$$
$$a(u)a(v) + a(v)a(u) = 0 = a^*(u)a^*(v) + a^*(v)a^*(u)$$

(1.1)

for all vectors $u, v \in H$. The map $u \mapsto a^*(u)$ is linear whereas $u \mapsto a(u)$ is antilinear.

Each polarization $H = H_+ \oplus H_-$ to a pair of infinite-dimensional subspaces defines an irreducible representation of CAR in a Hilbert space (the *Fock space*) $\mathcal{F} = \mathcal{F}(H_+ \oplus H_-)$. The (equivalence class of the) representation is uniquely defined by the requirement of existence of a *vacuum vector* ψ_0 such that

$$a^*(u)\psi_0 = 0 = a(v)\psi_0$$

(1.2)

for all $u \in H_-$ and $v \in H_+$. Any two representations defined by polarizations H_+, H'_+ are equivalent if and only if the projection operators P, P' to these subspaces differ by a Hilbert–Schmidt operator.

A unitary operator $S : H \to H$ can be promoted to a unitary operator $\hat{S} : \mathcal{F} \to \mathcal{F}$ such that

$$\hat{S}a(u)\hat{S}^{-1} = a(Su) \quad \forall u \in H,$$

and similarly for the creation operators $a^*(u)$, if and only if the off-diagonal blocks of S (in the given polarization) are Hilbert–Schmidt operators. Let us denote the group of unitaries S of this type as $U_{\text{res}} = U_{\text{res}}(H_+ \oplus H_-)$, [PS]. Note that the operator $\hat{S}$ is only defined up to a phase factor. The group of quantum operators $\hat{S}$ forms a central extension $\hat{U}_{\text{res}}$ of U_{res},

$$1 \to S^1 \to \hat{U}_{\text{res}} \to U_{\text{res}} \to 1.$$

Likewise, a bounded linear operator $X : H \to H$ can be 'second quantized' as a linear operator $\hat{X}$ in $\mathcal{F}$ such that

$$[\hat{X}, a^*(u)] = a^*(Xu), \quad [\hat{X}, a(u)] = -a(X^*u)$$

(1.3)

for all $u \in H$ if and only if the off-diagonal blocks of X are Hilbert–Schmidt. In this case the operator $\hat{X}$ is uniquely defined modulo an additive constant. The operators $\hat{X}$

form a Lie algebra, a central extension of the Lie algebra of the complex group GL_{res}, with commutation relations

$$[\hat{X}, \hat{Y}] = \widehat{[X, Y]} + c(X, Y),$$

where the complex valued 2-cocycle c depends on certain choices (physically, the choice of normal ordering in the Fock space), but its cohomology class is represented by, [Lu],

$$c(X, Y) = \frac{1}{4}\text{tr}\,\epsilon[\epsilon, X][\epsilon, Y],$$

where $\epsilon = P_{H_+} - P_{H_-}$.

In quantum field theory problems the polarization arises as a splitting of the 1-particle Hilbert space into positive and negative energy subspaces with respect to a (in general unbounded) self-adjoint Hamilton operator (e.g. a Dirac operator).

Consider next a case when we have a parametrized family of Hamilton operators. Let X be some manifold and let for each $x \in X$ a self-adjoint operator D_x (in a dense domain of) H be given, such that D_x depends smoothly on the parameter x in some appropriate topology. Tentatively, we would like to construct a family of Fock spaces $\mathcal{F}_x$ defined by the polarizations $H = H_+(D_x) \oplus H_-(D_x)$ to positive and negative spectral subspaces. However, in general this is not possible in a smooth way because of the spectral flow: each time an eigenvalue $\lambda_n(x)$ of D_x crosses the zero mode $\lambda = 0$ we have a discontinuity in the polarization and thus in the construction of the Fock spaces.

The potential resolution to the above problem lies in the fact that one is really interested only in the equivalence class of the CAR representation. Therefore one is happy with a choice of a function $x \mapsto P_x$, where P_x is a projection operator which differs from the projection operator onto $H_+(D_x)$ by a Hilbert–Schmidt operator. However, there can be an obstruction to the existence of the function P_x which depends on the K-theory class of the mapping $x \mapsto D_x$.

Recall the operator theoretic meaning of $K^1(X)$. Let Fred_* be the space of self-adjoint Fredholm operators in H with both negative and positive essential spectrum. Then $K^1(X)$ can be identified as the space of homotopy classes of maps $X \to \text{Fred}_*$. In particular, a family D_x of Dirac type operators defines an element in $K^1(X)$. Up to torsion, $K^1(X)$ is parametrized by the odd de Rham cohomology classes in $H^*(X, \mathbb{Z})$. It turns out that for the existence of the family P_x only the 3-form part is relevant.

As a concrete example in quantum field theory consider the case of Dirac operators coupled to vector potentials. Let $\mathcal{A}$ be the space of $\mathfrak{g}$ valued 1-forms on a compact odd dimensional spin manifold M where $\mathfrak{g}$ is the Lie algebra of a compact group G. Each $A \in \mathcal{A}$ defines a Dirac operator D_A in the space H of square integrable spinor fields twisted by a representation of G. The 'free' Dirac operator D_0 defines a background polarization $H = H_+ \oplus H_-$. Each potential A and a real number λ defines a spectral subspace $H_+(A, \lambda)$ corresponding to eigenvalues of D_A strictly bigger than λ.

Let $\mathrm{Gr}_p(H_+)$ be the Grassmann manifold consisting of all closed subspaces $W \subset H$ such that $P_W - P_{H_+}$ is in the Schatten ideal L_p of operators T with $|T|^p$ a trace-class operator. One can show that each $H_+(A, \lambda)$ belongs to $\mathrm{Gr}_p(H_+)$ when $p > \dim M$. All the Grassmannians Gr_p for $p \geq 1$ are homotopy equivalent and they are also homotopy equivalent to the space Fred of all Fredholm operators in H. For this reason Gr_p is a classifying space in K-theory. In particular, the connected components of Gr_p are labelled by the Fredholm index of the projection $W \to H_+$ and each component is simply connected. The second integral cohomology (and second homotopy) of each component is equal to $\mathbb{Z}$. For this reason the complex line bundles are generated by a single element DET_p. In the case $p \leq 2$ the curvature of the dual line bundle DET_p^* is particularly simple; it is given as 2π times the normalized 2-form

$$\omega = \frac{1}{16\pi} \mathrm{tr}\, P d P d P. \tag{1.4}$$

We can cover $\mathcal{A}$ with the open sets $U_\lambda = \{A \in \mathcal{A} | \lambda \notin Spec(D_A)\}$. On each U_λ the map $A \mapsto H_+(A, \lambda) \in \mathrm{Gr}_p$ is smooth. For this reason we may pull back the line bundle DET_p to a line bundle $DET_{p,\lambda}$ over U_λ. We shall not go into the explicit construction of DET_p here, [MR]. Instead, the difference bundles $DET_{p,\lambda\lambda'} = DET_{p,\lambda} \otimes DET_{p,\lambda'}^*$ over $U_{\lambda\lambda'} = U_\lambda \cap U_{\lambda'}$ are easy to describe. The fiber of $DET_{p,\lambda\lambda'}$ is simply the top exterior power of the spectral subspace of D_A for $\lambda < D_A < \lambda'$. By construction, we have a canonical identification

$$DET_{p,\lambda\lambda'} \otimes DET_{p,\lambda'\lambda''} = DET_{p,\lambda\lambda''} \tag{1.5}$$

on the triple overlaps, for $\lambda < \lambda' < \lambda''$. We set $DET_{p,\lambda\lambda'} = DET_{p,\lambda'\lambda}^*$ for $\lambda > \lambda'$. A system of complex line bundles with the cocycle property (1.5) defines a *gerbe*. In this case we have a trivial gerbe, it is generated by local line bundles $DET_{p,\lambda}$ over the open sets U_λ. However, we may push things down to the space of gauge orbits $X = \mathcal{A}/\mathcal{G}$, where $\mathcal{G}$ is the group of *based gauge transformations*, i.e., the group of smooth maps $f : M \to G$ such that $f(p) = 1$ at a base point $p \in M$; here we assume for simplicity that M is connected.

The gauge group acts covariantly on $\mathcal{A}$ and on the eigenvectors of D_A and therefore we may mod out by the $\mathcal{G}$ action to manufacture complex line bundles over $V_{\lambda\lambda'} = U_{\lambda\lambda'}/\mathcal{G}$. These line bundles (which we also denote by $DET_{p,\lambda\lambda'}$) satisfy the same cocycle property (1.5) as the original bundles over $U_{\lambda\lambda'}$. Thus we obtain a gerbe over X. Generically, this gerbe is nontrivial. In general, there is an obstruction to the trivialization of the gerbe given as a *Dixmier–Douady class* in $H^3(X, \mathbb{Z})$.

The Dixmier–Douady class can be computed as follows. Let $\omega_{\lambda\lambda'}$ be the curvature form of $DET_{\lambda\lambda'}$. These satisfy

$$\omega_{\lambda\lambda'} + \omega_{\lambda'\lambda''} = \omega_{\lambda\lambda''} \tag{1.6}$$

on the triple overlaps. Let $\{\rho_\lambda\}$ be a partition of unity subordinate to the covering by the open sets V_λ. The closed 3-forms

$$\omega_\lambda = \sum_{\lambda'} d\rho_{\lambda'} \omega_{\lambda\lambda'}$$

satisfy $\omega_\lambda = \omega_{\lambda'}$ on $V_{\lambda\lambda'}$ and therefore can be glued together to produce a global closed 3-form ω. This is easily seen to be integral. The class $[\omega] \in H^3(X, \mathbb{Z})$ is the DD class of the gerbe.

Let us again consider the case of the trivial gerbe over $\mathcal{A}$. The trivialization by the local line bundles $DET_{p,\lambda}$ resolves the problem of defining a continuous family of CAR representations parametrized by $\mathcal{A}$. Over each set U_λ we can define a CAR algebra representation in $\mathcal{F}'(A, \lambda) = \mathcal{F}(A, \lambda) \otimes DET^*_{p,\lambda}$ by the action $a^*(u) \otimes 1$ and $a(u) \otimes 1$ on the fibers. The crux is that the spaces $\mathcal{F}'(A, \lambda)$ are canonically isomorphic for different values of λ, and hence we have a well-defined family of spaces $\mathcal{F}'(A)$ for all $A \in \mathcal{A}$, [Mi]. In physics terminology, an isomorphism between $\mathcal{F}(A, \lambda)$ and $\mathcal{F}(A, \lambda')$ for $\lambda < \lambda'$ is obtained by 'filling the Dirac sea' from the vacuum level λ up to the level λ'. The filling is canonically defined up to a unitary rotation of the eigenvectors of D_A in the spectral interval $[\lambda, \lambda']$; a rotation of the basis by R leads to a phase factor $\det R$ in the filling, which is exactly compensated by the inverse phase factor in the isomorphism between the dual determinant lines $DET^*_{p,\lambda}(A)$ and $DET^*_{p,\lambda'}(A)$.

2 K-theory aspects of canonical quantization

Let X be a parameter space for a family of self-adjoint operators with both positive and negative essential spectrum, that is, we have a map $X \to \text{Fred}_*$ or in other words we have an element in $K^1(X)$.

As shown in [AS] the space Fred_* is homotopy equivalent to the group of unitaries g in the complex Hilbert space H such that $g - 1$ is compact. According to Palais this group is homotopy equivalent to the group $U^{(p)}$ of unitaries g such that $g - 1 \in L_p$ for any $p \geq 1$. The choice $G = U^{(1)}$ is the most convenient one since it allows to write in a simple way the generators for $H^*(G, \mathbb{Z})$. The cohomology is generated by the odd elements

$$c_{2k+1} = N_k \text{tr}\, (dg g^{-1})^{2k+1}$$

for $k = 0, 1, 2, \ldots$; $N_k = -(1/2\pi i)^{k+1} \frac{k!}{(2k+1)!}$ is a normalization constant.

The infinite-dimensional group U_{res} is of interest to us. Let $\mathcal{H} = L^2(S^1, H)$. Then the group ΩG of smooth based loops in G acts naturally in $\mathcal{H}$. The space $\mathcal{H}$ has a natural polarization $\mathcal{H} = \mathcal{H}_+ \oplus \mathcal{H}_-$ to positive (resp. zero and negative) Fourier modes. The ΩG orbit of $\mathcal{H}_+$ lies in the Hilbert–Schmidt Grassmannian

$Gr_2(\mathcal{H}_+)$. Thus we have $\Omega G \subset U_{\mathrm{res}}(\mathcal{H}_+ \oplus \mathcal{H}_-)$. Actually, the inclusion is a homotopy equivalence (a consequence of Bott periodicity), [CM1].

There is a universal ΩG bundle P over G, with total space the set of smooth paths $f : [0, 1] \to G$ starting from the unit element and such that $f^{-1}df$ is periodic at the end points. Replacing ΩG by U_{res} we obtain an universal U_{res} bundle over G. Thus G is a classifying space for U_{res} bundles and we have:

Proposition. $K^1(X)$ *is isomorphic to the group of equivalence classes of* U_{res} *bundles over* X.

The group structure in $K^1(X)$ comes from the representation of elements in $K^1(X)$ as homotopy classes of maps $g : X \to G$. The product is the pointwise multiplication of maps.

In canonical quantization, it is the U_{res} bundle aspect of $K^1(X)$ which is seen more directly. As we discussed earlier, the Fock representations of the CAR algebra are determined by polarizations of the 1-particle space. A family D_x of self-adjoint operators in Fred_* defines a principal U_{res} bundle over X. The fiber of the bundle at x is the set of unitaries g in $H = H_+ \oplus H_-$ such that the projection onto gH_+ differs from the spectral projection to the subspace $D_x > 0$ by a Hilbert–Schmidt operator. Since $\mathrm{Fred}_* \simeq G$ this bundle is the pull-back of the universal bundle P over G by the map $x \mapsto D_x$.

The second quantization may be viewed as a prolongation problem for a U_{res} bundle P_D over X. We want to construct a vector bundle $\mathcal{F}$ over X such that the fibers are Fock spaces carrying representations of the CAR algebra. The Fock bundle is an associated bundle, not to P_D because U_{res} does not act in the Fock spaces, but to an $\hat{U}_{\mathrm{res}}$ bundle $\hat{P}_D$ which is a prolongation of P_D by the center S^1 of $\hat{U}_{\mathrm{res}}$. The following was proven in [CM1]:

Theorem 1. *There is an obstruction for prolonging* P_D *to* $\hat{P}_D$ *given by the Dixmier–Douady class which is in the 3-form part of the K-theory class* $[D] \in K^1(X)$.

In many cases the DD class can be computed using index theory, see [CMM1,2], [CM1,2], but here we shall discuss a bit more the calculation based on the homotopy equivalence $\mathrm{Fred}_* \simeq G$.

First, one can construct a homotopy equivalence from Fred_* to the space of *bounded* self-adjoint operators with essential spectrum at ± 1, [AS]. For Dirac type operators on a compact manifold we could take for example the map $D \mapsto F_D = D/(|D| + \exp(-D^2))$. This is the approximate sign operator of D. The next step is to map the F_D's to unitary operators by $F_D \mapsto g_D = -\exp(i\pi F_D)$. It is not difficult to see that the difference $g_D - 1$ is trace-class. Note that for Dirac operators on a compact manifold of dimension n we could take $F_D = D/(|D| + 1)$, for simplicity, but then we would have the weaker condition $g_D - 1 \in L_p$ for $p > n$.

Formally, the pull-back of the universal DD class

$$c_3 = \frac{1}{24\pi^2} \text{tr} \, (dgg^{-1})^3$$

with respect to the mapping $F \mapsto g = -\exp(i\pi F)$ can be computed as $\omega = d\theta$ with

$$\theta = \frac{1}{8} \text{tr} \, dF h(\text{ad}_{i\pi F}) dF \tag{2.1}$$

where $h(x) = (\sinh(x) - x)/x^2$ and $\text{ad}_F(z) = [F, z]$. However, in general the expression after 'tr' is not trace-class (and for this reason ω is not exact). There are interesting cases when the above formula makes sense. When D is a Dirac operator on a manifold of dimension n then one can show using standard estimates on pseudodifferential symbols that dF is in L_p for any $p > n$. It follows that the right-hand-side of (2.1) is well-defined for a Dirac operator on a circle. The case of dimension $n = 3$ is a limiting case. In three dimensions the trace is logarithmically divergent (when defined as a conditional trace in a basis where F is diagonal) and subtracting the logarithmic divergence one obtains a finite expression which can be used to define θ.

The case $F^2 = 1$ is also interesting; these points define a Grassmann manifold since $(F + 1)/2$ is a projection onto an infinite-dimensional subspace. Again, in the case of sign operators defined by Dirac operators one can show that F defines a point in Gr_p for $p > n$. Assuming that the trace in (2.1) converges, one obtains a specially simple formula for the form θ,

$$\theta = \frac{1}{16\pi} \text{tr} \, F dF dF \quad \text{for } F^2 = 1 \, .$$

Not surprisingly, this is (mod a factor 2π) the curvature formula for the determinant bundle DET_2 over Gr_2.

3 Twisted k-theory and QFT

There has been an extensive discussion of twisted K-theory in the recent string theory literature, inspired by suggestions in [Wi]. I will not discuss any of the string theory applications here. Instead, I want to point out that twistings in K-theory are related to some very basic constructions in standard QFT.

Twisted K-theory was introduced in [Ro] as a generalization of algebraic K-theory on C^*-algebras. Today there are several equivalent definitions of twisted K, see e.g. [BCMMS]. For QFT problems I find it most convenient to use the topologist definition of twisted K-theory groups.

Twisted K-theory elements arise from principal $PU(H)$ bundles. Here $PU(H) = U(H)/S^1$ is the projective unitary group in a complex Hilbert space. Since $U(H)$ is contractible by Kuiper's theorem, the homotopy type of $PU(H)$ is simple: The only nonzero homotopy group is $\pi_2(PU(H)) = \mathbb{Z}$. For this reason $H^2(PU(H), \mathbb{Z}) = \mathbb{Z}$. On the Lie algebra level, the basic central extension $1 \to S^1 \to U(H) \to \text{PU}(H) \to 1$ of $\text{PU}(H)$ is given as follows. Let $\psi_0 \in H$ be a fixed vector of unit length and $\mathfrak{g}_0$ the subspace in the Lie algebra $\hat{\mathfrak{g}}$ of $U(H)$ consisting of operators x such that $(\psi_0, x\psi_0) = 0$. For each y in the Lie algebra $\mathfrak{g} = \hat{\mathfrak{g}}/i\mathbb{R}$ of $\text{PU}(H)$ there is a unique element $\hat{y} \in \mathfrak{g}_0$ such that $\pi(\hat{y}) = y$, where π is the canonical projection. We can write $\hat{\mathfrak{g}} = \mathfrak{g} \oplus i\mathbb{R}$ and the commutation relations in $\hat{\mathfrak{g}}$ can be written as

$$[(x, \alpha), (y, \beta)] = ([x, y]_\mathfrak{g}, c(x, y))$$

where the cocycle c is given by $c(x, y) = [\hat{x}, \hat{y}] - \widehat{[x, y]_\mathfrak{g}}$.

Given a principal $\text{PU}(H)$ bundle P over X we can construct an associated Fred_* bundle $Q(P)$ over X as $P \times_{\text{PU}(H)} \text{Fred}_*$ where $\text{PU}(H)$ acts on Fred_* through conjugation. Again, using the homotopy equivalence $\text{Fred}_* \simeq G$ we might consider G bundles as well; but then it is important to keep in mind that these are not principal G bundles. The twisted K^1 of X, to be denoted by $K^1(X, [P])$, is then defined as the set of homotopy classes of sections of the bundle $Q(P)$. A similar definition is used for the twisted K^0 group, the space Fred_* is then replaced by the space of all Fredholm operators in H, or alternatively, we can use the model $U_{\text{res}} \simeq \text{Fred}$.

Quantum field theory provides concrete examples of twisted bundles $Q(P)$ and its sections. As we have seen, a family of Dirac type hamiltonians parametrized by X is an element of $K^1(X)$ or equivalently, an equivalence class of U_{res} bundles over X. The basic observation is that the group U_{res} can be viewed as a subgroup of $\text{PU}(\mathcal{F})$ where $\mathcal{F}$ is a Fock space carrying a representation of the central extension $\hat{U}_{\text{res}}$. Therefore we can extend the U_{res} bundle to a $\text{PU}(\mathcal{F})$ bundle P over X. A section of the associated bundle $Q(P)$ becomes now a function f from P to the space of operators of type Fred_* in the Fock space $\mathcal{F}$ such that $f(pg) = g^{-1} f(p)g$ for all $g \in \text{PU}(\mathcal{F})$ and $p \in P$. Since our $\text{PU}(\mathcal{F})$ reduces to a U_{res} bundle P_0 we might as well construct a section of $Q(P)$ from an equivariant function on P_0 with values in Fred_*.

Let U_i be a family of open sets covering X equipped with local trivializations of the U_{res} bundle. Let $g_{ij} : U_i \cap U_j \to U_{\text{res}}$ be the corresponding transition functions. Then a section of $Q(P)$ can be given by a family of maps

$$a_i : U_i \to \text{Fred}_*(\mathcal{F})$$

such that

$$a_j(x) = \hat{g}_{ij}(x)^{-1} a_i(x) \hat{g}_{ij}(x) \quad \text{for } x \in U_{ij}.$$

Here $\hat{g}$ is the quantum operator, acting in $\mathcal{F}$, corresponding to the '1-particle operator' g.

In general, twisted K^1 is not easy to compute. Let us consider a simple example.

Example. Let $X = S^3$ which can be identified as the group $\text{SU}(2)$ of unitary 2×2 matrices of determinant $= 1$. A twisted $\text{PU}(H)$ bundle P over S^3 is constructed as follows. First, the space of smooth vector potentials A on a circle with values in the Lie algebra of $\text{SU}(2)$ can be viewed as a principal $\Omega\,\text{SU}(2)$ bundle over S^3. The group of based loops $\Omega\,\mathrm{SU}(2)$ acts on the vector potentials through gauge transformations $A \mapsto g^{-1}Ag + g^{-1}dg$. A vector potential modulo based gauge transformations is parametrized by the holonomy around the circle, giving an element in $S^3 = \mathrm{SU}(2)$.

The loop group $\Omega\,\mathrm{SU}(2)$ acts in the space H of square-integrable $\mathbb{C}^2$ valued functions on the unit circle, giving an embedding $\Omega\,\mathrm{SU}(2) \subset U_{\mathrm{res}}(H_+ \oplus H_-)$, the polarization being given by the splitting to negative and nonnegative Fourier modes. Thus we can extend the $\Omega\,\mathrm{SU}(2)$ bundle $\mathcal{A}$ over S^3 to a principal U_{res} bundle P_0. This extends, as explained above, to a principal $\mathrm{PU}(\mathcal{F})$ bundle over S^3.

All principal $\mathrm{PU}(\mathcal{F})$ bundles over S^3 are classified by the homotopy classes of transition functions $S^2 \to \mathrm{PU}(\mathcal{F})$, that is, by the elements $n \in \pi_2(\mathrm{PU}(\mathcal{F})) = \mathbb{Z}$. The construction above gives the basic bundle with $n = 1$. The higher bundles are obtained by taking tensor powers (and their duals) of the Fock space representations of the central extension of the loop group $\Omega\,\mathrm{SU}(2)$. The K^1-theory twist in this case is fixed by a choice of the integer n. An element of the twisted $K^1(S^3, n)$ is now given by the homotopy class of pairs of functions $h_\pm : S^3_\pm \to \mathrm{Fred}_*$ such that on the equator $S^2 \sim S^3_+ \cap S^3_-$

$$h_+(x) = g^{-1}(x)h_-(x)g(x),$$

where $g : S^2 \to \mathrm{PU}(H)$ is can be given explicitly, using the embedding $\Omega\,\mathrm{SU}(2) \subset \mathrm{PU}(\mathcal{F})$ and the fact that $\pi_2\Omega\,\mathrm{SU}(2) = \mathbb{Z}$, see [CMM2] for details.

All classes in $K^1(\mathrm{SU}(2), n)$ can actually be given in a simpler way. We can use the homotopy equivalence $\mathrm{Fred}_* \simeq U^{(1)} = G$. Choose then $h_+ : S^3_+ \to G$ such that it is equal to the unit element on the overlap $S^3_+ \cap S^3_-$ and take h_- as the constant function on S^3_- taking the value $1 \in G$. Then clearly $h_\pm$ are intertwined by the transition function g on the overlap. Since h_+ is constant on the boundary of S^3_+, it can be viewed as a map $g_+ : S^3 \to G$. The winding number of this map in $\pi_3(G) = \mathbb{Z}$, modulo n, determines the class in $K^1(S^3, n) = \mathbb{Z}/n\mathbb{Z}$.

The transition function in the present example, being a map from S^2 to U_{res} (which is a classifying space for K^0), can also be viewed as an element of $K^0(S^2) = \mathbb{Z} \oplus \mathbb{Z}$. This is an essential part of the computation of $K^1(S^3, n)$, based on the Mayer–Vietoris theorem in K-theory, see [BCMMS] for details, or the original computation in [Ro].

4 An example: supersymmetric WZW model

Let H_b be a complex Hilbert space carrying an irreducible unitary highest weight representation of the central extension $\widehat{LG}$ of the loop group LG of level k; here G is assumed to be compact and simple, $\dim G = N$. The level satisfies $2k/\theta^2 = 0, 1, 2, \ldots$, where θ is the length of the longest root of G.

Let H_f be a fermionic Fock space for the CAR algebra generated by elements ψ_n^a with $n \in \mathbb{Z}$ and $a = 1, 2, \ldots, N = \dim G$,

$$\psi_n^a \psi_m^b + \psi_m^b \psi_n^a = 2\delta_{n,-m} \delta_{a,b}. \tag{4.1}$$

The Fock vacuum is a subspace of H_f of dimension $2^{[N/2]}$ (here $[p]$ denotes the integral part of a real number p). The vacuum subspace carries an irreducible representation of the Clifford algebra generated by the ψ_0^a's and in addition any vector in the vacuum subspace is annihilated by all ψ_n^a's with $n < 0$.

The tensor product space $H = H_f \otimes H_b$ carries a tensor product representation of $\widehat{LG}$. The fermionic part of the representation is determined by the requirement

$$T_f(g)\psi(\alpha)T_f(g)^{-1} = \psi(g \cdot \alpha), \tag{4.2}$$

where α is a $\mathbb{C}^N$ valued smooth function on the unit circle and $\psi(\alpha) = \sum \psi_n^a \alpha_{-n}^a$, where the α_n^a's are the Fourier components of the vector valued function α. The action of $g \in LG$ on α is the point-wise adjoint action on the Lie algebra of the loop group.

The Lie algebra of $\widehat{LG}$ acting in H_b is generated by the Fourier modes T_n^a subject to the commutation relations

$$[T_n^a, T_m^b] = \lambda_{abc} T_{n+m}^c + kn\delta_{n,-m}\delta_{a,b}, \tag{4.3}$$

where the λ_{abc}'s are the structure constants of the Lie algebra $\mathfrak{g}$ in a basis T^a which is orthonormal with respect to the Killing form $\langle X, Y \rangle = -\text{tr}(\text{ad}_X \cdot \text{ad}_Y)$. There is, up to a phase factor, a unique normalized vector $x_\lambda \in H_b$ such that $T_n^a x_\lambda = 0$ for $n < 0$ and is a highest weight vector of weight λ for the finite-dimensional Lie algebra $\mathfrak{g}$.

We denote the loop algebra generators acting in the fermionic Fock space H_f by K_n^a. They satisfy the commutation relations

$$[K_n^a, K_m^b] = \lambda_{abc} K_{n+m}^c + \frac{1}{2}n\delta_{n,-m}\delta_{a,b}. \tag{4.4}$$

Explicitly, the generators are given by

$$K_n^a = -\frac{1}{4}\lambda_{abc} : \psi_{n+m}^b \psi_{-m}^c : . \tag{4.5}$$

The normal ordering $::$ is defined as the rule to write the operators with negative momentum index to the right of those with positive index. Actually, since our λ_{abc}'s are totally antisymmetric, the normal ordering in (4.5) is irrelevant.

We denote by S_n^a the generators of the tensor product representation in $H = H_b \otimes H_f$. They satisfy the relations

$$[S_n^a, S_m^b] = \lambda_{abc} S_{n+m}^c + \left(k + \frac{1}{2}\right)n\delta_{ab}\delta_{n,-m}. \tag{4.6}$$

The free hamilton operator is

$$h = h_b \otimes 1 + 1 \otimes (2k+1)h_f + \frac{N}{24}(1 \otimes 1)$$

where

$$h_b = - : T_n^a T_{-n}^a : \quad \text{and} \quad h_f = \frac{1}{4} : n \psi_n^a \psi_{-n}^a : \tag{4.7}$$

We use the conventions

$$(\psi_n^a)^* = \psi_{-n}^a \quad \text{and} \quad (T_n^a)^* = -T_{-n}^a. \tag{4.8}$$

As seen by a direct computation, the supercharge Q satisfies $Q^2 = h$ and is defined by

$$Q = i \psi_{-n}^a T_n^a - \frac{i}{12} \lambda_{abc} \psi_n^a \psi_m^b \psi_{-m-n}^c. \tag{4.9}$$

For a detailed description of the whole super current algebra, see [KT], or in somewhat different language, [La]. Again, by antisymmetry of the structure constants, no normal ordering is necessary in the last term on the right. The general structure of Q has similarities with Kostant's cubic Dirac operator, [Ko], (containing a cubic term in the 'gamma matrices' ψ_n^a); another variant of this operator has been discussed in conformal field theory context in [B-R]. Restricting to zero momentum modes, the operator Q in fact becomes Kostant's operator

$$\mathcal{K} = i \gamma^a T^a - \frac{i}{12} \lambda_{abc} \gamma^a \gamma^b \gamma^c, \tag{4.10}$$

where $\psi_0^a = \gamma^a$ are the Euclidean gamma matrices in dimension N. By the antisymmetry of the structure constants the last term is totally antisymmetrized product of gamma matrices.

The supercharge is a hermitean operator in a dense domain of the Fock space H, including all the states which are finite linear combinations of eigenvectors of h.

There is a difference between the cases $N = \dim G$ is odd or even. In the even case we can define a grading operator Γ which anticommutes with Q. It is given as $\Gamma = (-1)^F \psi_0^{N+1}$, where F is the fermion number operator, $\psi_n^a F + F \psi_n^a = \frac{n}{|n|} \psi_n^a$ for $n \neq 0$, and ψ_0^{N+1} is the chirality operator on the even dimensional group manifold G, with eigenvalues ± 1.

We can couple the supercharge to an external $\mathfrak{g}$ valued vector potential A on the circle by setting, with $\tilde{k} = k + \frac{1}{2}$,

$$Q_A = Q + \tilde{k} \psi_n^a i A_{-n}^a \tag{4.11}$$

where the Fourier components of the Lie algebra valued vector potential A satisfy $(A_n^a)^* = -A_{-n}^a$. By a direct computation,

$$[S_n^a, Q_A] = i\tilde{k}(n \psi_n^a + \lambda_{abc} \psi_m^b A_{n-m}^c). \tag{4.12}$$

This implies that for a finite gauge transformation $f \in LG$

$$S(f) Q_A S(f)^{-1} = Q_{A^f}, \tag{4.13}$$

where $A^f = f^{-1} A f + f^{-1} df$.

Theorem 2. *The family Q_A of hermitean operators in H defines an element of the twisted K-theory group $K^1(G, k')$ where the twist is $k' = (2k + 1)/\theta^2$ times the generator of $H^3(G, \mathbb{Z})$.*

Proof. As pointed out in [BCMMS], twisted K-theory classes over G can be thought of as equivariant maps $f : P \to \text{Fred}_*$, where P is a principal $PU(H)$ bundle over G with a given Dixmier–Douady invariant $\omega \in H^3(G, \mathbb{Z})$. The equivariance condition is $f(pg) = g^{-1}f(p)g$ for $g \in PU(H)$. In the case at hand, the principal bundle P is obtained by embedding of the loop group $LG \subset PU(H)$ through the projective representation of LG of level $k + \frac{1}{2}$. As we saw in (4.13), the family Q_A is equivariant with respect to the (projective) loop group action. Finally, the Dixmier–Douady class determined by the level $k+1/2$ of the projective representation is k' times the generator $\frac{1}{24\pi^2}\text{tr}(g^{-1}dg)^3$ on $G = \mathcal{A}/\Omega G$. $\square$

Note that in the even case the family Q_A gives necessarily a trivial element in K^1. This follows from the existence of the operator Γ which anticommutes with the hermitean operators Q_A. Thus there is no net spectral flow for this family of operators, which is an essential feature in odd K-theory.

However, in the even case we can define elements in K^0 by the standard method familiar from the theory of ordinary Dirac operators: We can split $Q_A = Q_A^+ + Q_A^-$, using the chiral projections $\frac{1}{2}(\Gamma \pm 1)$, where $(Q_A^+)^* = Q_A^-$ is a pair of nonhermitean operators with nontrivial index theory. Either of the families $Q_A^\pm$ can be used to define an element of $K^0(G, k')$. Again, we use the observation that elements in $K^0(G, k')$ can be viewed as equivariant maps from the total space P of a principal $PU(H)$ bundle over G to the set Fred of all Fredholm operators in H.

The operator Q is also of interest in cyclic cohomology. It can be used to construct the entire cyclic cocycle of Jaffe, Lesniewski, and Osterwalder [JLO] (they considered the case of abelian Wess–Zumino model). The key fact is that the operator $\exp(-sQ^2)$ is a trace class operator for any real $s > 0$; in fact, there is an explicit formula for the trace, it is equal to the product of Kac character formulas for two highest weight representations of the loop group, one in the bosonic Fock space and the second in the fermionic Fock space.

The second ingredient in cyclic cohomology is an associative algebra $\mathcal{B}$ acting in the Hilbert space such that each $[Q, a]$ is a bounded operator for $a \in \mathcal{B}$. This is the case for the elements $S(f) = f_n^a S_{-n}^a$ in the current algebra for each smooth function f on the unit circle. However, the operators $S(f)$ are not bounded. This should not be a serious problem since the norm of the restriction of $S(f)$ to a finite energy subspace is growing polynomially in energy, whereas $\text{tr} e^{-sQ^2}$ is decreasing exponentially in energy. Recall that the even entire JLO cocycle is composed of terms

$$\int_{s_i > 0, \sum s_i = 1} \text{tr}\, \Gamma a_0 e^{-s_0 Q^2}[Q, a_1]e^{-s_1 Q^2} \ldots [Q, a_n]e^{-s_n Q^2} ds_0 \ldots ds_{n-1}$$

with $a_i \in \mathcal{B}$. This is finite for elements a_i in the current algebra. The above formula can be used also in the odd case by setting $\Gamma = 1$.

Since the twisted K-theory classes above are labelled by the irreducible highest weight representations of an affine Kac–Moody algebra, it is natural to ask what is the relation of the twisted K-theory on G to the Verlinde algebra of G, on a given level k. Actually, D. Freed, M. Hopkins and C. Teleman have announced that there is a product in $K_G(G, k)$ (the G equivariant version of $K(G, k)$) which makes it isomorphic to the Verlinde algebra, [F], [FHT]. It would be interesting to understand the relation of their geometric construction to the algebraic construction based on the supersymmetric WZW model.

References

[AS]   M. F. Atiyah and I. Singer: Index theory for skew-adjoint Fredholm operators, *Inst. Hautes Études Sci. Publ. Math.* 37 (1969), 305.

[Ar]   H. Araki, Bogoliubov automorphisms and Fock representations of canonical anti-commutation relations, in *Operator algebras and mathematical physics*, Contemp. Math. 62, Amer. Math. Soc., Providence, R.I., 1987, 23.

[BCMMS]   P. Bouwknegt, Alan L. Carey, Varghese Mathai, Michael K. Murray, and Danny Stevenson, Twisted K-theory and K-theory of bundle gerbes, *Commun. Math. Phys.* 228 (2002), 17.

[Br]   J.-L. Brylinski, *Loop Spaces, Characteristic Classes, and Geometric Quantization*, Progr. Math. 107, Birkhäuser, Boston, Mass., 1993.

[B-R]   L. Brink and P. Ramond, Dirac equations, light cone supersymmetry, and super-conformal algebras, in *The many faces of the superworld*, World. Sci. Publ., 2000, 398.

[CMM1]   A. L. Carey, J. Mickelsson, and M. K. Murray, Index theory, gerbes, and hamiltonian quantization, *Commun. Math. Phys.* 183 (1997), 707.

[CMM2]   A. L. Carey, J. Mickelsson, and M. K. Murray, Bundle gerbes applied to quantum field theory, *Rev. Math. Phys.* 12 (2000), 65.

[CM1]   A. L. Carey and J. Mickelsson, A gerbe obstruction to quantization of fermions on odd-dimensional manifolds with boundary, *Lett. Math. Phys.* 51 (2000), 145.

[CM2]   A. L. Carey and J. Mickelsson, The universal gerbe, Dixmier-Douady class, and gauge theory, *Lett. Math. Phys.* 59 (2002), 47.

[F]   D. Freed, Twisted K-theory and loop groups, math.AT/0206237, in *Proceedings of the international congress of mathematicians* (ICM 2002, Beijing, China) Vol. III: Invited lectures, Higher Education Press, Beijing 2002, 419–430.

[FHT]   D. Freed, M. Hopkins, and C. Teleman, Twisted equivariant K-theory with complex coefficients, math.AT/0206257.

[GR]   K. Gawedzki and N. Reis, WZW branes and gerbes, *Rev. Math. Phys.* 14 (2002), 1281.

[JLO]   A. Jaffe, A. Lesniewski, K. Osterwalder, Quantum K theory. 1. The Chern character, *Commun. Math. Phys.* 118 (1988), 1.

[KT]   V. Kac and I. Todorov, Superconformal current algebra and their unitary representations, *Commun. Math. Phys.* 102 (1985), 337.

[Ko]   B. Kostant, A cubic Dirac operator and emergence of Euler multiplets of representations for equal rank subgroups, *Duke Math. J.* 100 (1999), 447.

[La]   G. Landweber, Multiplicities of representations and Kostants Dirac operator for equal rank loop groups, *Duke Math. J.* 110 (1) (2001), 121–160.

[Lu]   L.-E. Lundberg, Quasi-free second quantization, *Commun. Math. Phys.* 50 (1976), 103.

[Mi]   J. Mickelsson, On the hamiltonian approach to commutator anomalies in $3 + 1$ dimensions, *Phys. Lett.* B 241 (1990), 70.

[MR]   J. Mickelsson and S. Rajeev, Current algebras in $d + 1$ dimensions and determinant bundles over infinite-dimensional Grassmannians, *Commun. Math. Phys.* 116 (1988), 365.

[Mu]   M. K. Murray., Bundle gerbes, *J. London Math. Soc.* (2) 54 (2) (1996), 403.

[PS]   A. Pressley and G. Segal, *Loop Groups,* Clarendon Press, Oxford 1986.

[Ro]   J. Rosenberg, Homological invariants of extensions of C^*-algebras, *Proc. Symp. in Pure Math.* 38 (1982), 35.

[Se]   G. Segal, Topological structures in string theory, *Phil. Trans. R. Soc. Lond.* A 359 (2001), 1389.

[Wi]   E. Witten, D-branes and K-theory, *J. High Energy Phys.* 12 (1998), 019; hep-th/9810188.

Current groups for non-compact manifolds and their central extensions

Karl-Hermann Neeb

Technische Universität Darmstadt Schlossgartenstrasse 7
64289 Darmstadt, Germany
email: neeb@mathematik.tu-darmstadt.de

Dedicated to Karl Heinrich Hofmann on the occasion
of his 70th birthday

Abstract. In this paper we study two types of groups of smooth maps from a non-compact manifold M into a Lie group K which may be infinite-dimensional: the group $C_c^\infty(M, K)$ of compactly supported maps and for a compact manifold M and a closed subset S the group $C^\infty(M, S; K)$ of those maps which vanish on S, together with all their derivatives. We study central extensions of these groups associated to Lie algebra cocycles of the form $\omega(\xi, \eta) = [\kappa(\xi, d\eta)]$, where $\kappa \colon \mathfrak{k} \times \mathfrak{k} \to Y$ is a symmetric invariant bilinear map on the Lie algebra $\mathfrak{k}$ of K and the values of ω lie in $\Omega^1(M; Y)/dC^\infty(M; Y)$. For such cocycles we show that a corresponding central Lie group extension exists if and only if this is the case for $M = \mathbb{S}^1$. If K is finite-dimensional semisimple, this implies the existence of a universal central Lie group extension of the identity component of the current groups.

2000 Mathematics Subject Classification: 22E65; 58D15, 57T20.

Introduction

If M is a compact manifold and K a Lie group (which may be infinite-dimensional), then the so called current groups $C^\infty(M; K)$, endowed with the group structure given by pointwise multiplication, are interesting infinite-dimensional Lie groups arising in many circumstances. If M is a non-compact manifold, the full group $C^\infty(M; K)$ seems to be far too large to carry a Lie group structure compatible with its natural group topology, so that it is natural to study subgroups of maps $f \colon M \to K$ that either vanish outside a compact subset or decay fast enough at infinity. In the present paper we investigate the following two types of current groups on a non-compact manifold M. The first class consists of the groups $C_c^\infty(M; K)$ of compactly supported smooth maps and the second class of the groups $C^\infty(M, S; K)$ of maps on a compact manifold M for which all partial derivatives vanish on the closed subset $S \subseteq M$. The groups

$C^\infty(M, S; K)$ have the advantage that they are Fréchet–Lie groups if K is a Fréchet–Lie group, the Lie algebra is given by $C^\infty(M, S; \mathfrak{k})$. We consider them as groups of smooth maps on the non-compact manifold $M \setminus S$ vanishing at infinity. The groups $C_c^\infty(M; K)$ are modeled on the space $C_c^\infty(M; \mathfrak{k})$ which is not metrizable in its natural direct limit topology, not even for $K = \mathbb{R}$.

The goal of the present paper is to understand central extensions of current groups G which are identity components of groups of the type $C_c^\infty(M; K)$ or $C^\infty(M, S; K)$. For an infinite-dimensional Lie group G not every Lie algebra cocycle $\omega \colon \mathfrak{g} \times \mathfrak{g} \to \mathfrak{z}$ defines a central extension of $\mathfrak{g}$ by $\mathfrak{z}$ which can be integrated to a Lie group. In [Ne02a] we show that there are two kinds of obstructions. The first one is an element of $\mathrm{Hom}(\pi_1(G), \mathrm{Lin}(\mathfrak{g}, \mathfrak{z}))$, and we will see in Theorem V.8 that it always vanishes for current groups. The second obstruction is that the image of a certain "period map" $\mathrm{per}_\omega \colon \pi_2(G) \to \mathfrak{z}$ need not be discrete. To illuminate the obstructions for the class of current groups, we need a good deal of information on the abelian group $\pi_2(G)$. This information is obtained in Appendix A where we show that the computation of the homotopy groups of G can be reduced to the computation of those of groups $C(X; K)$ of continuous maps, where X is a compact manifold with boundary.

The Lie algebra cocycles we are interested in are those of *product type*, i.e., cocycles $\omega \colon \mathfrak{g} \times \mathfrak{g} \to \mathfrak{z}$ for which there exists a sequentially complete locally convex space Y and an invariant continuous symmetric bilinear form $\kappa \colon \mathfrak{k} \times \mathfrak{k} \to Y$ such that $\omega(\xi, \eta) = [\kappa(\xi, d\eta)]$ defines a cocycle with values in $\mathfrak{z} := \mathfrak{z}_{M,c}(Y) := \Omega_c^1(M; Y)/dC_c^\infty(M; Y)$ for $\mathfrak{g} = C_c^\infty(M; \mathfrak{k})$, and $\mathfrak{z} := \mathfrak{z}_{(M,S)}(Y) := \Omega^1(M, S; Y)/dC^\infty(M, S; Y)$ for $\mathfrak{g} = C^\infty(M, S; \mathfrak{k})$. We systematically use forms with values in an infinite-dimensional vector space to incorporate in particular the universal invariant symmetric bilinear form $\kappa \colon \mathfrak{k} \times \mathfrak{k} \to V(\mathfrak{k})$.

The main steps in our analysis of these cocycles and their period maps are as follows. In Section IV we show that the image of the period map always lies in the subspace of $\mathfrak{z}$ coming from the closed 1-forms. Then the problem is to determine the period group $\Pi_\omega := \mathrm{im}(\mathrm{per}_\omega) \subseteq \mathfrak{z}$ and to see if it is discrete. For the case $\mathfrak{g} = C_c(M; \mathfrak{k})$ it is quite hard to get information on the discreteness of a subgroup of $\mathfrak{z} = \mathfrak{z}_{M,c}(Y)$, resp., $H_{\mathrm{dR},c}^1(M; Y)$ because $\mathfrak{z}$ is a direct limit of spaces on which the topology is given by explicit seminorms. We address this problem by approximating the non-compact manifold M by suitably chosen submanifolds X_n with boundary in such a way that

$$H_{\mathrm{dR},c}^1(M; Y) = \varinjlim H_{\mathrm{dR}}^1(X_n, \partial X_n; Y)$$

(Section III). From this relation we then derive the existence of a countable set B so that

$$H_{\mathrm{dR},c}^1(M; Y) \cong Y^{(B)},$$

is a locally convex direct sum, where the projections are given by integrals over singular cycles or over piecewise smooth proper maps $\mathbb{R} \to M$. In Section IV this information permits us to see that Π_ω is discrete for each M if and only if this is the case for the circle $M = \mathbb{S}^1$. In the latter case $\pi_2(C^\infty(\mathbb{S}^1, K)) \cong \pi_2(K) \times \pi_3(K)$, the period map vanishes on $\pi_2(K)$, and $\mathfrak{Z}_{\mathbb{S}^1}(Y) \cong Y$, so that we arrive at a map $\pi_3(K) \to Y$ which depends only on the bilinear form κ. For finite-dimensional groups K we can now use information from [MN02] to see that the period group is discrete if κ is the universal invariant symmetric bilinear form. This is used in Section VI to construct for a finite-dimensional reductive Lie group K with simply connected center a universal central extension of the groups $C_c^\infty(M; K)_e$ and $C^\infty(M, S; K)_e$. In both cases there are many examples where the period group has infinite rank. A simple example with $M = \mathbb{S}^2$ and S a sequence with limit point is discussed in detail in Example II.12. All the concrete examples of central extensions of infinite-dimensional Lie groups which have been dealt with so far in the literature have finitely generated period groups. In this sense we provide new and concrete examples, where this is not the case.

The class of current groups most extensively studied is the class of loop groups ($M = \mathbb{S}^1$ and K compact) which is completely covered by Pressley and Segal's monograph [PS86]. The main point of the present paper is to see which Lie algebra cocycles of product type can be integrated to a central Lie group extension. These central extensions occur naturally in mathematical physics, where the problem to integrate projective representations of groups to representations of central extensions is at the heart of quantum mechanics ([Mic87], [LMNS98], [Wu01]). The central extensions of current groups are often constructed via representations by pulling back central extensions of certain operator groups ([Mic89]). It is our philosophy that one should try to understand the central extensions of a Lie group G first, and then construct representations of these central extensions. In this context certain discreteness conditions for Lie algebra cocycles appear naturally because they ensure that the corresponding central Lie algebra extension integrates to a central Lie group extension ([Ne02a]). We think of these discreteness conditions as an abstract version of the discreteness of quantum numbers in quantum physics. As an outcome of our analysis, we will see that for our general results we do not have to impose any restriction on the group K. It may be any infinite-dimensional Lie group. This permits in particular iterative constructions based on relations like $C^\infty(M \times N; K) \cong C^\infty(M, C^\infty(N; K))$ for compact manifolds M and N.

The content of the paper is as follows. In Section I we introduce the two kinds of Lie groups we are dealing with: $C_c^\infty(M; K)$ for M non-compact, and $C^\infty(M, S; K)$ for M compact and $S \subseteq M$ closed.

The main result of Section II is that the group $H^1_{\mathrm{dR}}(M, S; \Gamma)$ of all de Rham cohomology classes modulo S for which all integrals over singular cycles modulo S are contained in a discrete subgroup Γ of Y is discrete (Theorem II.7). In Section V this is used to prove the discreteness of period groups for cocycles of product type for the groups $C^\infty(M, S; K)$.

Our strategy to get a description of the spaces $\mathfrak{Z}_{M,c}(Y)$ and $H^1_{\mathrm{dR},c}(M; Y)$ for a non-compact manifold M is to describe M as a union of certain compact submanifolds with boundary $(X_n)_{n \in \mathbb{N}}$ with $X_n \subseteq X^0_{n+1}$. To get information on the space $H^1_{dR,c}(M; Y)$, we will need detailed information on the natural maps

$$H^1_{dR}(X_n, \partial X_n; Y) \to H^1_{dR}(X_{n+1}, \partial X_{n+1}; Y)$$

which is obtained in Theorem III.6. This result is used in Theorem IV.7 to obtain the isomorphism $H^1_{dR,c}(M; Y) \cong Y^{(B)}$ mentioned above. As a corollary, we show that if Γ is discrete, then $H^1_{dR,c}(M; \Gamma)$ is discrete.

In Section V we first explain the general setup for central extensions of Lie groups. The main question arising in the integration of Lie algebra cocycles ω to central extensions of Lie groups is whether the corresponding period group Π_ω is discrete. We then show that for cocycles of product type for the groups $C^\infty_c(M; K)_e$ and $C^\infty(M, S; K)_e$ the period group $\Pi_{M,\kappa}$ is discrete if and only if this is the case for $\Pi_{\mathbb{S}^1,\kappa}$. This reduces the discreteness problem to the case of loop groups, which is known for K compact, and therefore for all finite-dimensional Lie groups (cf. [PS86], [MN02]). We further show that $\Pi_{M,\kappa} = H^1_{dR,c}(M; \Pi_{\mathbb{S}^1,\kappa})$ for each non-compact manifold M and each κ.

In Section VI we finally turn to universal central extensions. For the special class of finite-dimensional semisimple Lie groups K, each Lie algebra cocycle $\omega \in Z^2_c(C^\infty_c(M, \mathfrak{k}), \mathfrak{z})$ is equivalent to a cocycle of product type ([Ma02], [Fe88]). This observation permits us to construct a universal central extension of the Lie algebra $\mathfrak{g} := C^\infty_c(M; \mathfrak{k})$, and we show that this construction can be globalized in our context, providing a universal central extension of the connected Lie group $C^\infty_c(M; K)_e$.

In Appendix A we deal with the topology of the groups $C^\infty_c(M; K)$ and $C^\infty(M, S; K)$. For our purposes it is of particular importance to know their homotopy groups. We write $C_0(M; K)$ for the group of continuous functions vanishing at infinity, endowed with the topology of uniform convergence. Information on homotopy groups is obtained by several approximation arguments showing that the inclusion maps

$$C^\infty_c(M; K) \hookrightarrow C_0(M; K) \quad \text{and} \quad C^\infty(M, S; K) \hookrightarrow C_0(M \setminus S; K)$$

are weak homotopy equivalences, i.e., induce isomorphisms of all homotopy groups. These results are motivated by the fact that it is usually much easier to deal with spaces of continuous maps than with spaces of differentiable maps. We also note that if K is a Banach-, resp., Fréchet–Lie group, then the same holds for the groups $C_0(M; K)$ and $C_0(M \setminus S; K)$.

Appendix B contains several results on direct limits of locally convex spaces. These are needed to deal with the spaces of compactly supported smooth functions or differential forms on a non-compact manifold. The difficulties with these spaces arise from the fact that they are not metrizable, which makes it harder to prove that a subgroup is discrete.

This paper contributes in particular to the program dealing with Lie groups G whose Lie algebras $\mathfrak{g}$ are *root graded* in the sense that there exists a finite irreducible root system Δ such that $\mathfrak{g}$ has a Δ-grading $\mathfrak{g} = \mathfrak{g}_0 \oplus \bigoplus_{\alpha \in \Delta} \mathfrak{g}_\alpha$, it contains the split simple Lie algebra $\mathfrak{k}$ corresponding to Δ as a graded subalgebra, and is generated, topologically, by the root spaces $\mathfrak{g}_\alpha, \alpha \in \Delta$. All Lie groups of the type $C_c^\infty(M; K)$, K simple complex, are of this type, and the same holds for their central extension. A different but related class of groups arising in this context are the Lie groups $SL_n(A)$ and their central extensions, where A is a continuous inverse algebra, i.e., a locally convex unital associative algebra with open unit group and continuous inversion ([Gl01c], [Ne03]).

In [Ne02b] we discuss the universal central extensions of the groups $SL_n(A)$, which are Lie group versions of the Steinberg groups $St_n(A)$. In [MN02, Rem. II.12] we have shown that for $K = SL_n(A)$, A a commutative continuous inverse algebra, the form $\kappa: \mathfrak{k} \times \mathfrak{k} \to A$, $\kappa(x, y) = \mathrm{tr}(xy)$ is universal, and that the image of the corresponding period map is discrete for the corresponding product type cocycle on the Lie algebra $C^\infty(M; \mathfrak{k})$ of the group $C^\infty(M; K)$. For non-commutative algebras the image of the period map is not always discrete ([Ne02b]).

Throughout this paper we will use the concept of an infinite-dimensional Lie group described in detail in [Mil83] (see also [Gl01a] for arguments showing that the completeness requirements made in [Mil83] are not necessary to define the concept). This means that a *Lie group* G is a smooth manifold modeled on a locally convex space $\mathfrak{g}$ for which the group multiplication and the inversion are smooth maps. We write $\lambda_g(x) = gx$, resp., $\rho_g(x) = xg$ for the left, resp., right multiplication on G. Let $e \in G$ be the identity element. Then each $X \in T_e(G)$ corresponds to a unique left invariant vector field X_l with $X_l(g) := d\lambda_g(\mathbf{1}).X$, $g \in G$. The space of left invariant vector fields is closed under the Lie bracket of vector fields, hence inherits a Lie algebra structure. In this sense we obtain on $\mathfrak{g} := T_e(G)$ a continuous Lie bracket which is uniquely determined by $[X, Y]_l = [X_l, Y_l]$.

All finite-dimensional manifolds M are assumed to be σ-compact which for connected manifolds is equivalent to requiring that M is paracompact or a second countable topological space. This excludes pathologies such as "long lines" which are one-dimensional smooth manifolds constructed from sets of countable ordinal numbers ([SS78, p.72]).

All topological vector spaces in this paper are assumed to be Hausdorff.

Acknowledgement. I am grateful to H. Biller and H. Glöckner for many extremely helpful suggestions to improve the exposition of this paper.

I Current groups on non-compact manifolds

In this section we introduce two classes of Lie groups of smooth maps: the group $C_c^\infty(M; K)$ of smooth maps with compact support on a non-compact manifold and the group $C^\infty(M, S; K)$ of smooth maps on a compact manifold M that together with all higher partial derivatives vanish on the closed subset S.

Compactly supported smooth maps

Definition I.1. For two topological spaces M and Y we write $C(M; Y)_c$ for the space $C(M; Y)$ of all continuous maps $M \to Y$ endowed with the *compact open topology*. The topology on this space is generated by the sets

$$W(C, O) := \{f \in C(M; Y): f(C) \subseteq O\},$$

where $C \subseteq M$ is compact and $O \subseteq Y$ is open.

(a) If M is locally compact and K is a topological group, then $C(M; K)_c$ is a topological group with respect to pointwise multiplication, and the topology coincides with the topology of uniform convergence on compact subsets of M ([Sch75, Satz II.4.5]). In particular the sets $W(C, U)$, where $C \subseteq M$ is compact and $U \subseteq K$ is an open identity neighborhood, form a basis of identity neighborhoods in $C(M; K)_c$.

For a function $f: M \to K$ let $\mathrm{supp}(f) := \overline{\{x \in M: f(x) \neq e\}}$ denote its *support*. Then for each compact subset $X \subseteq M$ the subset

$$C_X(M; K) := \{f \in C(M; K): \mathrm{supp}(f) \subseteq X\}$$

is a closed subgroup of $C(M; K)_c$ on which the subspace topology coincides with the topology of uniform convergence.

If M is a discrete set, then $C(M; K)_c \cong K^M$ as a topological group.

(b) If M is a locally compact space and Y is a locally convex space, then (a) implies that $C(M; Y)_c$ is a locally convex space, where the topology is defined by the seminorms

$$p_{X,q}(f) := \sup_{x \in X} q(f(x)),$$

where q is a continuous seminorm on Y and $X \subseteq M$ a compact subset.

If Y is a Fréchet space and M is σ-compact, then the topology is defined by a countable family of seminorms turning $C(M; Y)_c$ into a Fréchet space.

(c) If M is locally compact, $X \subseteq M$ compact, and Y is a locally convex space, then for each open 0-neighborhood $U \subseteq Y$ the subset

$$\{f \in C_X(M; Y)_c: f(M) \subseteq U\} = W(X, U) \cap C_X(M; Y)$$

is open in $C_X(M; Y)_c$.   □

Definition I.2. Let M be a smooth finite-dimensional σ-compact manifold. If Y is a locally convex space, then each smooth map $f: M \to Y$ defines a sequence of maps

$$d^n f: T^n M \to Y, \quad n \in \mathbb{N}.$$

We endow $C^\infty(M; Y)$ with the topology obtained from the embedding

$$C^\infty(M; Y) \hookrightarrow \prod_{n \in \mathbb{N}_0} C(T^n M, Y)_c.$$

turning $C^\infty(M; Y)$ into a locally convex space. If $X \subseteq M$ is a compact subset, we consider on $C^\infty_X(M; Y) \subseteq C^\infty(M; Y)$ the subspace topology.

(a) If K is a Lie group, then $C^\infty_X(M; K)$ is a group with respect to pointwise multiplication. It is shown in [Gl01b, 3.18] that it carries a Lie group structure which is uniquely determined by the property that for each open identity neighborhood $U \subseteq K$ and each chart $\phi \colon U \to \mathfrak{k}$ with $\phi(e) = 0$ there exists an open identity neighborhood $U_0 \subseteq U$ such that the map

$$\{f \in C^\infty_X(M; K) \colon f(M) \subseteq U_0\} \to \{h \in C^\infty_X(M; \mathfrak{k}) \colon h(M) \subseteq \phi(U_0)\}, \quad f \mapsto \phi \circ f$$

is a diffeomorphism onto an open subset of the locally convex space $C^\infty_X(M; \mathfrak{k})$. The Lie algebra of this group is the locally convex space $C^\infty_X(M; \mathfrak{k})$ with the pointwise Lie bracket, where $\mathfrak{k}$ is the Lie algebra of K ([Gl01b, 3.19]).

(b) For a locally convex space Y we endow the space

$$C^\infty_c(M; Y) := \{f \in C^\infty(M; Y) \colon \mathrm{supp}(f) \text{ compact}\} = \bigcup_X C^\infty_X(M; Y),$$

where X runs through all compact subsets of M, with the locally convex direct limit topology. This means that a seminorm on $C^\infty_c(M; Y)$ is continuous if and only if its restrictions to all the subspaces $C^\infty_X(M; Y)$ are continuous with respect to the topology defined above.

In M there exists an increasing sequence $(X_n)_{n \in \mathbb{N}}$ of compact subsets X_n with $X_n \subseteq X^0_{n+1}$ and $M = \bigcup_n X_n$. Then each compact subset $X \subseteq M$ is contained in some X_n, and each space $C^\infty_{X_n}(M; Y)$ is a closed subspace of $C^\infty_{X_{n+1}}(M; Y)$. Therefore

$$C^\infty_c(M; Y) = \varinjlim C^\infty_{X_n}(M; Y)$$

is a *strict inductive limit* of the locally convex spaces $C^\infty_{X_n}(M; Y)$ in the sense of [He89, Prop. 1.5.3]. In particular each bounded subset of $C^\infty_c(M; Y)$ is contained in one of the subspaces $C^\infty_{X_n}(M; Y)$. Moreover, $C^\infty_c(M; Y)$ is Hausdorff and the continuous maps $C^\infty_{X_n}(M; Y) \hookrightarrow C^\infty_c(M; Y)$ are embeddings, which in turn implies that all the inclusions

$$C^\infty_X(M; Y) \hookrightarrow C^\infty_c(M; Y)$$

are embeddings (cf. [Kö69, p. 222]).

If Y is a Fréchet space, this topology turns $C^\infty_c(M; Y)$ into an LF-space ([Gl01b, 4.6]). It is shown in [Gl01b, 4.18] that for each Lie group K the group $C^\infty_c(M; K)$ carries a Lie group structure, hence in particular the structure of a Hausdorff topological group. In the same way as for the groups $C^\infty_X(M; K)$, the Lie group structure is uniquely determined by the property that for each open identity neighborhood $U \subseteq K$ and each chart $\phi \colon U \to \mathfrak{k}$ with $\phi(e) = 0$ there exists an open identity neighborhood $U_0 \subseteq U$ such that the map

$$\{f \in C^\infty_c(M; K) \colon f(M) \subseteq U_0\} \to \{h \in C^\infty_c(M; \mathfrak{k}) \colon h(M) \subseteq \phi(U_0)\}, \quad f \mapsto \phi \circ f$$

is a diffeomorphism onto an open subset of the locally convex space $C_c^\infty(M; \mathfrak{k})$. The Lie algebra of this group is the locally convex space $C_c^\infty(M; \mathfrak{k})$ with the pointwise Lie bracket.

Remark I.3. From the fact that $C_c^\infty(M; \mathfrak{k})$ is a strict inductive limit of spaces $C_X^\infty(M; \mathfrak{k})$ and the description of the natural charts of the Lie group $C_c^\infty(M; K)$, we see that for each compact subset $X \subseteq M$ the inclusion map $C_X^\infty(M; K) \hookrightarrow C_c^\infty(M; K)$ is a topological embedding. $\square$

Remark I.4. If K is a Lie group with Lie algebra $\mathfrak{k}$, then the tangent bundle of K is a Lie group isomorphic to $\mathfrak{k} \rtimes K$, where K acts on $\mathfrak{k}$ by the adjoint representation (cf. [Ne01b]). Iterating this procedure, we obtain a Lie group structure on all iterated higher tangent bundles $T^n K$ which are diffeomorphic to $\mathfrak{k}^{2^n-1} \times K$.

It follows in particular that for each finite-dimensional manifold M and each $n \in \mathbb{N}_0$ we obtain topological groups $C(T^n M, T^n K)_c$ (Definition I.1(a)). Therefore the canonical inclusion map

$$C^\infty(M; K) \hookrightarrow \prod_{n \in \mathbb{N}} C(T^n M, T^n K)_c$$

leads to a natural topology on $C^\infty(M; K)$ turning it into a topological group.

If M is compact, then it is not hard to see that this procedure leads to the same topology as the Lie group structure defined in Definition I.2. A similar statement holds for $C_X^\infty(M; K)$ if $X \subseteq M$ is a compact subset.

We cannot expect for a general non-compact manifold M that $C^\infty(M; K)$ carries a natural Lie group structure. In the example $M = \mathbb{N}$ the group $C^\infty(\mathbb{N}; K) = C(\mathbb{N}; \mathbb{K}) \cong K^\mathbb{N}$ is the topological direct product group. As the example $K = \mathbb{T}$ already shows, the groups $K^\mathbb{N}$ need not be manifolds because they need not be locally contractible.

If M is connected, then the situation seems to be much better, but this needs to be investigated ([NW03]). One can show in particular that for each Banach–Lie group K the group $C^\infty(\mathbb{R}, K)$ is a Fréchet–Lie group with respect to its natural topology of uniform convergence of all derivatives on compact subsets of $\mathbb{R}$. Likewise, for each simply connected non-compact complex curve Σ and each complex Banach–Lie group K the group $\mathrm{Hol}(\Sigma, K)$ of all holomorphic maps $\Sigma \to K$ is a Lie group. $\square$

Fréchet current groups defined by vanishing conditions

In this subsection M denotes a connected finite-dimensional manifold and $S \subseteq M$ a closed subset. Mostly we will assume that M is compact.

Remark I.5. Let U be an open subset of a locally convex space X and Y another locally convex space. If for a smooth function $f: U \to Y$ its value together with all derivatives up to order k vanish in a point $p \in U$, then the formula for the Taylor expansion of compositions trivially implies that the same holds for all compositions

$f \circ \phi$ in q, where $\phi \colon V \to U$ is a C^k-map with $\phi(q) = p$. It follows in particular that for a smooth function on a manifold it makes sense to say that *all partial derivatives up to order k vanish in a point p*.   □

Definition I.6. Let M be a manifold with boundary and $S \subseteq M$ a closed subset. For a Lie group K we write $C^\infty(M, S; K)$ for the group of all those smooth maps for which their value together with all derivatives vanish on S. It clearly suffices that for each point $s \in S$ there exists one chart in which all partial derivatives vanish in s.

If M is compact and K is a (Fréchet-)Lie group, then also $C^\infty(M, S; K)$ is a Fréchet–Lie group, where we use the same charts as for $C^\infty(M; K)$ and observe that they restrict to charts of the subgroup $C^\infty(M, S; K)$. In particular $C^\infty(M, S; \mathbb{R})$ is a real Fréchet algebra. For non-compact M we consider $C^\infty(M, S; K)$ only as a topological subgroup of $C^\infty(M; K)$ in the sense of Remark I.4.   □

Remark I.7. Let us consider the category $\mathcal{P}$ whose objects are pairs (M, S), where M is a (finite-dimensional) manifold and S is a closed subset. A morphism $(M, S) \to (M', S')$ is a smooth map $\phi \colon M \to M'$ with $\phi(S) \subseteq S'$. Remark I.5 implies that the assignment $(M, S) \mapsto C^\infty(M, S; K)$ defines a contravariant functor from $\mathcal{P}$ to the category of topological groups. Here we use that for a morphism $\phi \colon (M, S) \to (M', S')$ the corresponding group homomorphism $C^\infty(M', S'; K) \to C^\infty(M, S; K)$, $f \mapsto f \circ \phi$ is continuous, which is an easy consequence of the definitions (cf. Lemma A.1.6).   □

Lemma I.8. *Let M be a finite-dimensional manifold, $X \subseteq M$ be a smooth submanifold with boundary, $\dim X = \dim M$, and Y a locally convex space. For a smooth function $f \colon X \to Y$ the extension by $f(M \setminus X) = \{0\}$ defines a smooth function $M \to Y$ if and only if f and all its derivatives vanish on ∂X.*

Proof. It clearly is a necessary condition that all derivatives of f vanish on ∂X. Suppose, conversely, that this condition is satisfied and extend f by 0 on $M \setminus X$ to a function $f_M \colon M \to Y$.

As the smoothness of f_M is equivalent to its weak smoothness (for this result of Grothendieck see [Wa72] or [KM97]), we may w.l.o.g. assume that $Y = \mathbb{R}$. Moreover, we may assume that $M = \mathbb{R}^n$ and that $X = \{x \in \mathbb{R}^n \colon x_n \leq 0\}$. Then it is clear that all partial derivatives of f extended by 0 on $M \setminus X$ yield continuous functions. Moreover, all partial derivatives of the extended function f_M exist and coincide with the extensions of the partial derivatives of f. This proves that f_M is a C^1-function. Iterating the argument shows that f_M is a C^k-function for each k, hence smooth.   □

Examples I.9. (a) Let X be a compact manifold with boundary and X^d the *double of X*. This is, by definition, a compact manifold without boundary containing X and a diffeomorphic copy $X^\sharp$ of X such that $X \cap X^\sharp = \partial X = \partial X^\sharp$ and $X \cup X^\sharp = X^d$. Then Lemma I.8 implies that

$$C^\infty(X, \partial X; K) \cong C_X^\infty(X^d; K)$$

and

$$C^\infty(X^d, \partial X; K) \cong C^\infty(X, \partial X; K) \times C^\infty(X^\sharp, \partial X; K) \cong C^\infty(X, \partial X; K)^2.$$

(b) We think of $C^\infty(M, S; K)$ as a group of smooth maps on the non-compact manifold $M \setminus S$. For $M = \mathbb{S}^n$ and $S = \{p\}$ have $M \setminus S \cong \mathbb{R}^n$, and hence a natural Lie group of smooth maps $\mathbb{R}^n \to K$ with a certain decay at infinity.

(c) Let $M = \mathbb{S}^1$. Then $M \setminus S$ is a countable union of intervals I_j, $j \in J$, and we thus obtain an inclusion

$$C^\infty(M, S; K) \hookrightarrow \prod_{j \in J} C^\infty(I_j, \partial I_j; K) \cong C^\infty(I, \partial I; K)^J,$$

where the right hand side does not carry the product topology but the l^∞-topology of uniform convergence of all derivatives uniformly in all components. □

II Relative de Rham cohomology

If M is a compact manifold, $S \subseteq M$ a compact subset, and Y a sequentially complete locally convex space (an s.c.l.c. space), then we consider the space $Z^1_{\mathrm{dR}}(M, S; Y)$ of all Y-valued closed smooth 1-forms that vanish, together with all their derivatives, on S. Integration of 1-forms with this property over singular cycles in M modulo S lead to the subgroup $Z^1_{\mathrm{dR}}(M, S; \Gamma)$ of those closed 1-forms for which all integrals over cycles have values in a subgroup Γ of Y. The main result of this section is Theorem II.7, saying that the image $H^1_{\mathrm{dR}}(M, S; \Gamma)$ of $Z^1_{\mathrm{dR}}(M, S; \Gamma)$ in $H^1_{\mathrm{dR}}(M, S; Y)$ is a discrete subgroup if Γ is discrete. In Examples II.11 and II.12 we see that these subgroups may have infinite rank, even for $Y = \mathbb{R}$.

We write $I := [0, 1]$ and assume that $S \neq \emptyset$ and that M is connected. Further, Y denotes an s.c.l.c. space, Γ is a subgroup of Y, and $T_\Gamma := Y/\Gamma$ the corresponding quotient group. If Γ is discrete, then the quotient topology turns T_Γ into a Lie group with Lie algebra Y. For some statements we do not have to assume that M is compact. If we assume compactness, we will mention it explicitly.

We write $\Omega^1(M; Y)$ for the space of smooth 1-forms on M with values in Y and endow this space with the natural topology corresponding in each chart to the uniform convergence of all derivatives on compact subsets mapping into coordinate charts (cf. [Gl01d]). For a subset $X \subseteq M$ we write $\Omega^1_X(M; Y)$ for the closed subspace of $\Omega^1(M; Y)$ consisting of those forms supported in X. We endow the space $\Omega^1_c(M; Y)$ with the locally convex direct limit topology with respect to the subspaces $\Omega^1_X(M; Y)$, where $X \subseteq M$ is a compact subset. For a closed subset $S \subseteq M$ we write $\Omega^1(M, S; Y) \subseteq \Omega^1(M; Y)$ for the subspace of all forms vanishing with all their partial derivatives on S.

The Lie group $C^\infty(M, S; T_\Gamma)$

Definition II.1. Let M be a smooth manifold and K a Lie group. For an element $f \in C^\infty(M; K)$ we write

$$\delta^l(f)(m) := d\lambda_{f(m)^{-1}}(f(m))df(m) : T_m(M) \to \mathfrak{k} \cong T_e(K)$$

for the *left logarithmic derivative of* f. This derivative can be viewed as a $\mathfrak{k}$-valued 1-form on M which we also write simply as $\delta^l(f) = f^{-1}.df$. We thus obtain a map

$$\delta^l : C^\infty(M; K) \to \Omega^1(M; \mathfrak{k})$$

satisfying the cocycle condition

$$\delta^l(f_1 f_2) = \mathrm{Ad}(f_2)^{-1}.\delta^l(f_1) + \delta^l(f_2).$$

We also have the *right logarithmic derivative* $\delta^r(f) = df.f^{-1}$ satisfying

$$\delta^r(f_1 f_2) = \delta^r(f_1) + \mathrm{Ad}(f_1).\delta^r(f_2).$$

(cf. [KM97, 38.1]). If K is abelian, then the cocycle condition shows that $\delta := \delta^l$ is a group homomorphism whose kernel consists of the locally constant maps.

The logarithmic derivatives $\delta^l(f)$, resp., $\delta^r(f)$, can also be defined as the pullbacks $f^*\theta_K^l$, resp., $f^*\theta_K^r$ or the left, resp., right *Maurer–Cartan form*, θ_K^l, resp., θ_K^r on K. $\qquad\square$

In Section V we will need the following continuity result for the logarithmic derivatives.

Lemma II.2. *For any Lie group K the maps $\delta^l, \delta^r : C_c^\infty(M; K) \to \Omega_c^1(M; \mathfrak{k})$ are smooth.*

Proof. In view of the cocycle relations

$$\delta^l(f_1 f_2) = \mathrm{Ad}(f_2)^{-1}.\delta^l(f_1) + \delta^l(f_2) \quad \text{and} \quad \delta^r(f_1 f_2) = \delta^r(f_1) + \mathrm{Ad}(f_1).\delta^r(f_2),$$

it suffices to prove the smoothness of δ^l and δ^r in an open identity neighborhood U of $C_c^\infty(M; K)$. Here we use that addition is continuous in $\Omega_c^1(M; \mathfrak{k})$, and that the continuity of the linear map $\mathrm{Ad}(f_1)$ on $\Omega_c^1(M; \mathfrak{k})$ follows from its continuity on the subspaces $\Omega_X^1(M; \mathfrak{k})$, $X \subseteq M$ compact. According to the definition of the Lie group structure on $C_c^\infty(M; K)$, we may assume that

$$U = \{f \in C_c^\infty(M; K) : f(M) \subseteq V_K\},$$

where $V_K \subseteq K$ is an open identity neighborhood for which there exists a diffeomorphism $\phi : V_\mathfrak{k} \to V_K$, where $V_\mathfrak{k}$ is an open subset of the locally convex space $\mathfrak{k}$. We now have to show that the map

$$D : C_c^\infty(M; V_\mathfrak{k}) \to \Omega_c^1(M; \mathfrak{k}), \quad f \mapsto \delta^l(\phi \circ f)$$

is smooth.

We think of D as a map between spaces of sections of vector bundles over M. Then the values of $D(f)$ in an open subset $O \subseteq M$ only depend on $f|_O$. This implies in particular that D is *local* in the sense of [Gl02, Def. 3.1]. Moreover, for each compact subset $X \subseteq M$ the map

$$D_X := D|_{C_X^\infty(M;K)} \colon C_X^\infty(M; K) \to \Omega_X^1(M; \mathfrak{k})$$

is smooth because the map

$$\delta^l \colon C_X^\infty(M; K) \to \Omega_X^1(M; \mathfrak{k})$$

is obviously smooth. Therefore the Smoothness Theorem 3.2 in [Gl02] implies that D is a smooth map and hence that δ^l is smooth. The smoothness of δ^r is shown similarly. $\qquad \square$

Lemma II.3. *If Γ is discrete, then*

$$\delta(C^\infty(M, S; T_\Gamma)) = \left\{ \beta \in \Omega^1(M, S; Y) \colon \left(\forall \alpha \in C^\infty((I, \partial I), (M, S)) \right) \int_\alpha \beta \in \Gamma \right\}.$$

Proof. If $\beta = \delta(f)$ for some $f \in C^\infty(M, S; T_\Gamma)$ and $\alpha \in C^\infty((I, \partial I), (M, S))$, then

$$f(\alpha(1)) - f(\alpha(0)) = \int_\alpha \beta + \Gamma$$

vanishes in $T_\Gamma = Y/\Gamma$, so that $\int_\alpha \beta \in \Gamma$.

Suppose, conversely, that $\beta \in \Omega^1(M, S; Y)$ satisfies

$$\int_\alpha \beta \in \Gamma \quad \text{for all } \alpha \in C^\infty((I, \partial I), (M, S)).$$

Pick $s_0 \in S$. Then all integrals of β over smooth loops based in s_0 are contained in Γ (here we need that Y is sequentially complete to ensure the existence of Y-valued Riemann integrals over curves), so that there exists a smooth function $f \colon M \to T_\Gamma$ with $\beta = \delta(f)$ and $f(s_0) = 0$ ([Ne02a, Prop. 3.9]). For each $s \in S$ there exists a smooth path $\alpha \in C^\infty((I, \partial I), (M, S))$ from s_0 to s, and we obtain

$$f(s) = f(s) - f(s_0) = \int_\alpha \beta + \Gamma \in \Gamma.$$

This means that $f|_S = 0$. As $\beta = \delta(f)$, all higher derivatives of f vanish on S, so that $f \in C^\infty(M, S; T_\Gamma)$. $\qquad \square$

Corollary II.4. *For each s.c.l.c. space Y we have*

$$dC^\infty(M, S; Y) = \left\{ \beta \in \Omega^1(M, S; Y) \colon \left(\forall \alpha \in C^\infty((I, \partial I), (M, S)) \right) \int_\alpha \beta = 0 \right\}.$$

In particular $dC^\infty(M, S; Y)$ is closed in $\Omega^1(M, S; Y)$. $\qquad \square$

Definition II.5. (a) In view of the closedness assertion in Corollary II.4, the quotient

$$\mathfrak{z}_{(M,S)}(Y) := \Omega^1(M, S; Y)/dC^\infty(M, S; Y)$$

carries a natural (Hausdorff) locally convex topology. Moreover, the subspace $Z^1_{dR}(M, S; Y)$ of closed forms in $\Omega^1(M, S; Y)$ is closed, which implies that

$$H^1_{dR}(M, S; Y) := Z^1_{dR}(M, S; Y)/dC^\infty(M, S; Y)$$

is a closed subspace of $\mathfrak{z}_{(M,S)}(Y)$. Let $q \colon \Omega^1(M, S; Y) \to \mathfrak{z}_{(M,S)}(Y)$ denote the quotient map.

We want to relate $H^1_{dR}(M, S; Y)$ to the singular Y-valued cohomology of M modulo S. The abelian group $Z_1(M, S)$ of singular 1-cycles modulo S is generated by those given by continuous maps $(I, \partial I) \to (M, S)$. Therefore $H_1(M, S)$ is generated by the image of the set $\pi_1(M, S) := [(I, \partial I), (M, S)]$ of homotopy classes of maps of pairs (see [Br93, VII.4.10] for more details on Hurewicz maps from homotopy groups to homology groups). Let $\beta \in Z^1_{dR}(M, S; Y)$. Then we can define for each singular 1-chain α the integral $\int_\alpha \beta$. According to Stoke's formula, these integrals vanish on boundaries and also on chains supported by S. We thus obtain a map

$$Z^1_{dR}(M, S; Y) \to H^1(M, S; Y) := \mathrm{Hom}(H_1(M, S); Y),$$

where $H_1(M, S)$ denotes the singular homology group with coefficients in $\mathbb{Z}$ and $H^1(M, S; Y)$ a relative singular cohomology group (cf. [Br93, V.7.2]).

The kernel of this map consists of all closed 1-forms β for which all the integrals of cycles in $Z_1(M, S)$ vanish, which means that $\beta = df$ for some $f \in C^\infty(M, S; Y)$ (Corollary II.4). Hence we obtain an embedding

$$\eta \colon H^1_{dR}(M, S; Y) \hookrightarrow H^1(M, S; Y). \tag{2.1}$$

As we will see in Example II.12 below, this map is not always surjective.
(b) For a subgroup $\Gamma \subseteq Y$ we define

$$Z^1_{dR}(M, S; \Gamma) := \left\{ \beta \in Z^1_{dR}(M, S; Y) \colon (\forall \alpha \in C^\infty((I, \partial I), (M, S))) \int_\alpha \beta \in \Gamma \right\}.$$

By Corollary II.4 we see that $dC^\infty(M, S; Y)$ is a closed subspace of $Z^1_{dR}(M, S; \Gamma)$, so that

$$H^1_{dR}(M, S; \Gamma) := Z^1_{dR}(M, S; \Gamma)/dC^\infty(M, S; Y)$$

carries a natural Hausdorff locally convex topology. We also define

$$Z^1_{dR}(M; \Gamma) := \left\{ \beta \in Z^1_{dR}(M; Y) \colon (\forall \alpha \in C^\infty(\mathbb{S}^1, M)) \int_\alpha \beta \in \Gamma \right\}$$

and $H^1_{dR}(M; \Gamma) := Z^1_{dR}(M; \Gamma)/dC^\infty(M; Y)$. $\qquad \square$

Remark II.6. Let M be a connected manifold.

(a) Assume that $\Gamma \subseteq Y$ is a discrete subgroup and let $T_\Gamma := Y/\Gamma$ denote the corresponding quotient Lie group and $q_\Gamma \colon Y \to T_\Gamma$ the quotient map. We consider the abelian topological group $G := C^\infty(M; T_\Gamma)$, the space $\mathfrak{g} := C^\infty(M; Y)$, and the exponential function

$$\exp_G \colon \mathfrak{g} \to G, \quad \xi \mapsto q_\Gamma \circ \xi.$$

The map

$$\delta \colon G = C^\infty(M; T_\Gamma) \to Z^1_{\mathrm{dR}}(M; Y), \quad f \mapsto \delta(f) = f^{-1}df$$

is a continuous group homomorphism whose kernel consists of the locally constant functions on M. If M is connected, then $\ker \delta$ consists only of the constant functions.

According to [Ne02a, Prop. 3.9], a closed 1-form in $Z^1_{\mathrm{dR}}(M; Y)$ can be written as $\delta(f)$ for some $f \in C^\infty(M; T_\Gamma)$ if and only if all integrals over closed piecewise smooth paths are contained in Γ. This means that

$$\mathrm{im}(\delta) = Z^1_{\mathrm{dR}}(M; \Gamma).$$

Using the decomposition $G \cong G_* \times T_\Gamma$ with $G_* := \{f \in G \colon f(x_M) = 0\}$, where $x_M \in M$ is a base point, it follows that

$$\delta \colon G_* \to Z^1_{\mathrm{dR}}(M; \Gamma)$$

is an isomorphism of groups. Here the subgroup $B^1_{\mathrm{dR}}(M; Y) \subseteq Z^1_{\mathrm{dR}}(M; \Gamma)$ corresponds to $\mathrm{im}(\exp_G)$, so that

$$G/\exp_G(\mathfrak{g}) \cong Z^1_{\mathrm{dR}}(M; \Gamma)/B^1_{\mathrm{dR}}(M; Y) = H^1_{\mathrm{dR}}(M; \Gamma).$$

If, in addition, M is compact, then G is a Lie group with Lie algebra $\mathfrak{g}$, $\exp_G$ is the universal covering map of G_e, and $\delta \colon G_* \to Z^1_{\mathrm{dR}}(M; \Gamma)$ is an isomorphism of Lie groups. This leads to

$$\pi_0(G) \cong G/\exp_G(\mathfrak{g}) \cong Z^1_{\mathrm{dR}}(M; \Gamma)/B^1_{\mathrm{dR}}(M; Y) = H^1_{\mathrm{dR}}(M; \Gamma).$$

(b) If M is compact and $S \subseteq M$ a non-empty closed subset, then we obtain with similar arguments as in (a) that the group $G := C^\infty(M, S; T_Y)$ is a Lie group and that $\exp_G$ is the universal covering map of the identity component G_e of G. The connectedness of M and $S \neq \emptyset$ imply $\ker \exp_G = \{0\}$. Therefore the exponential function $\exp_G$ induces a diffeomorphism

$$\exp_G \colon \mathfrak{g} = C^\infty(M, S; Y) \to G_e.$$

Moreover, δ is an injective homomorphism of Lie groups with $\delta(G) = Z^1_{\mathrm{dR}}(M, S; \Gamma)$ (Lemma II.3), where G_e corresponds to the subspace $dC^\infty(M, S; Y)$, so that

$$\pi_0(G) \cong H^1_{\mathrm{dR}}(M, S; \Gamma).$$

(c) The set $M \setminus S$ is an open subset of M, hence a non-compact manifold. We have inclusions

$$\Omega^1_c(M \setminus S; Y) \hookrightarrow \Omega^1(M, S; Y) \quad \text{and} \quad Z^1_{\mathrm{dR},c}(M \setminus S; Y) \hookrightarrow Z^1_{\mathrm{dR}}(M, S; Y).$$

Moreover,

$$dC_c^\infty(M \setminus S; Y) \subseteq Z_{\mathrm{dR},c}^1(M \setminus S; Y) \cap dC^\infty(M, S; Y)$$

and if, conversely, $\zeta = df \in \Omega_c(M \setminus S; Y)$ with $f \in C^\infty(M, S; Y)$, then df vanishes in a neighborhood of S, so that $f^{-1}(0)$ is an open neighborhood of S. If M is compact, then it follows that f has compact support, and therefore that

$$dC_c^\infty(M \setminus S; Y) = Z_{\mathrm{dR},c}^1(M \setminus S; Y) \cap dC^\infty(M, S; Y).$$

This means that we also obtain an inclusion

$$\phi: H_{\mathrm{dR},c}^1(M \setminus S; Y) \hookrightarrow H_{\mathrm{dR}}^1(M, S; Y).$$

If X is a compact manifold with boundary, $M = X \cup X^\sharp$ as in Example I.9, and $\mathrm{int}(X) = M \setminus S$, we claim that

$$H_{\mathrm{dR},c}^1(\mathrm{int}(X); Y) \cong H_{\mathrm{dR}}^1(X, \partial X; Y) := H_{\mathrm{dR}}^1(M, M \setminus \mathrm{int}(X); Y). \qquad (2.2)$$

In fact, if $\zeta \in Z_{\mathrm{dR}}^1(X, \partial X; Y)$, then the restriction of ζ to ∂X vanishes. Moreover, there exists a tubular neighborhood U of ∂X diffeomorphic to $\partial X \times I$, so that the inclusion $\partial X \hookrightarrow U$ induces an isomorphism $\pi_1(\partial X) \to \pi_1(U)$. We conclude that all periods of $\zeta|_U$ vanish, and hence that there exists a smooth function $f \in C^\infty(U, \partial X; Y)$ with $df = \zeta|_U$. Let $\chi \in C^\infty(X; \mathbb{R})$ be constant 1 in a neighborhood of ∂X and 0 on $X \setminus U$. Then $\zeta - d(\chi f) \in Z_{\mathrm{dR},c}^1(\mathrm{int}(X); Y)$ has the same cohomology class as ζ. This proves (2.2).

From [Br97, Prop. II.12.3, Th. III.1.1, Cor. III.4.12] applied to the paracompactifying family Φ of closed subsets of $X \setminus \partial X$, we derive that for singular cohomology we have

$$H^1(X, \partial X; Y) \cong H_c^1(\mathrm{int}(X); Y).$$

Further the general version of de Rham's Theorem with values in sheaves ([Br97, §III.3]) yields an isomorphism

$$H_c^1(\mathrm{int}(X); Y) \cong H_{\mathrm{dR},c}^1(\mathrm{int}(X); Y).$$

Therefore

$$H_{\mathrm{dR}}^1(X, \partial X; Y) \cong H_{\mathrm{dR},c}^1(\mathrm{int}(X); Y) \cong H_c^1(\mathrm{int}(X); Y)$$

$$\cong H^1(X, \partial X; Y) \cong \mathrm{Hom}(H_1(X, \partial X); Y). \qquad \square$$

The following theorem on the discreteness of the group $H_{\mathrm{dR}}^1(M, S; \Gamma)$ is the main result of the present section.

Theorem II.7. *Let S be a non-empty closed subset of the compact manifold M and $\Gamma \subseteq Y$ a discrete subgroup. Then the subgroup*

$$H_{\mathrm{dR}}^1(M, S; \Gamma) = \left\{ [\beta] \in \mathfrak{z}_{(M,S)}(Y) : (\forall \alpha \in C^\infty((I, \partial I), (M, S))) \int_\alpha \beta \in \Gamma \right\}$$

of $\mathfrak{z}_{(M,S)}(Y)$ *is discrete.*

Proof. Let $Z^1_{\mathrm{dR}}(M; Y) \subseteq \Omega^1(M; Y)$ denote the closed subspace of closed 1-forms. As $\pi_1(M)$ is finitely generated (cf. Proposition III.1 below) and Γ is discrete,

$$dC^\infty(M; Y) = \left\{ \beta \in Z^1_{\mathrm{dR}}(M; Y) : (\forall [\alpha] \in \pi_1(M)) \int_\alpha \beta = 0 \right\}$$

is an open subgroup of

$$Z^1_{\mathrm{dR}}(M; \Gamma) = \left\{ \beta \in Z^1_{\mathrm{dR}}(M; Y) : (\forall \alpha \in C^\infty(\mathbb{S}^1, M)) \int_\alpha \beta \in \Gamma \right\}.$$

That $H^1_{\mathrm{dR}}(M, S; \Gamma)$ is a discrete subgroup of the quotient space $\mathfrak{z}_{(M,S)}(Y)$ is equivalent to $dC^\infty(M, S; Y)$ being an open subgroup of $Z^1_{\mathrm{dR}}(M, S; \Gamma)$. As a consequence of what we have just seen, the group $Z^1_{\mathrm{dR}}(M, S; \Gamma) \cap dC^\infty(M; Y)$ is open in $Z^1_{\mathrm{dR}}(M, S; \Gamma)$. Therefore it suffices to verify that $dC^\infty(M, S; Y)$ is an open subgroup of $Z^1_{\mathrm{dR}}(M, S; \Gamma) \cap dC^\infty(M; Y)$.

Fix a point $x_M \in S$. We consider the map

$$\Phi : Z^1_{\mathrm{dR}}(M, S; \Gamma) \to C(M; T_\Gamma), \quad \Phi(\beta)(x) := \int_{x_M}^x \beta + \Gamma \in T_\Gamma.$$

Then

$$\Phi(Z^1_{\mathrm{dR}}(M, S; \Gamma)) \subseteq C^\infty(M; T_\Gamma), \quad d(\Phi(\beta)) = \beta, \quad \Phi(\beta)|_S = 0,$$

and Φ is continuous with respect to the topology of uniform convergence on compact subsets of M. Hence

$$\Phi^{-1}(C(M; T_\Gamma)_e) = \Phi^{-1}(\exp(C(M; Y))) = dC^\infty(M, S; Y)$$

is an open subgroup of $Z^1_{\mathrm{dR}}(M, S; \Gamma)$ because $C(M; T_\Gamma)$ is a Lie group (Remark II.6). $\qquad\square$

Lemma II.8. *Let* $I = [0, 1]$. *The integration maps*

$$I_{\mathbb{R}} : \Omega^1_c(\mathbb{R}; Y) = Z^1_{\mathrm{dR},c}(\mathbb{R}; Y) \to Y, \quad \beta \mapsto \int_{\mathbb{R}} \beta, \tag{2.3}$$

$$I_I : \Omega^1(I, \partial I; Y) = Z^1_{\mathrm{dR}}(I, \partial I; Y) \to Y, \quad \beta \mapsto \int_I \beta, \tag{2.4}$$

and

$$I_{\mathbb{S}^1} : \Omega^1(\mathbb{S}^1; Y) = Z^1_{\mathrm{dR}}(\mathbb{S}^1; Y) \to Y, \quad \beta \mapsto \int_{\mathbb{S}^1} \beta \tag{2.5}$$

induce topological isomorphisms

$$H^1_{\mathrm{dR},c}(\mathbb{R}; Y) \to Y, \quad H^1_{\mathrm{dR}}(I, \partial I; Y) \to Y \quad \text{and} \quad H^1_{\mathrm{dR}}(\mathbb{S}^1; Y) \to Y.$$

Proof. We have a continuous map $\Omega_c^1(\mathbb{R}; Y) \to Y$, $\beta \mapsto \int_{\mathbb{R}} \beta$, and it is easy to see that this map is surjective because there exists a smooth real-valued 1-form γ with compact support and $\int_{\mathbb{R}} \gamma = 1$. Since the map $Y \to \Omega_c^1(\mathbb{R}; Y)$, $v \mapsto \gamma \cdot v$ is continuous, the integration map splits linearly. Further its kernel coincides with the space of exact forms, which proves (2.3). The other two assertions follow by similar arguments.   □

Remark II.9. (a) For each smooth map $\alpha: (I, \partial I) \to (M, S)$ of pairs we obtain a natural map

$$I_\alpha : \mathfrak{z}_{(M,S)}(Y) \to Y \cong \mathfrak{z}_{(I,\partial I)}(Y)$$

which is given on the equivalence class of a Y-valued 1-form β by

$$I_\alpha([\beta]) = \int_\alpha \beta := \int_I \alpha^* \beta$$

(cf. Lemma II.8). The description of $dC^\infty(M, S; Y)$ in Lemma II.3 implies that the maps $I_\alpha : \mathfrak{z}_{(M,S)}(Y) \to Y$ separate points.

(b) For $(M, S) = (I, \partial I)$ the set $\pi_1(I, \partial I)$ consists of 4 elements. In fact, if $f: I \to I$ is a continuous function with $f(\partial I) \subseteq \partial I$, then the convexity of I implies that f is homotopy equivalent to the affine interpolation of the restriction $f|_{\partial I}$, and there are precisely four different maps $\partial I \to \partial I$.   □

Lemma II.10. *The subspace $H_{\mathrm{dR}}^1(M, S; Y)$ of $\mathfrak{z}_{(M,S)}(Y)$ coincides with those elements $[\beta]$ for which all the integrals $I_\alpha([\beta])$ only depend on the homotopy class of $\alpha \in C^\infty((I, \partial I), (M, S))$ in $\pi_1(M, S)$. In particular*

(1) $H_{\mathrm{dR}}^1(M, S; Y)$ is a closed subspace of $\mathfrak{z}_{(M,S)}(Y)$, and

(2) if Γ is discrete, then

$$Z_{\mathrm{dR}}^1(M, S; \Gamma) = \left\{ \beta \in \Omega^1(M, S; Y) \colon (\forall \alpha \in C^\infty((I, \partial I), (M, S))) \int_\alpha \beta \in \Gamma \right\}.$$

Proof. Fix a point $x_M \in S$. Then we have a natural inclusion $C((I, \partial I), (M, x_M)) \to C((I, \partial I), (M, S))$ inducing the map $\pi_1(M, x_M) \to \pi_1(M, S)$.

Let $\beta \in \Omega^1(M, S; Y)$. Suppose first that for $\alpha \in C^\infty((I, \partial I), (M, S))$ the integrals $\int_\alpha \beta$ only depend on the homotopy class. This implies in particular that the integrals over loops in $C^\infty((I, \partial I), (M, x_M)) \subseteq C_*^\infty(\mathbb{S}^1, M) := \{ f \in C^\infty(\mathbb{S}^1, M) \colon f(1) = x_M \}$ in x_M only depend on the homotopy class. Here we use $1 \in \mathbb{S}^1 \subseteq \mathbb{C}$ as a base point. Let $q_M : \widetilde{M} \to M$ denote the universal covering manifold. That the integrals of β over loops in x_M only depend on the homotopy class implies that there exists a smooth function $f: \widetilde{M} \to Y$ with $df = q_M^* \beta$, hence in particular that $d\beta = 0$, and therefore that $[\beta] \in H_{\mathrm{dR}}^1(M, S; Y)$.

Suppose, conversely, that $[\beta] \in H_{\mathrm{dR}}^1(M, S; Y)$, i.e., that β is closed. Then integrals over continuous maps $I \to M$ are well-defined. Then $q_M^* \beta$ is exact ([Ne02a, Th. 3.6]), and there exists a smooth function $f \in C^\infty(\widetilde{M}; Y)$ with $df = q_M^* \beta$. It follows in particular that all integrals of β over contractible loops vanish. Let $\alpha: I \times I \to M$ be a continuous map such that the maps $\alpha_t := \alpha(t, \cdot): I \to M$ satisfy $\alpha_t(\{0, 1\}) \subseteq S$. We have to show that $\int_{\alpha_0} \beta = \int_{\alpha_1} \beta$. We define

$$\tilde{\alpha}: I \times [0, 3] \to M, \quad \tilde{\alpha}(t, s) := \begin{cases} \alpha(st, 0) & \text{for } 0 \leq s \leq 1 \\ \alpha(t, s-1) & \text{for } 1 \leq s \leq 2 \\ \alpha((3-s)t, 1) & \text{for } 2 \leq s \leq 3 \end{cases}$$

and observe that $\tilde{\alpha}$ is continuous and that the curves $\tilde{\alpha}_t := \tilde{\alpha}(t, \cdot)$ start in $\alpha_0(0)$ and end in $\alpha_0(1)$, where $s \mapsto \tilde{\alpha}_0(3s)$ is homotopic to α_0. We conclude that for each $t \in I$ we have

$$0 = \int_{\tilde{\alpha}_t} \beta - \int_{\tilde{\alpha}_0} \beta = \int_{\tilde{\alpha}_t} \beta - \int_{\alpha_0} \beta = \int_1^2 \tilde{\alpha}_t^* \beta - \int_{\alpha_0} \beta = \int_{\alpha_t} \beta - \int_{\alpha_0} \beta.$$

Here we use that the vanishing of β on S implies that the integrals $\int_0^1 \tilde{\alpha}_t^* \beta$ and $\int_2^3 \tilde{\alpha}_t^* \beta$ vanish. For $t = 1$ we obtain $\int_{\alpha_0} \beta = \int_{\alpha_1} \beta$, and hence the homotopy characterization of the subspace $H^1_{dR}(M, S; Y)$ of $\mathfrak{z}_{(M,S)}(Y)$.

This implies in particular that $H^1_{dR}(M, S; Y)$ is closed, because it is defined as the intersection of the kernels of the continuous linear maps

$$[\beta] \mapsto \int_{\alpha_1} \beta - \int_{\alpha_0} \beta, \quad \alpha_i \in C^\infty((I, \partial I), (M, S))$$

from above.

Assume now that $\Gamma \subseteq Y$ is a discrete subgroup. Then the requirement $\int_\alpha \beta \in \Gamma$ for each map $\alpha \in C^\infty((I, \partial I), (M, S))$ together with the continuous dependence of the integral from α implies that $\int_\alpha \beta$ only depends on the homotopy class of α in $\pi_1(M, S)$. If all these integrals are contained in the discrete subgroup Γ, it follows from the first part of the proof that β is closed.   □

Example II.11. We consider the closed subset

$$S = \{0\} \cup \left\{ \tfrac{1}{n} : n \in \mathbb{N} \right\} \subseteq \mathbb{R}.$$

We claim that

$$H^1_{dR}(\mathbb{R}, S; \mathbb{R}) \cong E := \{(\lambda_n)_{n \in \mathbb{N}} : (\forall k \in \mathbb{N}_0) \lim_{n \to \infty} n^k \lambda_n = 0\}$$

as Fréchet spaces, where the topology on E is given by the seminorms $p_k(\lambda) := \sup_{n \in \mathbb{N}} n^k |\lambda_n|$ for $k \in \mathbb{N}$.

As $\dim \mathbb{R} = 1$, we have $Z^1_{dR}(M, S; \mathbb{R}) = \Omega^1(\mathbb{R}, S; \mathbb{R})$, and each element on this space can be written as the differential of a unique function $f \in C^\infty(\mathbb{R}; \mathbb{R})$ with $f(0) = 0$. We have to study the possible restrictions $f|_S$ because they give as the values of $[df]$ on the relative 1-cycles in $Z_1(\mathbb{R}, S)$.

First we derive necessary conditions. As $f^{(k)}(0) = 0$ for each $k \in \mathbb{N}$ and

$$f^{(k)}(0) = \lim_{x \to 0} \frac{k! f(x)}{x^k} = \lim_{n \to \infty} k! f\left(\tfrac{1}{n}\right) n^k, \tag{2.6}$$

we obtain for each $k \in \mathbb{N}$ the condition $\lim_{n \to \infty} f(\frac{1}{n}) n^k = 0$. ·

Let $(\lambda_n)_{n \in \mathbb{N}}$ satisfy $\lim_{n \to \infty} n^k \lambda_n = 0$ for each $k \in \mathbb{N}_0$. We are looking for a smooth function f in $C^\infty(\mathbb{R}; \mathbb{R})$ with $f' \in C^\infty(\mathbb{R}, S; \mathbb{R})$ and $f(\frac{1}{n}) = \lambda_n$ for each n. Let $\psi \in C_c^\infty(\mathbb{R}; \mathbb{R})$ be a function with $\mathrm{supp}(\psi) = [-1, 1]$, $\mathrm{im}(\psi) \subseteq [0, 1]$ and equal to 1 on a neighborhood of 0. Then we obtain for each $a \in \mathbb{R}$ and $\varepsilon > 0$ a smooth function $\psi_{a,\varepsilon}(x) := \psi(\varepsilon^{-1}(x - a))$ supported by $[a - \varepsilon, a + \varepsilon]$ which is constant 1 in a neighborhood of a. We define $\psi_n := \psi_{\frac{1}{n}, \frac{1}{4n(n+1)}}$. Then ψ_n is a function constant 1 in a neighborhood of $\frac{1}{n}$ with support contained in

$$]\tfrac{1}{2}(\tfrac{1}{n} + \tfrac{1}{n+1}), \tfrac{1}{2}(\tfrac{1}{n} + \tfrac{1}{n-1})[.$$

In particular the supports of the functions ψ_n are pairwise disjoint. We claim that

$$f := \sum_{n=1}^{\infty} \lambda_n \psi_n$$

defines a function in $C^\infty(\mathbb{R}; \mathbb{R})$ with $f' \in C^\infty(\mathbb{R}, S; \mathbb{R})$. This will be achieved by showing that all derivatives of the sequence defining f are uniformly convergent. In fact, for $k \in \mathbb{N}_0$ we have

$$\|\psi_n^{(k)}\|_\infty \le (4n(n+1))^k \|\psi^{(k)}\|_\infty \le c_k n^{2k}$$

for some positive constant c_k. Therefore

$$\sum_n |\lambda_n| \|\psi_n^{(k)}\|_\infty \le \sum_n |\lambda_n| c_k n^{2k} \le c_k \sum_n |\lambda_n| n^{2k} < \infty.$$

We conclude that the series $f = \sum_n \lambda_n \psi_n$ defines a smooth function. It follows directly from the construction that f is constant λ_n in a neighborhood of $\frac{1}{n+1}$ and that all derivatives of f vanish in 0 because f vanishes on $]-\infty, 0[$.

This proves that the map

$$\Phi: Z_{\mathrm{dR}}^1(\mathbb{R}, S; \mathbb{R}) \to E, \quad h(t) dt \mapsto \left(\int_0^{\frac{1}{n}} h(\tau) \, d\tau \right)_{n \in \mathbb{N}},$$

is surjective. Formula (2.6) easily implies that Φ is continuous, hence a quotient map by the Open Mapping Theorem. This proves that the induced map $H_{\mathrm{dR}}^1(\mathbb{R}, S; \mathbb{R}) \to E$ is a topological isomorphism.   □

In the next example we take a convergent sequence out of the sphere. This aims at an example of a Fréchet–Lie group $C^\infty(M, S; K)$ where the period group $\Pi_{(M,S)}$ (cf. Definition III.7) is discrete but not finitely generated (see Proposition VII.16).

Example II.12 (Removing a convergent sequence from the sphere). Let $M := \mathbb{S}^2 \subseteq \mathbb{R}^3$ and $S = \{x_n : n \in \mathbb{N}\} \cup \{(0, 0, 1)\}$, where

$$x_n = \left(\tfrac{1}{n}, 0, \sqrt{1 - \tfrac{1}{n^2}} \right).$$

As $\pi_1(M)$ is trivial, there exists for each $n \in \mathbb{N}$ a path $\gamma_n \colon [0, 1] \to M$ from $x_0 :=$ $(0, 0, 1)$ to x_n such that the group $H_1(M, S)$ is generated by the classes $[\gamma_n]$, $n \in \mathbb{N}$.

For each $n \in \mathbb{N}$ there exists a smooth function $f_n \in C^\infty(M, S; \mathbb{R})$ which is constant 1 in a neighborhood of x_n and vanishes in a neighborhood of $S \setminus \{x_n\}$. Then $df_n \in \Omega^1(M, S; \mathbb{R})$ and we have

$$\int_{\gamma_m} df_n = f_n(\gamma_m(1)) - f_n(\gamma_m(0)) = \delta_{mn}.$$

It follows in particular that the classes $[\gamma_n]$ are linearly independent over $\mathbb{Z}$, so that we obtain

$$H_1(M, S) = \bigoplus_{n \in \mathbb{N}} \mathbb{Z}[\gamma_n] \cong \mathbb{Z}^{(\mathbb{N})}$$

and therefore that the map

$$H^1(M, S; \mathbb{R}) \to \mathbb{R}^{\mathbb{N}}, \qquad f \mapsto (f([\gamma_n]))_{n \in \mathbb{N}}$$

is bijective.

We want to determine the subgroup $H^1_{\mathrm{dR}}(M, S; \mathbb{R})$ in $H^1(M, S; \mathbb{R})$. Let $\zeta \in Z^1_{\mathrm{dR}}(M, S; \mathbb{R})$. Since $H^1_{\mathrm{dR}}(\mathbb{S}^2; \mathbb{R})$ is trivial, there exists a smooth function $f \colon \mathbb{S}^2 \to \mathbb{R}$ with $f(x_0) = 0$ and $df = \zeta$. Then

$$\int_{\gamma_n} \zeta = \int_{\gamma_n} df = f(x_n) - f(x_0) = f(x_n),$$

and the question is how to characterize those sequences in $\mathbb{R}^{\mathbb{N}}$ which arise as $(f(x_n))_{n \in \mathbb{N}}$ for such a function f. We obtain a natural chart around x_0 via

$$\phi \colon U := \{x \in \mathbb{R}^2 \colon \|x\|_2 < 1\} \to \mathbb{S}^2, \qquad \phi(x) = \left(x_1, x_2, \sqrt{1 - x_1^2 - x_2^2}\right).$$

Each of the functions constructed in Example II.11 may be extended to a smooth compactly supported function on a neighborhood of S in $\mathbb{R}^2$ in such a way that it does not depend on the second variable x_2 in a neighborhood of S. Then we may use the chart ϕ to obtain a function in $C^\infty(M, S; \mathbb{R})$. We thus obtain

$$H^1_{\mathrm{dR}}(M, S; \mathbb{R}) \cong \{(\lambda_n)_{n \in \mathbb{N}} \colon (\forall k \in \mathbb{N})\, \lambda_n n^k \to 0\} \subseteq H^1(M, S; \mathbb{R}) \cong \mathbb{R}^{\mathbb{N}},$$

i.e., that $H^1_{\mathrm{dR}}(M, S; \mathbb{R})$ corresponds to the space of rapidly decreasing sequences with its usual topology.

A function f yields an element in the group $H^1_{\mathrm{dR}}(M, S; \mathbb{Z})$ if and only if all its values in the x_n are integral, so that $H^1_{\mathrm{dR}}(M, S; \mathbb{Z}) \cong \mathbb{Z}^{(\mathbb{N})}$ corresponds to the integer-valued functions with finite support. In particular $H^1_{\mathrm{dR}}(M, S; \mathbb{Z})$ is a discrete subgroup of $H^1_{\mathrm{dR}}(M, S; \mathbb{R})$ (cf. Theorem IV.7). $\qquad\square$

We conclude this section with some additional remarks on the relation between the two spaces $H^1_{\mathrm{dR}}(M, S; Y)$ and $H^1_{\mathrm{dR},c}(M \setminus S; Y)$.

Remark II.13. We recall from Remark II.6 (c) the injection

$$\phi: H^1_{dR,c}(M \setminus S; Y) \hookrightarrow H^1_{dR}(M, S; Y).$$

(a) If S is a compact submanifold of M, then ϕ is surjective. In fact, if $\zeta \in Z^1_{dR}(M, S; Y)$, then $\zeta \mid_S = 0$. Let U be a tubular neighborhood of S diffeomorphic to $S \times \mathbb{R}$. Then $\zeta \mid_U$ is exact, and there exists $f \in C^\infty(U; Y)$ with $df = \zeta \mid_U$. Now there exists a function $f_1 \in C^\infty(M; Y)$ which coincides with f on a neighborhood of S, and then $\zeta - df$ vanishes in a neighborhood of S. This proves that $[\zeta] = [\zeta - df_1] \in \text{im}(\phi)$.

(b) If $H^1_{dR,c}(M \setminus S; \mathbb{R})$ is infinite-dimensional, then ϕ is not surjective. In fact, then the space $H^1_{dR,c}(M \setminus S; \mathbb{R})$ is a countable direct limit of finite-dimensional spaces, hence of countable dimension (cf. Theorem IV.16). On the other hand $H^1_{dR}(M, S; \mathbb{R})$ is a quotient of the Fréchet space $Z^1_{dR}(M, S; \mathbb{R})$ by a closed subspace, hence a Fréchet space. As ϕ is injective, this space is infinite-dimensional, so that the Baire property implies that it is not countably dimensional. Hence ϕ is not surjective.

(c) If $H_0(S)$ is finitely generated, i.e., S has only finitely many arc-components, then the exact homology sequence of the pair (M, S) implies that $H_1(M, S)$ is finitely generated, which in turn implies that

$$H^1(M, S; \mathbb{R}) \cong \text{Hom}(H_1(M, S), \mathbb{R})$$

is finite-dimensional. Therefore $H^1_{dR}(M, S; \mathbb{R})$ is also finite-dimensional (cf. Definition II.5).

Conversely, every locally constant function $S \to \mathbb{Z}$ can be extended to a smooth function $f: M \to \mathbb{R}$ (it suffices to consider functions $S \to \{0, 1\}$) which is locally constant in a neighborhood of the compact set S. Then $df \in Z^1_{dR,c}(M \setminus S; \mathbb{Z})$. The class of $[df]$ in $H^1_{dR,c}(M \setminus S; \mathbb{Z})$ is non-zero if $f \mid_S$ is not constant. Therefore $H^1_{dR,c}(M \setminus S; \mathbb{Z})$ has infinite rank if $C(S, \mathbb{Z})$ has infinite rank. Note that this condition is weaker than the requirement that S has only finitely many arc-components.   □

III   Compact manifolds with boundary

Our strategy to get a better description of the spaces $\mathfrak{z}_M(Y)$ and $H^1_{dR,c}(M; Y)$ for a non-compact manifold is to describe M as a union of certain compact submanifolds with boundary $(X_n)_{n \in \mathbb{N}}$ with $X_n \subseteq X^0_{n+1}$ (cf. Section IV). To get information on the space $H^1_{dR,c}(M; Y)$, we will need detailed information on the natural maps $H^1_{dR}(X_n, \partial X_n; Y) \to H^1_{dR}(X_{n+1}, \partial X_{n+1}; Y)$. To obtain this information is the main goal of the present section (Theorem III.6). In this section we only deal with compact manifolds with boundary, and in Section IV we describe the approximation of non-compact manifolds.

In the following we write for a topological space X simply $H_*(X) := H_{\mathrm{sing},*}(X; \mathbb{Z})$ for the *singular homology groups* with coefficients in $\mathbb{Z}$. We likewise write $H_*(X, A)$ for the singular homology groups for space pairs (X, A).

Proposition III.1. *Let X be a compact manifold with boundary ∂X. Then the following assertions hold:*

(i) *The singular homology groups $H_*(X)$ are finitely generated.*

(ii) *All homotopy groups $\pi_k(X), k \in \mathbb{N}_0$, are finitely generated.*

(iii) *For each commutative ring R the cohomology groups $H^*(X, R)$ are finitely generated R-modules.*

(iv) *The relative homology groups $H_*(X, \partial X)$ are finitely generated.*

(v) *The inclusion $\mathrm{int}(X) \hookrightarrow X$ is a homotopy equivalence.*

Proof. There exists a compact manifold X^d, the *double of X*, in which X embeds. In particular Whitney's Embedding Theorem implies that X^d and hence X embeds smoothly into $\mathbb{R}^{2d+1}$, where $d = \dim X$. From the proof of Corollary E.5 in [Br93] we derive that there exists a finite CW-complex $K \subseteq \mathbb{R}^{2n+1}$ such that K is a neighborhood of X and there exists a retraction $r \colon K \to X$. The inclusion $j \colon X \hookrightarrow K$ satisfies $r \circ j = \mathrm{id}_X$.

(i) We immediately derive that the spaces $H_*(X)$ are direct summands in $H_*(K)$, hence in particular finitely generated abelian groups.

(ii) We likewise see that for each $k \in \mathbb{N}_0$ we have $\pi_k(K) \cong \ker \pi_k(r) \rtimes \pi_k(X)$. As $\pi_k(K)$ is finitely generated, the same holds for the group $\pi_k(X) \cong \pi_k(K)/\ker \pi_k(r)$.

(iii) In view of [Fu70, Th. 52.2], we have for abelian groups A and $C_j, j \in J$:

$$\mathrm{Ext}(\oplus_{j \in J} C_j, A) = \prod_{j \in J} \mathrm{Ext}(C_j, A).$$

As $\mathrm{Ext}(\mathbb{Z}, A) \cong \mathbf{0}$ and $\mathrm{Ext}(\mathbb{Z}/n\mathbb{Z}, A) \cong A/nA$, we conclude that for every commutative ring R and every finitely generated abelian group Γ the group $\mathrm{Ext}(\Gamma, R)$ is a finitely generated R-module. Therefore the Universal Coefficient Theorem implies that for every compact manifold with boundary the groups $H^*(X, R)$ are finitely generated R-modules.

(iv) In view of [Br93, Th. IV.6.15], we further have an exact sequence

$$H_*(\partial X) \to H_*(X) \to H_*(X, \partial X) \to H_{*-1}(\partial X).$$

The fact that $H_{*-1}(\partial X)$ and $H_*(X)$ are finitely generated groups implies that the groups $H_*(X, \partial X)$ are finitely generated.

(v) Using the collar construction for a compact manifold with boundary, we obtain inclusions $\mathrm{int}(X) \hookrightarrow X \hookrightarrow \mathrm{int}(X) \hookrightarrow X$, where the compositions of two successive ones are homotopic to the identity on $\mathrm{int}(X)$, resp., X. Therefore the inclusion $\mathrm{int}(X) \hookrightarrow X$ is a homotopy equivalence. $\square$

Lemma III.2. *For each compact manifold X with boundary the space $H^1_{\mathrm{dR}}(X, \partial X; \mathbb{R})$ is finite-dimensional.*

Proof. In Definition II.5 we have described an embedding

$$H^1_{\mathrm{dR}}(X, \partial X; \mathbb{R}) \hookrightarrow H^1(X, \partial X; \mathbb{R}).$$

Hence the assertion follows from Proposition III.1 which implies that $H^1(X, \partial X; \mathbb{R})$ is finite-dimensional. ☐

We take a closer look at the embedding

$$H^1_{\mathrm{dR}}(X, \partial X; \mathbb{R}) \hookrightarrow H^1(X, \partial X; \mathbb{R}) \cong \mathrm{Hom}(H_1(X, \partial X); \mathbb{R})$$

introduced in Definition II.5. The injectivity of this embedding implies that the integration maps

$$I_\gamma : H^1_{\mathrm{dR}}(X, \partial X; \mathbb{R}) \to \mathbb{R}, \quad [\zeta] \mapsto \int_\gamma \zeta$$

for singular cycles $\gamma \in Z_1(X, \partial X)$ separate points. We are interested in a nice set of such cycles for which the integration maps form a basis of the dual space of the finite-dimensional vector space $H^1_{\mathrm{dR}}(X, \partial X; \mathbb{R})$.

We recall the part

$$H_1(\partial X) \to H_1(X) \xrightarrow{\iota} H_1(X, \partial X) \to H_0(\partial X) \xrightarrow{s} H_0(X)$$

of the long exact homology sequence of the pair $(X, \partial X)$ ([Br93, Th. IV.6.15]). Let $\iota : H_1(X) \to H_1(X, \partial X)$ be the natural map and choose piecewise smooth loops $\alpha_1, \ldots, \alpha_a$ in X for which the images $\iota([\alpha_i]) \in H_1(X, \partial X)$ form a $\mathbb{Z}$-basis of the image $\iota(H_1(X))$ modulo torsion. Let $b := \mathrm{rk}\, H_0(\partial X) - 1$ and choose a minimal system of piecewise smooth arcs $\beta_1, \ldots, \beta_b$ in $Z_1(X, \partial X)$ connecting the boundary components of ∂X. Since there are $b+1$ boundary components, b arcs suffice and less would not be enough. Then the images of the classes $[\beta_i]$ in $H_0(\partial X)$ form a $\mathbb{Z}$-basis of the kernel of the summation map $s : H_0(\partial X) \cong \mathbb{Z}^{b+1} \to H_0(X) \cong \mathbb{Z}$.

Since the classes $[\beta_j]$ form a basis of the image of $H_1(X, \partial X)$ in $H_0(\partial X)$, and the classes $\iota([\alpha_i])$ generate the kernel of the map $H_1(X, \partial X) \to H_0(\partial X)$ modulo torsion, the classes $\iota([\alpha_i])$ and $[\beta_j]$ form a $\mathbb{Z}$-basis of the abelian group $H_1(X, \partial X)$ modulo torsion.

The bijectivity of the map η in the following proposition (see also (2.1)) can alternatively be derived from the discussion in Remark II.6(c), which implies that the real vector spaces $H^1_{\mathrm{dR}}(X, \partial X; \mathbb{R})$ and $\mathrm{Hom}(H_1(X, \partial X); \mathbb{R})$ have the same dimension, so that the injectivity of η implies that it is bijective. We will see that Proposition III.3 provides more concrete information which is needed later on.

Proposition III.3. *The integration functionals $I_{\alpha_i}, i = 1, \ldots, a$, and $I_{\beta_j}, j = 1, \ldots, b$, form a basis of the dual space of $H^1_{\mathrm{dR}}(X, \partial X; \mathbb{R})$. In particular, the natural homo-*

morphism

$$\eta: H^1_{dR}(X, \partial X; \mathbb{R}) \to \mathrm{Hom}(H_1(X, \partial X); \mathbb{R}), \quad \eta([\zeta])([\gamma]) = \int_\gamma \zeta$$

from Definition II.5 (2.1) is bijective.

Proof. Since the classes $\iota([\alpha_i])$ and $[\beta_j]$ generate $H_1(X, \partial X)$ modulo torsion and η is injective (Definition II.5), the integration maps I_{α_i} and I_{β_j} separate the points of $H^1_{dR}(X, \partial X; \mathbb{R})$, hence span its dual space.

Let $\chi_0: H_1(X, \partial X) \to \mathbb{R}$ be a homomorphism and $\chi: H_1(X) \to \mathbb{R}$ its pull-back to $H_1(X)$. Then χ vanishes on the image of $H_1(\partial X)$ in $H_1(X)$, so that there exists a closed 1-form α on X with

$$\int_\gamma \alpha = \chi(\gamma), \quad \gamma \in H_1(X).$$

This can be proved as [Ne02a, Prop. 3.8]. The main idea is to associate to χ, viewed as a homomorphism $\pi_1(X) \to \mathbb{R}$, an affine $\mathbb{R}$-bundle over X and then to use partitions of unity to obtain a smooth global section s, whose differential can be taken as α. Since χ vanishes on the image of $H_1(\partial X)$ in $H_1(X)$, we can think of it as a homomorphism of the image $\iota(H_1(X))$ of $H_1(X)$ in $H_1(X, \partial X)$ to $\mathbb{R}$.

Let C be a connected component of ∂X, $I := [0, 1]$ and $\hat{C}$ be a neighborhood of C in X diffeomorphic to $I \times C$ in such a way that $\{0\} \times C$ corresponds to C. Then the homomorphism $H_1(C) \to H_1(\partial X) \to \mathbb{R}$ induced by the 1-form α vanishes, so that there exists a smooth function $g_0: \hat{C} \to \mathbb{R}$ with $\alpha \mid_{\hat{C}} = dg$. If $\phi: I \to \mathbb{R}$ is smooth with $\phi = 1$ in a neighborhood of 0 and 0 in a neighborhood of 1, then $\hat{\phi}: (t, x) \mapsto \phi(t)$ yields a smooth function on X vanishing in a neighborhood of $X \setminus \hat{C}$ and taking the value 1 on a neighborhood of C. Hence $\hat{\phi} \cdot g$ can be viewed as a smooth function $X \to \mathbb{R}$ whose differential coincides with dg in a neighborhood of C. Now $\alpha - d(\hat{\phi} \cdot g_0)$ defines the same homomorphism $\pi_1(X) \to \mathbb{R}$ but, in addition, this 1-form vanishes in a neighborhood of C. Repeating this construction for the other connected components of ∂X yields a closed 1-form $\alpha' \in \Omega^1(X, \partial X; \mathbb{R})$ vanishing in a neighborhood of ∂X for which α' represents χ on $H_1(X)$. We conclude that $\chi_0 - \eta([\alpha'])$ vanishes on $\iota(H_1(X))$ in $H_1(X, \partial X)$, so that it remains to see that each homomorphisms $\chi: H_1(X, \partial X) \to \mathbb{R}$ vanishing on the image of $H_1(X)$ is contained in $\mathrm{im}(\eta)$. Let $r: H_1(X, \partial X) \to H_0(X)$ denote the boundary map. Then $\chi_0 = \chi' \circ r$ for some $\chi': H_0(\partial X) \cong \mathbb{Z}^{b+1} \to \mathbb{R}$.

Let $C \subseteq \partial X$ be a connected component. Using the collar construction, we obtain a smooth function $f_C: X \to \mathbb{R}$ which is 1 in a neighborhood of C and 0 in a neighborhood of all other connected components of ∂X. Then $df_C \in Z^1_{dR}(X, \partial X; \mathbb{R})$ and because the form df_C is exact, it vanishes on all cycles in $\iota(H_1(X))$. Moreover, the function f_C defines a homomorphism

$$F_C: H_0(\partial X) \to \mathbb{Z}, \quad C' \mapsto f_C(C') = \delta_{C,C'},$$

and, as a homomorphism $H_1(X, \partial X) \to \mathbb{R}$, the integration of df_C over cycles modulo ∂X is obtained by pulling F_C back via the natural map $H_1(X, \partial X) \to H_0(\partial X)$. As the F_C form a $\mathbb{Z}$-basis of $\operatorname{Hom}(H_0(\partial X), \mathbb{R})$, we conclude that χ' lies in the span of the $\eta([df_C])$, hence is contained in the image of η. This completes the proof of the surjectivity of η. □

Lemma III.4. *For any s.c.l.c. space Y the exactness of a closed 1-form $\zeta \in \Omega^1(X, \partial X; Y)$ is equivalent to the vanishing of all integrals $\int_{\alpha_i} \zeta$ and $\int_{\beta_j} \zeta$.*

Proof. If $\zeta \in \Omega^1(X, \partial X; Y)$ is exact, then clearly all integrals $\int_\gamma \zeta$ vanish for $\gamma \in Z_1(X, \partial X)$. Suppose, conversely, that all integrals $\int_{\alpha_i} \zeta$ and $\int_{\beta_j} \zeta$ vanish. For each continuous linear functional $\lambda \in Y'$ we then obtain

$$\int_{\alpha_i} \lambda \circ \zeta = \lambda\left(\int_{\alpha_i} \zeta\right) = \lambda\left(\int_{\beta_j} \zeta\right) = \int_{\beta_j} \lambda \circ \zeta = 0$$

for each i and j. Since Y' separates points of Y, all integrals of ζ on $Z_1(X, \partial X)$ are trivial, and therefore ζ is exact. □

Remark III.5. Let $[\alpha_i^*], [\beta_j^*] \in H_{dR}^1(X, \partial X; \mathbb{R})$ be a basis dual to the integrals I_{α_i} and I_{β_j} from above. Then the map

$$\Phi_X : H_{dR}^1(X, \partial X; Y) \to Y^{a+b}, \quad \Phi_X([\zeta]) := \left(\int_{\alpha_i} \zeta, \int_{\beta_j} \zeta\right)_{i=1,\dots,a;\, j=1,\dots,b}$$

is continuous and injective (Lemma III.4). Moreover, it is surjective and its inverse is given by

$$\Phi_X^{-1}(y_1, \dots, y_{a+b}) := \sum_{i=1}^{a} [\alpha_i^* \cdot y_i] + \sum_{j=1}^{b} [\beta_j^* \cdot y_{a+j}].$$

It follows in particular that Φ_X^{-1} is continuous, and therefore that Φ_X is an isomorphism of topological vector spaces. The extension of Φ_X to a map

$$\widetilde{\Phi}_X : \mathfrak{z}_{(X,\partial X)}(Y) \to Y^{a+b}, \quad \Phi_X([\zeta]) := \left(\int_{\alpha_i} \zeta, \int_{\beta_j} \zeta\right)_{i=1,\dots,a;\, j=1,\dots,b}$$

is continuous and surjective. Therefore its kernel is a closed complement to $H_{dR}^1(X, \partial X; Y)$ and the corresponding projection onto $H_{dR}^1(X, \partial X; Y)$ is given by

$$p_X : [\zeta] \mapsto \sum_{i=1}^{a} \left[\alpha_i^* \cdot \int_{\alpha_i} \zeta\right] + \sum_{j=1}^{b} \left[\beta_j^* \cdot \int_{\beta_j} \zeta\right]. \qquad □$$

Theorem III.6. *Let Z be a compact connected manifold with boundary and $X \subseteq \operatorname{int}(Z)$ a compact connected equidimensional submanifold with boundary. We assume that each connected component of $Z \setminus X$ intersects ∂Z. Then the following assertions hold:*

(1) *The inclusion $Z^1_{dR}(X, \partial X; Y) \hookrightarrow Z^1_{dR}(Z, \partial Z; Y)$ obtained by extension by 0 on $Z \setminus X$ induces an injective map*

$$H^1_{dR}(X, \partial X; Y) \hookrightarrow H^1_{dR}(Z, \partial Z; Y).$$

(2) *The continuous projection p_X extends to a continuous projection p_Z, so that we obtain the commutative diagram*

$$
\begin{array}{ccc}
\mathfrak{z}_{(X,\partial X)}(Y) & \xrightarrow{p_X} & H^1_{dR}(X, \partial X; Y) \\
\downarrow & & \downarrow \\
\mathfrak{z}_{(Z,\partial Z)}(Y) & \xrightarrow{p_Z} & H^1_{dR}(Z, \partial Z; Y).
\end{array}
$$

Proof. Let α_i, $i = 1, \ldots, a$ and β_j, $j = 1, \ldots, b$ be as in Proposition III.3. Then the integration functionals $I_{\alpha_1}, \ldots, I_{\alpha_a}, I_{\beta_1}, \ldots, I_{\beta_b}$ form a basis of the dual space of $H^1_{dR}(X, \partial X; \mathbb{R})$.

(1) We claim that

$$dC^\infty(Z, \partial Z; Y) \cap Z^1_{dR}(X, \partial X; Y) = dC^\infty(X, \partial X; Y).$$

The inclusion "$\supseteq$" is trivial. Conversely, let $f \in C^\infty(Z, \partial Z; Y)$ and suppose that $df \in Z^1_{dR}(X, \partial X; Y)$, i.e., that df vanishes on $Z \setminus X$. Then f is constant on all connected components of $Z \setminus X$. By our initial assumptions, all connected components of $Z \setminus X$ intersect ∂Z, which implies that f vanishes on all these components, hence that $f \in C^\infty(X, \partial X; Y)$. This proves (1).

(2) Next we want to choose integration maps $H^1_{dR}(Z, \partial Z; Y) \to Y$ in such a way that those which are additional to the ones needed for X are supported by $Z \setminus \mathrm{int}(X)$, hence vanish on $Z^1_{dR}(X, \partial X; Y)$.

We have to modify the curves β_i so that they represent elements on $Z_1(Z, \partial Z)$. Since every connected component of $Z \setminus X$ meets ∂Z, we can extend every piecewise smooth curve β_i to a piecewise smooth curve $\tilde{\beta}_i$ connecting two boundary components of Z. For this we may w.l.o.g. assume that we have parametrizations $\beta_j : [0, 1] \to X$ and $\tilde{\beta}_j : [-1, 2] \to Z$ with $\tilde{\beta}_j |_{[0,1]} = \beta_j$ and $[0, 1] = \tilde{\beta}_j^{-1}(X)$. In particular we have for each 1-form ζ supported by X the relation

$$\int_{\beta_i} \zeta = \int_{\tilde{\beta}_i} \zeta.$$

Next we choose piecewise smooth closed curves $\gamma_1, \ldots, \gamma_c$ in $Z \setminus X$ connecting those connected components of ∂Z lying in the same connected component of $Z \setminus X$. We further need closed curves on $\delta_1, \ldots, \delta_d$ in $Z \setminus \mathrm{int}(X)$ whose homology classes generate $H_1(Z \setminus \mathrm{int}(X); \mathbb{R})$ modulo the image of $H_1(\partial Z; \mathbb{R})$. We will show below that the classes of $\alpha_i, \tilde{\beta}_j, \gamma_k$ and δ_l generate $H_1(Z, \partial Z)$ modulo torsion by showing that the corresponding integrals separate points on $H^1_{dR}(Z, \partial Z; \mathbb{R})$.

Let $\zeta \in Z^1_{dR}(Z, \partial Z; Y)$ be such that all integrals over the $\alpha_i, \tilde{\beta}_j, \gamma_k$ and δ_l vanish. We claim that ζ is exact. In particular all integrals coming from $H_1(\partial X)$ vanish, so that there exists an open neighborhood $U \cong I \times \partial X$ of ∂X on which ζ is exact. Let $f \in$

$C^\infty(U; Y)$ with $df = \zeta\,|_U$. Multiplying f with a smooth function χ on U of the form $(t, x) \mapsto \phi(t)$, where $\phi \in C^\infty(I; \mathbb{R})$ is 1 on a neighborhood of 0 and vanishes outside some interval $[-\varepsilon, \varepsilon]$, we obtain a smooth function $\widetilde{f} := \chi \cdot f \in C^\infty(Z, \partial Z; Y)$ with $d\widetilde{f} = \zeta$ in a neighborhood of ∂X. Replacing ζ by $\zeta - d\widetilde{f}$, we may assume that ζ vanishes on a neighborhood of ∂X. Then $\zeta\,|_X \in Z^1_{\mathrm{dR}}(X, \partial X; Y)$ is exact because the integrals over the α_i vanish. Likewise $\zeta\,|_{Z \setminus X}$ is exact because all integrals over the δ_i vanish. Let $f_1 \in C^\infty(X; Y)$ with $df_1 = \zeta\,|_X$ and $f_2 \in C^\infty(Z \setminus \mathrm{int}(X); Y)$ with $df_2 = \zeta\,|_{Z \setminus X}$. We normalize f_2 by the condition that it vanishes on ∂Z. That this is possible follows from the vanishing of all integrals of ζ over the γ_i. We further normalize f_1 such that on one boundary point $x \in \partial X$ we have $f_1(x) = f_2(x)$. In a neighborhood of ∂X both functions f_1 and f_2 are locally constant, hence constant on all connected components of ∂X. It remains to show that $f_1\,|_{\partial X} = f_2\,|_{\partial X}$, so that both combine to a function $f \in C^\infty(Z, \partial Z; Y)$ with $df = \zeta$.

Let β_i be such that either its end or starting point lies in the same connected component of ∂X as x. We recall the parametrizations $\beta_i : [0, 1] \to X$ from above. We further observe that $f_1(x) = f_1(\beta_i(0)) = f_2(x) = f_2(\beta_i(0))$ because $f_1 = f_2$ is constant on the whole component of X containing x. We also recall the parameterization of $\widetilde{\beta}_j$ on $[-1, 2]$ from above and put $y := \beta_i(1) \in \partial X$. Let $p := \widetilde{\beta}_i(-1)$ and $q := \widetilde{\beta}_i(2)$. Then

$$f_1(y) - f_2(y) = \left(f_1(x) + \int_{\beta_i} \zeta \right) + \underbrace{f_2(q) - f_2(y)}_{=0} = f_2(x) + \int_{\beta_i} \zeta + f_2(q) - f_2(y)$$

$$= \int_{-1}^{0} \widetilde{\beta}_i^* \zeta + \int_0^1 \widetilde{\beta}_i^* \zeta + \int_1^2 \widetilde{\beta}_i^* \zeta = \int_{\widetilde{\beta}_i} \zeta = 0.$$

This proves $f_1(y) = f_2(y)$. Using the other paths $\widetilde{\beta}_i$, we conclude inductively that $f_1 = f_2$ holds on all connected components of ∂X, and this completes the proof of the exactness of ζ.

Hence the integration maps I_{α_i}, $I_{\widetilde{\beta}_j}$, I_{γ_k} and I_{δ_l} separate points on $H^1_{\mathrm{dR}}(Z, \partial Z; \mathbb{R})$. Since the maps I_{α_i}, $i = 1, \dots, a$, and $I_{\widetilde{\beta}_j}$, $j = 1, \dots, b$, are linearly independent on the subspace $H^1_{\mathrm{dR}}(X, \partial X; \mathbb{R})$, by omitting some of the γ_k and δ_l, we may w.l.o.g. assume that the whole collection is linearly independent.

We recall the maps Φ_X and p_X from Remark III.5. Then we see that

$$\Phi_Z : H^1_{\mathrm{dR}}(Z, \partial Z; Y) \to Y^{a+b+c+d},$$

$$\Phi_Z([\zeta]) := \left(\int_{\alpha_i} \zeta, \int_{\beta_j} \zeta, \int_{\gamma_k} \zeta, \int_{\delta_l} \zeta \right)_{i=1,\dots,a;\, j=1,\dots,b;\, k=1,\dots,c;\, l=1,\dots,d}$$

is a topological isomorphism. The corresponding projection

$$p_Z : \mathfrak{z}_{(Z, \partial Z)}(Y) \to H^1_{\mathrm{dR}}(Z, \partial Z; Y)$$

is given by

$$p_Z : [\zeta] \mapsto \sum_{i=1}^{a} \left[\alpha_i^* \cdot \int_{\alpha_i} \zeta\right] + \sum_{j=1}^{b} \left[\beta_j^* \cdot \int_{\tilde{\beta}_j} \zeta\right] + \sum_{k=1}^{c} \left[\gamma_k^* \cdot \int_{\gamma_k} \zeta\right] + \sum_{l=1}^{d} \left[\delta_l^* \cdot \int_{\delta_l} \zeta\right].$$

Since the integrals over the γ_k and δ_l vanish for $\zeta \in \Omega^1(X, \partial X; Y)$, and the integrals over β_j and $\tilde{\beta}_j$ are the same for these 1-forms, we obtain $p_Z \mid_{\mathfrak{z}(X, \partial X)(Y)} = p_X$. $\square$

Example III.7 (Oriented surfaces). Let X be an oriented compact connected surface with boundary. All the boundary components are diffeomorphic to the circle. Collapsing each boundary component to a point leads to an oriented compact surface Σ. Let $g := g(X) := g(\Sigma)$ denote the *genus* of Σ and $p := p(X)$ be the number of boundary components.

We recall the part

$$\cdots \to H_2(X) \to H_2(X, \partial X) \to H_1(\partial X) \xrightarrow{\alpha} H_1(X)$$
$$\to H_1(X, \partial X) \to H_0(\partial X) \to H_0(X)$$

of the long exact homology sequence of the pair $(X, \partial X)$. Then $H_0(\partial X) \cong H_1(\partial X) \cong \mathbb{Z}^P$. According to Proposition III.1(v), the inclusion $\mathrm{int}(X) \hookrightarrow X$ is a homotopy equivalence, so that $H_1(X) \cong H_1(\mathrm{int}(X))$. On the other hand $\mathrm{int}(X) \cong \Sigma \setminus P$, where P is the image of ∂X in Σ.

Let $\hat{P}$ be a disjoint union of open discs in Σ around each point of P. Then $\Sigma = \mathrm{int}(X) \cup \hat{P}$ is a union of two open subsets, and the exact Mayer–Vietoris Sequence ([Br93, Th. IV.18.1]) yields an exact sequence

$$\cdots \to H_2(\mathrm{int}(X)) \oplus H_2(\hat{P}) \to H_2(\Sigma) \to H_1(\mathrm{int}(X) \cap \hat{P})$$
$$\to H_1(\mathrm{int}(X)) \oplus H_1(\hat{P})$$
$$\to H_1(\Sigma) \to H_0(\mathrm{int}(X) \cap \hat{P})$$
$$\to H_0(\mathrm{int}(X)) \oplus H_0(\hat{P}) \to H_0(\Sigma).$$

We have $H_0(\hat{P}) \cong \mathbb{Z}^P$, $H_1(\hat{P}) = H_2(\hat{P}) = \mathbf{0}$, $H_0(\mathrm{int}(X)) \cong \mathbb{Z}$, $H_0(\mathrm{int}(X) \cap \hat{P}) \cong \mathbb{Z}^P$, $H_1(\mathrm{int}(X) \cap \hat{P}) \cong \mathbb{Z}^P$, and $H_2(\mathrm{int}(X)) = \mathbf{0}$ because $\mathrm{int}(X)$ is not compact. Therefore we obtain an exact sequence

$$H_2(\Sigma) \cong \mathbb{Z} \hookrightarrow \mathbb{Z}^P \to H_1(\mathrm{int}(X)) \to H_1(\Sigma) \cong \mathbb{Z}^{2g} \xrightarrow{0} \mathbb{Z}^P \hookrightarrow \mathbb{Z} \oplus \mathbb{Z}^P \to \mathbb{Z}.$$

The vanishing of the homomorphism in the middle follows from the injectivity of the map $H_0(\mathrm{int}(X) \cap \hat{P}) \to H_0(\hat{P})$. This implies that the sequence

$$\mathbb{Z} \hookrightarrow \mathbb{Z}^P \to H_1(\mathrm{int}(X)) \to \mathbb{Z}^{2g} \to 0$$

is exact. As $\pi_1(\mathrm{int}(X))$ is a free group [tD00, Satz II.8.8], the homology group $H_1(\mathrm{int}(X)) \cong \pi_1(\mathrm{int}(X))/[\pi_1(\mathrm{int}(X)), \pi_1(\mathrm{int}(X))]$ is a free abelian group, which leads to

$$H_1(X) \cong H_1(\text{int}(X)) \cong \mathbb{Z}^{2g(X)+p(X)-1}.$$

Now we obtain with $H_2(X) \cong H_2(\text{int}(X)) = 0$ for $H_1(X, \partial X)$ the exact sequence

$$H_2(X, \partial X) \hookrightarrow H_1(\partial X) \cong \mathbb{Z}^p \xrightarrow{\alpha} H_1(X) \cong \mathbb{Z}^{2g+p-1} \to H_1(X, \partial X) \to \mathbb{Z}^p \to \mathbb{Z}.$$

The image of α in $H_1(X)$ corresponds to the image of $H_1(\text{int}(X) \cap \hat{P})$ in $H_1(\text{int}(X))$ in the exact Mayer–Vietoris Sequence, and is isomorphic to $\mathbb{Z}^{p-1}$. The cokernel of α is isomorphic to $\mathbb{Z}^{2g}$. The map $H_0(\partial X) \cong \mathbb{Z}^p \to H_0(X) \cong \mathbb{Z}$ is the summation map, so that its kernel is isomorphic to $\mathbb{Z}^{p-1}$. We thus obtain a short exact sequence

$$\text{coker}(\alpha) \cong \mathbb{Z}^{2g} \hookrightarrow H_1(X, \partial X) \twoheadrightarrow \mathbb{Z}^{p-1},$$

and finally

$$H_1(X, \partial X) \cong \mathbb{Z}^{2g(X)+p(X)-1}. \qquad \square$$

Example III.8 (Non-orientable surfaces). Let X be a non-orientable compact connected surface with boundary and proceed as in Example III.7. Then Σ is non-orientable. We define $g(X)$ and $p(X)$ as in Example III.8.

For the finite subset $P \subseteq \Sigma$ we now obtain with the exact Mayer–Vietoris sequence:

$$\cdots \to H_2(\Sigma) = 0 \to H_1(\text{int}(X) \cap \hat{P}) \cong \mathbb{Z}^p \to H_1(\text{int}(X)) \oplus H_1(\hat{P})$$
$$\to H_1(\Sigma) \cong \mathbb{Z}^g \oplus \mathbb{Z}_2 \to H_0(\text{int}(X) \cap \hat{P}) \cong \mathbb{Z}^p$$
$$\to H_0(\text{int}(X)) \oplus H_0(\hat{P}) \cong \mathbb{Z}^{p+1} \to H_0(\Sigma) \cong \mathbb{Z}.$$

This leads to an exact sequence

$$\mathbb{Z}^p \hookrightarrow H_1(\text{int}(X)) \to H_1(\Sigma) \cong \mathbb{Z}^g \oplus \mathbb{Z}_2 \to \mathbb{Z}^p \hookrightarrow \mathbb{Z}^{p+1},$$

and further to

$$\mathbb{Z}^p \hookrightarrow H_1(\text{int}(X)) \twoheadrightarrow \mathbb{Z}^g \oplus \mathbb{Z}_2.$$

As $H_1(\text{int}(X))$ is a free abelian group, it follows that

$$H_1(X) \cong H_1(\text{int}(X)) \cong \mathbb{Z}^{g(X)+p(X)}.$$

Now we obtain with the long exact homology sequence of the pair $(X, \partial X)$:

$$\cdots \to H_2(X) \to H_2(X, \partial X) \to H_1(\partial X) \xrightarrow{\alpha} H_1(X)$$
$$\to H_1(X, \partial X) \to H_0(\partial X) \to H_0(X)$$

and hence

$$\mathbb{Z}^p \xrightarrow{\alpha} \mathbb{Z}^{g+p} \to H_1(X, \partial X) \to \mathbb{Z}^p \xrightarrow{s} \mathbb{Z}.$$

The image of α in $H_1(X)$ corresponds to the image of $H_1(\text{int}(X) \cap P)$ in $H_1(\text{int}(X))$, hence is isomorphic to $\mathbb{Z}^p$, and $\text{coker}(\alpha) \cong \mathbb{Z}^g$. Here $s : H_0(\partial X) \cong \mathbb{Z}^p \to H_0(X) \cong$

$\mathbb{Z}$ is the summation map, so that its kernel is isomorphic to $\mathbb{Z}^{p-1}$. We thus obtain a short exact sequence

$$\operatorname{coker}(\alpha) \cong \mathbb{Z}^g \hookrightarrow H_1(X, \partial X) \twoheadrightarrow \mathbb{Z}^{p-1} = \ker s,$$

which leads to

$$H_1(X, \partial X) \cong \mathbb{Z}^{g(X)+p(X)-1}. \qquad \qquad \square$$

IV Approximating non-compact manifolds by compact ones

In this section M denotes a connected σ-compact finite-dimensional manifold. We call a submanifold X of M *equidimensional* if $\dim X = \dim M$. In this section we first prove the existence of well behaved sequences $(X_n)_{n \in \mathbb{N}}$ of equidimensional compact submanifolds with boundary exhausting M (Lemma IV.4). The main result of this section is Theorem IV.16 providing a topological isomorphism

$$\Phi_M \colon H^1_{\mathrm{dR},c}(M; Y) \to Y^{(B)}$$

for a certain set B which might be infinite. The components of Φ_M are given by integration over singular cycles in M or over curves obtained from proper maps $\mathbb{R} \to M$. Here we make heavy use of Theorem III.6 about the cohomology of compact manifolds with boundary to construct the set B in such a way that Φ_M becomes an isomorphism. As a corollary, we show that if Γ is discrete, then $H^1_{\mathrm{dR},c}(M; \Gamma) \cong \Gamma^{(B)}$ is discrete.

Saturated exhaustive sequences

Lemma IV.1. *For each compact equidimensional submanifold $X \subseteq M$ with boundary the number of connected components of $M \setminus X$ is finite.*

Proof. As every connected component of $M \setminus X$ contains some component of ∂X in its closure, and the number of components of the compact manifold ∂X is finite, the assertion follows. $\qquad \square$

Definition IV.2. Let $X \subseteq M$ be an equidimensional compact submanifold with boundary. We observe that each connected component of ∂X is contained in the closure of exactly one connected component of $M \setminus X$. We write $\hat{X}$ for the union of X with all those components of $M \setminus X$ which are relatively compact. As the number of these components is finite (Lemma IV.1), $\hat{X}$ is compact, because for each component $C \subseteq M \setminus X$ the boundary ∂C is a union of connected components of ∂X. This argument further shows that $\hat{X}$ is a compact submanifold with boundary in M. $\qquad \square$

Lemma IV.3. *For two equidimensional submanifolds with boundary $X_1, X_2 \subseteq M$ with $X_1 \subseteq X_2^0$ we have $\hat{X}_1 \subseteq \hat{X}_2^0$.*

Proof. Let $C \subseteq M \setminus X_1$ be a relatively compact connected component. Then $C \setminus X_2$ is also relatively compact in M, hence contained in $\hat{X}_2$. Therefore $\hat{X}_1 \subseteq \hat{X}_2$. If $p \in \partial \hat{X}_2$ is a boundary point, then it is in particular a boundary point of X_2, hence not contained in X_1, and therefore not in ∂X_1. If the connected component of $M \setminus X_2^0$ containing p is non-compact, then this is likewise true for the connected component of $M \setminus X_1$ containing p, which shows that it is not contained in $\hat{X}_1$. This proves $\hat{X}_1 \subseteq \hat{X}_2^0$. $\square$

For the case of surfaces the following lemma can also be found in [tD00, Satz 7.3].

Lemma IV.4. *There exists a sequence X_n of compact connected manifolds with boundary in M such that*

(E1) $X_n \subseteq X_{n+1}^0$,

(E2) $\bigcup_n X_n = M$,

(E3) $\hat{X}_n = X_n$, *i.e., each connected component of $M \setminus X_n$ is not relatively compact in M.*

Proof. Let $\phi \colon M \to \mathbb{R}$ be a proper smooth function which is bounded from below. Such a function can be obtained from an embedding $\iota \colon M \hookrightarrow \mathbb{R}^n$ as $\phi(x) := \|x\|_2^2$. Then Sard's Theorem implies that there exists an increasing sequence $(r_n)_{n \in \mathbb{N}}$ of regular values of ϕ with $r_n \to \infty$. Then each $Y_n := \{x \in M \colon \phi(x) \leq r_n\}$ is a compact equidimensional submanifold with boundary. Pick $x_0 \in Y_1$. We define Z_n to be the connected component of Y_n containing x_0 and $X_n := \hat{Z}_n$. From $r_n < r_{n+1}$ we derive $Y_n \subseteq Y_{n+1}^0$, so that $Z_n \subseteq Z_{n+1}^0$, and Lemma IV.3 implies (E1). From $r_n \to \infty$ we get $\bigcup_n Y_n = M$. Each $x \in M$ can be connected to x_0 by an arc, which lies in some Y_n, whence $x \in Z_n$, and (E2) follows. Eventually (E3) follows from the definition of $\hat{Z}_n$. $\square$

We call a sequence $(X_n)_{n \in \mathbb{N}}$ as in Lemma IV.4 a *saturated exhaustive sequence* of M.

Lemma IV.5. *For each $x \in M$ there exists a proper smooth map $\gamma \colon \mathbb{R}^+ := [0, \infty[\to M$ with $\gamma(0) = x$. If $X = \hat{X}$ is an equidimensional compact submanifold with boundary and $x \in \partial X$, then there exists a γ as above with $\gamma(]0, \infty[) \subseteq M \setminus X$.*

Proof. Pick a saturated exhaustive sequence $(X_n)_{n \in \mathbb{N}}$ of M and choose points $x_n \in \partial X_n$ such that x_{n+1} lies in the connected component of $M \setminus X_n$ containing x_n in its boundary. Since this component is not relatively compact in M, it intersects ∂X_{n+1}. Then there exists a smooth curve $\gamma \colon \mathbb{R}^+ \to M$ with $\gamma(0) = x$, $\gamma(n) = x_n$ for all $n \in \mathbb{N}$, and $\gamma([n, n+1]) \subseteq X_{n+1} \setminus X_n^0$. The latter condition implies that γ is proper.

If $x \in \partial X$ holds for an equidimensional compact submanifold with boundary X, then $X \subseteq X_N$ for N sufficiently large, and we can proceed as above by connecting first x in $X_N \setminus X_1$ to a point in the boundary of X_N, then to a point in X_{N+1} etc. We thus obtain γ with the required properties. $\square$

Lemma IV.6. *For $x, y \in M$ there exists a proper smooth map $\gamma : \mathbb{R} \to M$ with $\gamma(0) = x$ and $\gamma(1) = y$.*

Proof. Using Lemma IV.5, we find a smooth map $\gamma : \mathbb{R} \to M$ with $\gamma(0) = x$ and $\gamma(1) = y$ such that the restrictions to $[1, \infty[$ and $] - \infty, 0]$ are proper. This implies that γ itself is proper.   □

The following lemma is obvious.

Lemma IV.7. *Let M be a topological space and $(M_j)_{j \in J}$ a directed family of open subsets of M with $M = \bigcup_j M_j$. Then $M = \varinjlim M_j$ holds in the category of topological spaces, each compact subset of M is contained in some M_j, and for each $x_M \in M$ and $k \in \mathbb{N}_0$ we have*

$$\pi_k(M, x_M) \cong \varinjlim \pi_k(M_j, x_M),$$

where $\{ j \in J : x_M \in M_j \}$ is cofinal in J.   □

Remark IV.8. The preceding lemma applies in particular to saturated exhaustions $(X_n)_{n \in \mathbb{N}}$ of a non-compact manifold M with $M_n = X_n^0$. Then we obtain with Proposition III.1(v):

$$\pi_k(M) \cong \varinjlim \pi_k(X_n^0) \cong \varinjlim \pi_k(X_n)$$ □

Proposition IV.9. *For each σ-compact connected finite-dimensional manifold M all homotopy groups are countable.*

Proof. This is a direct consequence of Lemma IV.7, Remark IV.8 and Proposition III.1 (ii).   □

De Rham cohomology with compact supports is a direct sum

If Y is a s.c.l.c. space and $(X_n)_{n \in \mathbb{N}}$ is a saturated exhaustive sequence of M, then $\Omega_c^1(M; Y)$ carries the locally convex direct limit topology of the spaces $\Omega_{X_n}^1(M; Y) \subseteq \Omega^1(M; Y)$ (cf. Section II). The differential $d : C_c^\infty(M; Y) \to \Omega_c^1(M; Y)$ is a continuous linear map because $C_c^\infty(M; Y)$ carries the locally convex direct limit topology of the subspaces $C_{X_n}^\infty(M; Y)$ on which d is continuous.

Lemma IV.10. *Let $X = \hat{X}$ be an equidimensional compact submanifold with boundary. Then $\Omega_X^1(M; Y) \cong \Omega^1(X, \partial X; Y)$ and*

$$\Omega_X^1(M; Y) \cap dC_c^\infty(M; Y) = dC_X^\infty(M; Y).$$

Proof. (cf. Step 1 in the proof of Theorem III.6) It is clear that $dC_X^\infty(M; Y)$ is contained in $\Omega_X^1(M; Y) \cap dC_c^\infty(M; Y)$. To prove the converse inclusion, let $\beta \in \Omega_X^1(M; Y)$

and $f \in C_c^\infty(M; Y)$ with $\beta = df$. Then f is constant on all connected components of $M \setminus X$. Since all these components are not relatively compact in M and f has compact support, it follows that $f(M \setminus X) = \{0\}$, and therefore $f \in C_X^\infty(M; Y)$. □

From the isomorphisms

$$\Omega_X^1(M; Y) \cong \Omega^1(X, \partial X; Y) \quad \text{and} \quad C_X^\infty(M; Y) \cong C^\infty(X, \partial X; Y)$$

obtained by extension on $M \setminus X$ by 0, we now derive

$$\Omega_X^1(M; Y)/\big(dC_c^\infty(M; Y) \cap \Omega_X^1(M; Y)\big)$$
$$\cong \Omega^1(X, \partial X; Y)/dC^\infty(X, \partial X; Y) = \mathfrak{z}_{(X, \partial X)}(Y).$$

Lemma IV.11. *For each s.c.l.c. space Y the subspace $B_{\mathrm{dR},c}^1(M; Y) = dC_c^\infty(M; Y)$ of $\Omega_c^1(M; Y)$ is closed.*

Proof. For each equidimensional compact submanifold $X = \hat{X}$ with boundary, Lemma IV.10 implies that $\Omega_X^1(M; Y) \cap dC_c^\infty(M; Y) = dC_X^\infty(M; Y)$, which corresponds to the subspace

$$dC^\infty(X, \partial X; Y) \subseteq \Omega^1(X, \partial X; Y)$$

whose closedness follows from Corollary II.4 which also applies to the pair $(X, \partial X)$, as it has the same space of smooth functions as the pair $(X^d, X^\sharp)$ (cf. Example I.9 (a)).

For each saturated exhaustive sequence $(X_n)_{n \in \mathbb{N}}$, the space $\Omega_c^1(M; Y)$ is the locally convex direct limit of the subspaces $\Omega_{X_n}^1(M; Y)$, so that the closedness of $dC_c^\infty(M; Y)$ follows from the closedness of the intersections with the spaces $\Omega_{X_n}^1(M; Y)$ (Lemma B.4 (ii)). □

Definition IV.12. As a consequence of Lemma IV.11, the space

$$\mathfrak{z}_{M,c}(Y) := \Omega_c^1(M; Y)/dC_c^\infty(M; Y)$$

carries a natural (Hausdorff) locally convex topology. It is isomorphic to

$$\varinjlim \Omega_{X_n}^1(M; Y)/\big(\Omega_{X_n}^1(M; Y) \cap dC_c^\infty(M; Y)\big) \cong \varinjlim \Omega_{X_n}^1(M; Y)/dC_{X_n}^\infty(M; Y)$$
$$= \varinjlim \mathfrak{z}_{(X_n, \partial X_n)}(Y)$$

(Lemmas B.4 and IV.10). We write $q: \Omega_c^1(M; Y) \to \mathfrak{z}_{M,c}(Y)$ for the quotient map. The cohomology space

$$H_{\mathrm{dR},c}^1(M; Y) := Z_{\mathrm{dR},c}^1(M; Y)/dC_c^\infty(M; Y)$$

is a closed subspace of $\mathfrak{z}_{M,c}(Y)$. For a compact subset $X \subseteq M$ we define

$$H_{\mathrm{dR},X}^1(M; Y) := Z_{\mathrm{dR},X}^1(M; Y)/\big(Z_{\mathrm{dR},X}^1(M; Y) \cap dC_c^\infty(M; Y)\big)$$

and observe that $H_{\mathrm{dR},c}^1(M; Y)$ is the union of the subspaces $H_{\mathrm{dR},X_n}^1(M; Y)$. □

Remark IV.13. For each compact equidimensional submanifold $X \subseteq M$ with $X = \hat{X}$, Lemma IV.10 implies that

$$H^1_{\mathrm{dR},X}(M; Y) = Z^1_{\mathrm{dR},X}(M; Y)/dC^\infty_X(M; Y) \cong Z^1_{\mathrm{dR}}(X, \partial X; Y)/dC^\infty(X, \partial X; Y)$$
$$= H^1_{\mathrm{dR}}(X, \partial X; Y).$$

Therefore Lemma III.2 implies that for $\dim Y < \infty$ these spaces are finite-dimensional[1]. $\square$

Lemma IV.14. *Let M be a non-compact finite-dimensional manifold, $(X_n)_{n\in\mathbb{N}}$ a saturated exhaustion of M and Y a Fréchet space. Then the following assertions hold:*

(i) $\Omega^1_c(M; \mathbb{R})$ *is a nuclear LF-space.*

(ii) $H^1_{\mathrm{dR},c}(M; Y)$ *is the locally convex direct limit of the subspaces $H^1_{\mathrm{dR}}(X_n, \partial X_n; Y)$.*

Proof. (i) $\Omega^1_c(M; \mathbb{R})$ is the direct limit of the Fréchet spaces $\Omega^1_{X_n}(M; \mathbb{R})$. Each space $\Omega^1_{X_n}(M; \mathbb{R})$ can be embedded into a product of finitely many spaces of the form $\Omega^1(U; \mathbb{R})$, where U is an open subset of $\mathbb{R}^d$, $d = \dim M$. As the spaces $\Omega^1(U; \mathbb{R})$ are nuclear, the spaces $\Omega^1_{X_n}(M; \mathbb{R})$ are nuclear, and the assertion follows ([Tr67, Prop. 50.1]).

(ii) First we verify that the pairs $X_n \subseteq X_{n+1}$ satisfy the assumptions of Theorem III.6. Let C be a connected component of $X_{n+1} \setminus X_n$. If C does not intersect ∂X_{n+1}, then it also is a connected component of $M \setminus X_n$. Further it is contained in the compact set X_{n+1}, so that $X_{n+1} = \hat{X}_{n+1}$ leads to a contradiction. Therefore all connected components of $X_{n+1} \setminus X_n$ are non-compact, Theorem III.6 applies, and we obtain inductively continuous projections

$$p_n \colon \mathfrak{z}_n := \mathfrak{z}_{(X_n, \partial X_n)}(Y) \to H^1_n := H^1_{\mathrm{dR}}(X_n, \partial X_n; Y)$$

which are compatible in the sense that $p_{n+1}|_{\mathfrak{z}_n} = p_n$. Since $\mathfrak{z}_{(M,c)}$ is the locally convex direct limit of the subspaces $\mathfrak{z}_n$ (Definition IV.12), there exists a continuous projection

$$p \colon \mathfrak{z}_{(M,c)}(Y) \to H^1_{\mathrm{dR},c}(M; Y)$$

with $p|_{\mathfrak{z}_n} = p_n$ for each $n \in \mathbb{N}$.

Now let $f_n \colon H^1_n \to E$ be continuous linear functions into a locally convex space E with

$$f_{n+1}|_{H^1_n} = f_n, \quad \text{for } n \in \mathbb{N}.$$

Then the functions $f_n \circ p_n \colon \mathfrak{z}_n \to E$ are continuous linear maps with $f_{n+1} \circ p_{n+1}|_{\mathfrak{z}_n} = f_n \circ p_n$, so that there exists a continuous linear map $F \colon \mathfrak{z}_{(M,c)}(Y) \to E$ with $F|_{\mathfrak{z}_n} =$

---

[1]There is some subtle point that one has to observe here. In general a closed subspace Y of an LF-space $X=\varinjlim X_n$ does not have to carry the LF-space topology defined by the subspaces $Y\cap X_n$ (cf. [Tr67, Rem. 13.2]).

$f_n \circ p_n$ for each $n \in \mathbb{N}$, and therefore the restriction $f := F|_{H^1_{dR,c}(M;Y)}$ is continuous. This proves the universal direct limit property of the locally convex space $H^1_{dR,c}(M;Y)$.

$\square$

Lemma IV.15. *If Y is a locally convex space and $\Gamma \subseteq Y$ a discrete subgroup, then the subgroup $\Gamma^{(\mathbb{N})}$ is discrete in the space $Y^{(\mathbb{N})}$ endowed with the locally convex direct limit topology of the finite products $Y^n = Y^{\{1,\dots,n\}}, n \in \mathbb{N}$.*

Proof. Let $U \subseteq Y$ be a convex 0-neighborhood with $U \cap \Gamma = \{0\}$. Then $U^{(\mathbb{N})}$ is a convex 0-neighborhood in $Y^{(\mathbb{N})}$ with $U^{(\mathbb{N})} \cap \Gamma^{(\mathbb{N})} = \{0\}$.

$\square$

Theorem IV.16. *Let Y be a s.c.l.c. space and M a non-compact connected manifold with a saturated exhaustion $(X_n)_{n\in\mathbb{N}}$. Then there exists a set $B = \bigcup_n B_n$ consisting of piecewise smooth cycles and of piecewise smooth proper maps $\mathbb{R} \to M$ such that:*

(1) *For each $n \in \mathbb{N}$ the subset B_n is finite, and the integration map*

$$\Phi_{X_n} : H^1_{dR}(X_n, \partial X_n; Y) \to Y^{B_n}, \quad [\zeta] \mapsto \left(\int_b \zeta\right)_{b\in B_n}$$

is a topological isomorphism.

(2) *The integration map*

$$\Phi_M : H^1_{dR,c}(M; Y) \to Y^{(B)} \cong \varinjlim Y^{B_n}, \quad [\zeta] \mapsto \left(\int_b \zeta\right)_{b\in B}$$

is a topological isomorphism.

Proof. Using the construction in the proof of Theorem III.6, we inductively obtain finite sets B_n of piecewise smooth cycles in X_n modulo ∂X_n such that $B_n \subseteq B_{n+1}$ holds in the sense that those cycles in B_n which are not cycles in X_{n+1} are "extended" to relative cycles modulo ∂X_{n+1} in X_{n+1}, and the set $B_{n+1} \setminus B_n$ consists of cycles supported in $X_{n+1} \setminus X_n$. Moreover, for each $n \in \mathbb{N}$ the integration map Φ_{X_n} is a topological isomorphism (Remark III.5) which, in addition, satisfies

$$\Phi_{X_{n+1}}|_{H^1_{dR}(X_n,\partial X_n;Y)} = \Phi_{X_n}.$$

Therefore Lemma IV.14(ii) leads to a topological isomorphism

$$\Phi : H^1_{dR,c}(M; Y) \to \varinjlim Y^{B_n} \cong Y^{(B)},$$

where $B := \bigcup_n B_n$, and the space $Y^{(B)} = \bigcup_n Y^{B_n}$ carries the locally convex direct limit topology.

$\square$

Discrete subgroups of de Rham cohomology

Remark IV.17. In the following we write $C^\infty_p(N, M)$ for the set of proper smooth maps from the manifold N to the manifold M.

Every smooth loop in $C^\infty(\mathbb{S}^1, M)$ is homotopic to a smooth loop α for which all derivatives vanish in the base point $1 \in \mathbb{S}^1$, where we consider $\mathbb{S}^1$ as a subset of $\mathbb{C}$. Then we can view it as a smooth map $[0, 1] \to M$ which extends to a proper smooth map $\tilde{\alpha} \colon \mathbb{R} \to M$ by using a smooth proper map $\gamma \colon \mathbb{R}^+ \to M$ with $\gamma(0) = \alpha(1)$ for which all derivatives vanish in 0 and then define $\tilde{\alpha}(t) := \gamma(t - 1)$ for $t \geq 1$ and $\tilde{\alpha}(t) := \gamma(-t)$ for $t \leq 0$ (cf. Lemma IV.5). For each compactly supported 1-form β we then have

$$\int_\alpha \beta = \int_{\tilde{\alpha}} \beta - \int_\gamma \beta + \int_\gamma \beta = \int_{\tilde{\alpha}} \beta. \qquad \square$$

Lemma IV.18. *Let* $X = \hat{X} \subseteq M$ *be an equidimensional compact submanifold with boundary. Then the following assertions hold:*

(i) *For* $x, y \in \partial X$ *there exists a smooth proper curve* $\alpha \colon \mathbb{R} \to M$ *with* $\alpha(0) = x$, $\alpha(1) = y$, *and* $[0, 1] = \alpha^{-1}(X)$. *For* $\zeta \in \Omega^1(X, \partial X; Y)$ *we then have*

$$\int_\alpha \zeta = \int_{\alpha|_{[0,1]}} \zeta.$$

(ii) *For* $\zeta \in Z^1_{\mathrm{dR}}(X, \partial X; Y)$ *the subgroup of* Y *generated by the set of all integrals* $\int_\alpha \zeta$, $\alpha \in C^\infty_p(\mathbb{R}, M)$, *coincides with the set of all integrals over elements in* $Z_1(X, \partial X)$.

Proof. (i) This follows from Lemma IV.6 and its proof.

(ii) From (i), Remark IV.17 and Proposition III.3 it follows that each integral over a cycle in $Z_1(X, \partial X)$ can also be written as a sum of integrals over proper smooth maps $\mathbb{R} \to M$.

Suppose, conversely, that $\alpha \colon \mathbb{R} \to M$ is smooth and proper. Then α is smoothly homotopic to a proper curve γ which is transversal to the compact submanifold ∂X of M ([BJ73, Satz 14.7; p.158]). Therefore $\gamma^{-1}(X)$ is a finite union of compact intervals $I_1, \ldots, I_m$, because it is locally connected and compact. Then

$$\int_\alpha \zeta = \int_\gamma \zeta = \sum_j \int_{\gamma|_{I_j}} \zeta,$$

and the restrictions $\gamma|_{I_j}$ can be interpreted as cycles in $Z_1(X, \partial X)$. $\qquad \square$

We conclude from Lemma IV.18 that for the sake of testing integrality conditions of 1-forms supported by X, we could either work with 1-cycles in X modulo ∂X or with proper smooth maps $\mathbb{R} \to M$. The latter approach has the advantage of being independent of X.

Definition IV.19. For a subgroup $\Gamma \subseteq Y$ let

$$Z^1_{\mathrm{dR},c}(M; \Gamma) := \left\{ \beta \in Z^1_{\mathrm{dR},c}(M; Y) \colon (\forall \alpha \in C^\infty_p(\mathbb{R}, M)) \int_\alpha \beta \in \Gamma \right\}$$

and observe that this equals $\{\beta \in \Omega_c^1(M; Y): (\forall \alpha \in C_p^\infty(\mathbb{R}, M)) \int_\alpha \beta \in \Gamma\}$ if Γ is discrete (cf. Lemma II.10(2)). We also define

$$H_{dR,c}^1(M; \Gamma) := Z_{dR,c}^1(M, \Gamma)/dC_c^\infty(M; Y). \qquad \square$$

Proposition IV.20. *Let* $\Gamma \subseteq Y$ *be a discrete subgroup and* $T_\Gamma := Y/\Gamma$. *Then* $\delta(C_c^\infty(M; T_\Gamma))$ *consists of those* 1-*forms whose integrals over all elements of* $C_p^\infty(\mathbb{R}, M)$ *are contained in* Γ. *In particular,*

$$H_{dR,c}^1(M; \Gamma) = \delta(C_c^\infty(M; T_\Gamma))/d(C_c^\infty(M; Y)).$$

Proof. For each closed 1-form $\delta(f)$, $f \in C_c^\infty(M; T_\Gamma)$, the integrals over elements of $C_p^\infty(\mathbb{R}, M)$ are obviously contained in Γ. If, conversely, $\zeta \in \Omega_c^1(M; Y)$ has this property, then we pick an equidimensional compact manifold $X = \hat{X}$ with boundary containing the support of ζ. Then Lemmas II.3 and IV.18 imply the existence of $f \in C^\infty(X, \partial X; T_\Gamma) \subseteq C_c^\infty(M; T_\Gamma)$ with $\beta = \delta(f)$. This proves that $\delta(C_c^\infty(M; T_\Gamma))$ consists of those 1-forms whose integrals over all elements of $C_p^\infty(\mathbb{R}, M)$ are contained in Γ. $\qquad \square$

For the following corollary we recall the set B from Theorem IV.16. For the case where Y is finite-dimensional, the following discreteness result can also be obtained from Proposition B.3, combined with Theorem II.7.

Corollary IV.21. *We have* $\Phi_M\left(H_{dR,c}^1(M; \Gamma)\right) = \Gamma^{(B)}$ *and in particular*

$$H_{dR,c}^1(M; \Gamma) \cong \Gamma^{(B)} \subseteq Y^{(B)} \cong H_{dR,c}^1(M; Y).$$

Moreover, for $H_{dR,c}^1(M; \mathbb{R}) \neq \{0\}$ *the group* Γ *is discrete if and only if* $H_{dR,c}^1(M; \Gamma)$ *is discrete.*

Proof. In view of Lemma IV.18 (ii), we have

$$Z_{dR,c}^1(M; \Gamma) = \bigcup_{n \in \mathbb{N}} Z_{dR}^1(X_n, \partial X_n; \Gamma),$$

and therefore $\Phi_M(H_{dR,c}^1(M; \Gamma)) \subseteq \Gamma^{(B)}$. On the other hand, we have for each n the restriction isomorphism

$$\Phi_{X_n} = \Phi_M|_{H_{dR}^1(X_n, \partial X_n; Y)}: H_{dR}^1(X_n, \partial X_n; Y) \to Y^{B_n} \subseteq Y^{(B)}.$$

Let $x_M \in X_1$ be a base point. If $\Phi_{X_n}([\zeta]) \in \Gamma^{B_n}$, then the construction of the set B_n (cf. Theorem III.6) implies that all integrals of ζ over cycles in $Z_1(X_n, \partial X_n)$ lie in Γ, and hence that all integrals over curves in $C_p^\infty(\mathbb{R}, M)$ lie in Γ (Lemma IV.18 (ii)). Therefore $\zeta \in Z_{dR,c}^1(M; \Gamma)$ and $\Phi_M([\zeta]) = \Phi_{X_n}([\zeta])$. We conclude that $\Phi_M(H_{dR,c}^1(M; \Gamma)) = \Gamma^{(B)}$.

Now we use Lemma IV.15 to see that for a non-empty set B the subgroup $\Gamma^{(B)}$ of the locally convex direct sum $Y^{(B)}$ is discrete if and only if Γ is discrete in Y. $\qquad \square$

For the following, we observe that we have a natural continuous multiplication map

$$\Omega^1(M; \mathbb{R}) \times Y \to \Omega^1(M; Y), \quad (\zeta, y) \mapsto \zeta \cdot y$$

which induces continuous bilinear maps

$$H^1_{\mathrm{dR}}(M; \mathbb{R}) \times Y \to H^1_{\mathrm{dR}}(M; Y) \quad \text{and} \quad H^1_{\mathrm{dR},c}(M; \mathbb{R}) \times Y \to H^1_{\mathrm{dR},c}(M; Y).$$

Corollary IV.22. *For each subgroup Γ of Y we have $H^1_{\mathrm{dR},c}(M; \mathbb{Z})\cdot\Gamma = H^1_{\mathrm{dR},c}(M; \Gamma)$.*

Proof. The inclusion $H^1_{\mathrm{dR},c}(M; \mathbb{Z}) \cdot \Gamma \subseteq H^1_{\mathrm{dR},c}(M; \Gamma)$ is trivial. For the converse, let $\zeta \in Z^1_{\mathrm{dR}}(X_n, \partial X_n; \Gamma)$. Then $\Phi_{X_n}([\zeta]) \in \Gamma^{B_n}$ (Lemma IV.18). Suppose that $B_n = \{b_1, \ldots, b_m\}$. Let $b_i^* \in Z^1_{\mathrm{dR}}(X_n, \partial X_n; \mathbb{R})$ be elements with $I_{b_i} b_j^* = \delta_{ij}$. Then $\int_b \zeta = 0$ for $b \in B \setminus B_n$ implies that $\zeta - \sum_{i=1}^m b_i^* \cdot \int_{b_i} \zeta$ is exact, so that

$$[\zeta] = \sum_{j=1}^n [b_i^*] \cdot \int_{b_i} \zeta \in H^1_{\mathrm{dR},c}(M; \mathbb{Z}) \cdot \Gamma$$

holds in $H^1_{\mathrm{dR},c}(M; Y)$. As B_n generates $Z_1(X_n, \partial X_n)$ modulo torsion, we get $b_i^* \in H^1_{\mathrm{dR},c}(M; \mathbb{Z})$ (Lemma IV.18). □

The following proposition will be helpful in understanding the assertion of Proposition V.12 below.

Proposition IV.23. *If S is a closed subset of the compact manifold M, then for each discrete subgroup $\Gamma \subseteq Y$ we have*

$$H^1_{\mathrm{dR}}(M, S; \mathbb{Z}) \cdot \Gamma = H^1_{\mathrm{dR}}(M, S; \Gamma).$$

Proof. The inclusion "$\subseteq$" is clear. It remains to show the converse. So let $\zeta \in Z^1_{\mathrm{dR}}(M, S; \Gamma)$. First we show that the group $\langle \zeta, H_1(M, S) \rangle \subseteq \Gamma$ is finitely generated.

Since $H_1(M)$ is finitely generated, $\Gamma_0 := \langle \zeta, H_1(M) \rangle$ is a finitely generated subgroup of Γ. Let $p: Y \to Y/\Gamma_0$ denote the quotient map. Then all periods of the 1-form $\zeta_0 := p \circ \zeta$ are trivial, and there exists a smooth function $f_0: M \to Y/\Gamma_0$ with $df_0 = \zeta_1$ and $f_0(S) \subseteq \Gamma/\Gamma_0$. Moreover, the function f_0 lifts to a smooth function $f_1: \widetilde{M} \to Y$, with $f_1(q_M^{-1}(S)) \subseteq \Gamma$, where $q_M: \widetilde{M} \to M$ is a universal covering of M. As Γ is discrete, the function f_1 is locally constant on $q_M^{-1}(S)$, and therefore f_0 is locally constant on S. Therefore $f_0(S)$ is finite. As $\langle \zeta, H_1(M, S) \rangle / \Gamma_0 \subseteq \langle f_0(S) \rangle$, it follows that $\langle \zeta, H_1(M, S) \rangle$ is finitely generated.

Moreover, there exists a smooth function $f_2: M \to Y$ locally constant on a neighborhood of S such that for each $s \in S$ we have $f_2(s) + \Gamma_0 = f_0(s)$. Then $df_2 \in H^1_{\mathrm{dR}}(M, S; \Gamma)$ lies in the image of

$$H^1_{\mathrm{dR},c}(M \setminus S; \Gamma) \cong H^1_{\mathrm{dR},c}(M \setminus S; \mathbb{Z}) \cdot \Gamma$$

(Corollary IV.22). For $\zeta_1 := \zeta - df_2$ we now have

$$\Gamma_0 = \langle \zeta, H_1(M) \rangle = \langle \zeta_1, H_1(M) \rangle = \langle \zeta_1, H_1(M, S) \rangle,$$

so that there exists some $f_3 \in C^\infty(M, S; Y/\Gamma_0)$ with $df_3 = \zeta_1$.

As Γ_0 is finitely generated, it spans a finite-dimensional subspace $Y_0 \subseteq Y$. Extending the identity map $Y_0 \to Y_0$ to a continuous linear map $Y \to Y_0$ using the Hahn–Banach Extension Theorem, we obtain a topological direct sum decomposition $Y \cong Y_0 \oplus Y_1$, where Y_1 is the kernel of the extension. Then $Y/\Gamma_0 \cong (Y_0/\Gamma_0) \times Y_1$ as Lie groups. Moreover, $\zeta_1 = \alpha_1 + \alpha_2$ with $\alpha_j \in Z^1_{dR}(M, S; Y_j)$, $j = 1, 2$, and $f_3 = h_1 + h_2$ with $h_1 \in C^\infty(M, S; Y_0/\Gamma_0)$, $h_2 \in C^\infty(M, S; Y_1)$, $\delta(h_1) = \alpha_1$ and $dh_2 = \alpha_2$. This proves that $[\zeta_1] = [\alpha_1]$. As Y_0/Γ_0 is a finite-dimensional torus, we can write it as $\mathbb{R}^d/\mathbb{Z}^d$ with $Y_0 \cong \mathbb{R}^d$ and $\mathbb{Z}^d \cong \Gamma_0$. This means that h_1 is a finite product of the d component functions $l_1, \ldots, l_d \in C^\infty(M, S; \mathbb{T})$. If $e_1, \ldots, e_d$ denote the canonical basis vectors in $\mathbb{R}^d$, this leads to

$$[\alpha_1] = \sum_{j=1}^{d} [dl_j] \cdot e_i \in H^1_{dR}(M, S; \mathbb{Z}) \cdot \Gamma.$$

Summing up, we obtain

$$H^1_{dR}(M, S; \Gamma) \subseteq H^1_{dR,c}(M \setminus S; \Gamma) + H^1_{dR}(M, S; \mathbb{Z}) \cdot \Gamma$$
$$= H^1_{dR,c}(M \setminus S; \mathbb{Z}) \cdot \Gamma + H^1_{dR}(M, S; \mathbb{Z}) \cdot \Gamma \subseteq H^1_{dR}(M, S; \mathbb{Z}) \cdot \Gamma. \quad \square$$

Example IV.24. Let $M := \mathbb{R}^2 \setminus P$, where P is a subset without cluster points. We want to get an explicit picture of $H^1_{dR,c}(M; \mathbb{R})$.

(a) First we consider on $\mathbb{R}^2 \setminus \{(0, 0)\}$ in polar coordinates the 1-form

$$\alpha(re^{i\phi}) := f(r)dr,$$

where $f :]0, \infty[\to \mathbb{R}$ has compact support and satisfies $\int_0^\infty f(r) \, dr = 1$. Then

$$d\alpha = f'(r)dr \wedge dr + \frac{\partial f}{\partial \phi} d\phi \wedge dr = 0,$$

and for each proper map $\gamma : \mathbb{R} \to \mathbb{R}^2$ with $\lim_{t \to -\infty} \gamma(t) = (0, 0)$ and $\lim_{t \to \infty} \gamma(t) = \infty$ we have

$$\int_\gamma \alpha = 1.$$

(b) To calculate $H^1_{dR,c}(M; \mathbb{R})$, we approximate M by compact submanifolds X_n which are obtained from closed discs D_n with $\partial D_n \cap P = \emptyset$ by removing open discs around the finitely many points in $D_n \cap P$. Note that the set P is countable, so that there exist arbitrarily large discs D_n whose boundaries do not intersect P.

Assume that $D := D_n$ contains k elements of P and put $X := X_n$. Then $\pi_1(X) \cong \pi_1(\mathrm{int}(X))$ is a free group of k generators. For each closed 1-form ζ with compact support in X^0 the integrals over the loops in X are trivial (make them very small around the points in P). Hence every such 1-form is exact. Let $\zeta = df$ with $f \in C^\infty(X; \mathbb{R})$. As ζ has compact support, f is constant on the connected complement of D, so that we may w.l.o.g. assume that $f = 0$ on the outer circle $\partial D \subseteq \partial X$. Then we connect ∂D by arcs $\gamma_1, \ldots, \gamma_k$ to the other boundary components. If all integrals of ζ over the γ_j vanish, then $\zeta \in dC^\infty(X, \partial X; \mathbb{R})$. If $\alpha_1, \ldots, \alpha_k \in Z^1_{\mathrm{dR}}(X, \partial X; \mathbb{R})$ are the 1-forms supported close to the elements of $P \cap D$ as in (a), we see that $\int_{\gamma_i} \alpha_j = \delta_{ij}$ for an appropriate normalization, so that $[\zeta] = \sum_j \int_{\gamma_j} \zeta \cdot [\alpha_j]$. Therefore

$$H^1_{\mathrm{dR}}(X, \partial X; \mathbb{R}) = \bigoplus_{p \in P \cap D} \mathbb{R}[\alpha_p],$$

and further

$$H^1_{\mathrm{dR,c}}(M; \mathbb{R}) = \varinjlim H^1_{\mathrm{dR}}(X_n, \partial X_n; \mathbb{R}) = \bigoplus_{p \in P} \mathbb{R}[\alpha_p] \cong \mathbb{R}^{(P)}.$$

The subgroup $H^1_{\mathrm{dR,c}}(M; \mathbb{Z})$ of integral elements in $H^1_{\mathrm{dR,c}}(M; \mathbb{R})$ consists of those cohomology classes whose integrals over all paths between elements of P are integers. For $p, q \in P$ we write $\gamma_{p,q}$ for an arc from p to q. Then

$$\int_{\gamma_{p,q}} \alpha_r = \delta_{p,r} - \delta_{q,r}.$$

This means that $\sum_r \lambda_r \alpha_r$ is integral if and only if all differences $\lambda_r - \lambda_s$ are integral. As only finitely many coefficients λ_r are non-zero, it follows that

$$H^1_{\mathrm{dR,c}}(M; \mathbb{Z}) = \sum_p \mathbb{Z}[\alpha_p] \cong \mathbb{Z}^{(P)}.$$

$\square$

V Central extensions of Lie groups and period maps

In this section we first explain the general setup for central extensions of infinite-dimensional Lie groups. The main question arising in the integration process of Lie algebra cocycles ω to central extensions of Lie groups is whether the corresponding period group Π_ω is discrete. In this section we show that for cocycles of product type for the groups $C^\infty_c(M; K)_e$ and $C^\infty(M, S; K)_e$ the period group is discrete for any M if and only if this is the case for $M = \mathbb{S}^1$. This reduces the discreteness problem to the case of loop groups, which is known for K compact, and therefore for all finite-dimensional Lie groups K.

Generalities on central Lie group extensions

Definition V.1. (a) Let $\mathfrak{z}$ be a topological vector space and $\mathfrak{g}$ a topological Lie algebra. A *continuous $\mathfrak{z}$-valued 2-cocycle* is a continuous skew-symmetric function $\omega \colon \mathfrak{g} \times \mathfrak{g} \to \mathfrak{z}$ with

$$\omega([x, y], z) + \omega([y, z], x) + \omega([z, x], y) = 0.$$

It is called a *coboundary* if there exists a continuous linear map $\alpha \in \mathrm{Lin}(\mathfrak{g}, \mathfrak{z})$ with $\omega(x, y) = \alpha([x, y])$ for all $x, y \in \mathfrak{g}$. We write $Z_c^2(\mathfrak{g}, \mathfrak{z})$ for the space of continuous $\mathfrak{z}$-valued 2-cocycles and $B_c^2(\mathfrak{g}, \mathfrak{z})$ for the subspace of coboundaries. We define the *second continuous Lie algebra cohomology space*

$$H_c^2(\mathfrak{g}, \mathfrak{z}) := Z_c^2(\mathfrak{g}, \mathfrak{z})/B_c^2(\mathfrak{g}, \mathfrak{z}).$$

(b) If ω is a continuous $\mathfrak{z}$-valued cocycle on $\mathfrak{g}$, then we write $\mathfrak{g} \oplus_\omega \mathfrak{z}$ for the topological Lie algebra whose underlying topological vector space is the product space $\mathfrak{g} \times \mathfrak{z}$, and the bracket is defined by

$$[(x, z), (x', z')] = \big([x, x'], \omega(x, x')\big).$$

Then $q \colon \mathfrak{g} \oplus_\omega \mathfrak{z} \to \mathfrak{g}$, $(x, z) \mapsto x$ is a central extension and $\sigma \colon \mathfrak{g} \to \mathfrak{g} \oplus_\omega \mathfrak{z}$, $x \mapsto (x, 0)$ is a continuous linear section of q.

If, conversely, a central Lie algebra extension $q \colon \hat{\mathfrak{g}} \to \mathfrak{g}$ with kernel $\mathfrak{z}$ has a continuous linear section $\sigma \colon \mathfrak{g} \to \hat{\mathfrak{g}}$, then it can be described by a continuous Lie algebra cocycle $\omega \in Z_c^2(\mathfrak{g}, \mathfrak{z})$ defined by $\omega(x, y) := [\sigma(x), \sigma(y)] - \sigma([x, y])$, because the map

$$\mathfrak{g} \oplus_\omega \mathfrak{z} \to \hat{\mathfrak{g}}, \quad (x, z) \mapsto \sigma(x) + z$$

is an isomorphism of topological Lie algebras. As two Lie algebra cocycles define equivalent central extensions if and only if they differ by a coboundary, we obtain an identification of the set of equivalence class of all central $\mathfrak{z}$-extensions of $\mathfrak{g}$ (with a continuous linear section) with the vector space $H_c^2(\mathfrak{g}, \mathfrak{z})$.   □

Definition V.2. (a) *Central extensions of Lie groups* are always assumed to have a smooth local section. Let $Z \hookrightarrow \hat{G} \twoheadrightarrow G$ be a central extension of the connected Lie group G by the abelian Lie group Z. We assume that the identity component Z_e of Z can be written as $Z_e = \mathfrak{z}/\pi_1(Z)$, where the Lie algebra $\mathfrak{z}$ of Z is a s.c.l.c. space. The group $(\mathfrak{z}, +)$ can be identified in a natural way with the universal covering group of Z_e, and Z_e is a quotient of $\mathfrak{z}$ modulo a discrete subgroup which can be identified with $\pi_1(Z)$. Since the quotient map $q \colon \hat{G} \to G$ has a smooth local section, the corresponding Lie algebra homomorphism $\hat{\mathfrak{g}} \to \mathfrak{g}$ has a continuous linear section $\sigma \colon \mathfrak{g} \to \hat{\mathfrak{g}}$, hence can be described by a continuous Lie algebra cocycle (Definition V.1).

(b) If G is a group and Z an abelian group, then we define the group

$$Z^2(G, Z) := \{f \colon G \times G \to Z \colon (\forall x, y, z \in G)$$
$$f(\mathbf{1}, x) = f(x, \mathbf{1}) = \mathbf{1}, \quad f(x, y)f(xy, z) = f(x, yz)f(y, z)\}$$

of Z-*valued 2-cocycles* and the subgroup

$$B^2(G, Z) := \{f: G \times G \to Z: (\exists h: G \to Z)\, h(\mathbf{1}) = \mathbf{1},$$

$$(\forall x, y \in G)\, f(x, y) = h(xy)h(x)^{-1}h(y)^{-1}\}$$

of Z-*valued 2-coboundaries*. In both cases the group structure is given by pointwise multiplication.

If G and Z are Lie groups, we write $Z_s^2(G, Z)$ for the subgroup of $Z^2(G, Z)$ consisting of those cocycles f which are smooth in a neighborhood of (e, e), and $B_s^2(G, Z)$ for the subgroup of all functions of the form $(g, g') \mapsto h(gg')h(g)^{-1}h(g')^{-1}$, where $h: G \to Z$ is smooth in an identity neighborhood. We recall from [Ne02a, Prop. 4.2] that central Lie group extensions as above can always be written as

$$\hat{G} \cong G \times_f Z \quad \text{with } (g, z)(g', z') = \big(gg', zz'\,f(g, g')\big),$$

for some $f \in Z_s^2(G, Z)$. Two cocycles f_1, f_2 define equivalent Lie group extensions if and only if $f_1 \cdot f_2^{-1} \in B_s^2(G, Z)$ (for $f_2^{-1}(x, y) := f_2(x, y)^{-1}$), and the quotient group $H_s^2(G, Z) := Z_s^2(G, Z)/B_s^2(G, Z)$ parametrizes the equivalence classes of central Z-extensions of G with smooth local sections ([Ne02a, Remark 4.4]). There is a natural map $H_s^2(G, Z) \to H_c^2(\mathfrak{g}, \mathfrak{z})$ induced by the map

$$D: Z_s^2(G, Z) \to Z_c^2(\mathfrak{g}, \mathfrak{z}),$$
$$D(f)(x, y) = d^2 f(e, e)((x, 0), (0, y)) - d^2 f(e, e)((y, 0), (x, 0)) \tag{5.1}$$

([Ne02a, Lemma 4.6]), where $d^2 f(e, e)$ is well-defined because $df(e, e)$ vanishes, which follows from $f(g, e) = f(e, g) = \mathbf{1}$. For more details on central extensions of Lie groups we refer to [Ne02a].   □

Definition V.3. If $\mathfrak{z}$ is a s.c.l.c. space, G a Lie group, and $\Omega \in \Omega^2(G, \mathfrak{z})$ a closed $\mathfrak{z}$-valued 2-form, then we obtain with [Ne02a, Lemma 5.7] a group homomorphism

$$\mathrm{per}_\Omega: \pi_2(G) \to \mathfrak{z}$$

called the *period map*. It is given on smooth representatives $\sigma: \mathbb{S}^2 \to G$ of classes in $\pi_2(G)$ by the integral

$$\mathrm{per}_\Omega([\sigma]) = \int_{\mathbb{S}^2} \sigma^* \Omega = \int_\sigma \Omega.$$

We recall that each homotopy class contains smooth representatives. Here we use the sequential completeness of $\mathfrak{z}$ to ensure that the integrals, which can be obtained as limits of Riemann sums, do exist. If Ω is exact, then the period map is trivial by Stoke's Theorem. The image $\Pi_\Omega := \mathrm{per}_\Omega(\pi_2(G))$ is called the *period group* of Ω.   □

Definition V.4. Let G be a connected Lie group with Lie algebra $\mathfrak{g}$ and $\omega \in Z_c^2(\mathfrak{g}, \mathfrak{z})$ a continuous Lie algebra cocycle with values in the s.c.l.c. space $\mathfrak{z}$. Let $\Gamma \subseteq \mathfrak{z}$ be a discrete subgroup and $Z := \mathfrak{z}/\Gamma$ the corresponding quotient Lie group. Further let

Ω be the corresponding left invariant closed $\mathfrak{z}$-valued 2-form on G. Then we define a homomorphism

$$P: H_c^2(\mathfrak{g}, \mathfrak{z}) \to \mathrm{Hom}(\pi_2(G), Z) \times \mathrm{Hom}(\pi_1(G), \mathrm{Lin}(\mathfrak{g}, \mathfrak{z}))$$

as follows. For the first component we take

$$P_1([\omega]) := q_Z \circ \mathrm{per}_\omega,$$

where $q_Z: \mathfrak{z} \to Z$ is the quotient map and $\mathrm{per}_\omega := \mathrm{per}_\Omega: \pi_2(G) \to \mathfrak{z}$ is the period map of ω. To define the second component, for each $X \in \mathfrak{g}$ we write X_r for the corresponding right invariant vector field on G. Then $i_{X_r}\Omega$ is a closed $\mathfrak{z}$-valued 1-form ([Ne02a, Lemma 3.11]) to which we associate a homomorphism $\pi_1(G) \to \mathfrak{z}$ via

$$P_2([\omega])([\gamma])(X) := \int_\gamma i_{X_r}\Omega.$$

We refer to [Ne02a, Sect. 7] for arguments showing that P is well-defined, i.e., that the right hand sides only depend on the Lie algebra cohomology class of ω.   □

The following theorem completely describes the obstructions for a Lie algebra cocycle to integrate to a central Lie group extension. It is the main result of [Ne02a].

Theorem V.5. *Let $\omega \in Z_c^2(\mathfrak{g}, \mathfrak{z})$ be a continuous Lie algebra cocycle. Then the central Lie algebra extension $\mathfrak{z} \hookrightarrow \hat{\mathfrak{g}} := \mathfrak{g} \oplus_\omega \mathfrak{z} \twoheadrightarrow \mathfrak{g}$ integrates to a central Lie group extension $Z \hookrightarrow \hat{G} \twoheadrightarrow G$ if and only if $P([\omega]) = 0$.*

Proof. [Ne02a, Th. 7.12].   □

Applications to current groups

Now we turn to central extensions of the two classes of current Lie groups given as the identity components of $C_c^\infty(M; K)$ and $C^\infty(M, S; K)$. The methods developed in this paper are well suited for the study of Lie algebra cocycles of product type introduced below. Here the main problem is to decide for a given cocycle if its period group is discrete (cf. Theorem V.5).

Definition V.6. Let $\mathfrak{k}$ be a locally convex topological Lie algebra, M a manifold and $\mathfrak{g} := C^\infty(M; \mathfrak{k})$. We consider a continuous invariant symmetric bilinear map $\kappa: \mathfrak{k} \times \mathfrak{k} \to Y$, where Y is a s.c.l.c. space. We then obtain a continuous $\mathfrak{z}_M(Y)$-valued cocycle on $\mathfrak{g}$ by

$$\omega_M(\xi, \eta) := \omega_{M,\kappa}(\xi, \eta) := [\kappa(\xi, d\eta)] \in \mathfrak{z}_M(Y),$$

where we view $\kappa(\xi, d\eta)$ as the element of $\Omega^1(M; Y)$ whose value in a tangent vector $v \in T_p(M)$ is given by $\kappa(\xi(p), d\eta(p)(v))$.

(a) On $C^\infty(M, S; \mathfrak{k})$ we obtain by restriction a continuous $\mathfrak{z}_{(M,S)}(Y)$-valued Lie algebra cocycle $\omega_{(M,S)}$. For a compact manifold M the group $C^\infty(M, S; K)$ has a natural Lie group structure (Definition I.6), so that we can define the period map

$$\mathrm{per}_{\omega_{(M,S)}} : \pi_2(C^\infty(M, S; K)) \to \mathfrak{z}_{(M,S)}(Y)$$

corresponding to the left invariant 2-form $\Omega_{(M,S)}$ on $C^\infty(M, S; K)$ with $\Omega_{(M,S),e} = \omega_{(M,S)}$. We write $\Pi_{(M,S)}$ for the corresponding period group.

(b) If ξ and η have compact support, then the same holds for $\kappa(\xi, \eta)$, so that we also obtain a Lie algebra cocycle

$$\omega_M \in Z_c^2(C_c^\infty(M; \mathfrak{k}), \mathfrak{z}_{M,c}(Y)), \quad \mathfrak{z}_{M,c}(Y) := \Omega_c^1(M; Y)/dC_c^\infty(M; Y).$$

The continuity of this cocycle follows from the continuity of the map

$$C_c^\infty(M; \mathfrak{k}) \times \Omega_c^1(M; \mathfrak{k}) \to \Omega_c^1(M; Y), \quad (f, \xi) \mapsto \kappa(f, \xi),$$

which in turn follows from [Gl01d, Th. 4.7] because it can be interpreted as a map on the level of compactly supported sections of vector bundles induced by the bundle map determined by the continuous map

$$\mathfrak{k} \times \mathrm{Lin}(T_p(M); \mathfrak{k}) \to \mathrm{Lin}(T_p(M); Y), \quad (x, \beta) \mapsto \kappa(x, \beta(\cdot))$$

on the fiber in $p \in M$.

(c) For any Lie group K we define $V(\mathfrak{k})$ as follows. We first endow $\mathfrak{k} \otimes \mathfrak{k}$ with the projective tensor product topology and define $V(\mathfrak{k})$ as the completion of the quotient of $V(\mathfrak{k})$ by the closure of the subspace spanned by all elements of the form

$$x \otimes y - y \otimes x \quad \text{and} \quad [x, y] \otimes z + y \otimes [x, z], \quad x, y, z \in \mathfrak{k}.$$

If $[z]$ denotes the image of $z \in \mathfrak{k} \otimes \mathfrak{k}$ in $V(\mathfrak{k})$, we obtain a continuous invariant bilinear map

$$\kappa : \mathfrak{k} \times \mathfrak{k} \to V(\mathfrak{k}), \quad \kappa(x, y) := [x \otimes y]$$

which leads to the cocycle $\omega = \omega_{\mathbb{S}^1, \kappa} \in Z_c^2(\mathfrak{g}, V(\mathfrak{k}))$ on $\mathfrak{g} := C^\infty(\mathbb{S}^1; \mathfrak{k})$ given by $\omega(\xi, \eta) := [\kappa(\xi, d\eta)]$. As $\pi_2(C^\infty(\mathbb{S}^1; K)) \cong \pi_3(K)$ (Corollary A.15), the period map per_ω yields a homomorphism

$$\mathrm{per}_K : \pi_3(K) \to V(\mathfrak{k}). \qquad \square$$

Proposition V.7. *Let* $\mathfrak{g} := C_c^\infty(M; \mathfrak{k})$ *and* $\kappa : \mathfrak{k} \times \mathfrak{k} \to Y$ *be a continuous invariant symmetric bilinear form. Then we obtain for the cocycle* $\omega(\xi, \eta) := [\kappa(\xi, d\eta)]$ *an automorphic action of the group* $C^\infty(M, K)$ *on* $\hat{\mathfrak{g}} := \mathfrak{g} \oplus_\omega \mathfrak{z}_M(Y)$ *by*

$$f.(\xi, z) := (\mathrm{Ad}(f).\xi, z - [\kappa(\delta^l(f), \xi)]). \tag{5.2}$$

The corresponding derived action is given by

$$\eta.(\xi, z) = [(\eta, 0), (\xi, z)] = ([\eta, \xi], \omega(\eta, \xi)). \tag{5.3}$$

Proof. The arguments can be taken over from [MN02, Prop. III.3]. Here we only have to add Lemma II.2 to see that δ^l is smooth.   □

Theorem V.8. *Let K be a connected Lie group, M a connected manifold, $G :=
C_c^\infty(M, K)_e$ and $\omega_{M,\kappa} \in Z_c^2(\mathfrak{g}, {}_{3M}(Y))$ as above. Suppose that the period group
$\Pi_{M,\kappa} \subseteq {}_{3M}(Y)$ is discrete. For $Z := {}_{3M}(Y)/\Pi_{\omega_{M,\kappa}}$ we then obtain a central Lie
group extension $Z \hookrightarrow \hat{G} \twoheadrightarrow G$ corresponding to the cocycle $\omega_{M,\kappa}$.*

Proof. In view of Theorem V.5, we only have to see that $P_2([\omega_{M,\kappa}]) = 0$. According
to [Ne02a, Prop. 7.6], this is equivalent to the existence of a smooth linear action of
G on $\hat{\mathfrak{g}}$ whose derived action is given by $\eta.(\xi, z) = ([\eta, \xi], \omega(\eta, \xi))$. Proposition V.7
implies that such a representation exists.   □

For the following theorem we recall that we can use the continuous bilinear form
$\kappa: \mathfrak{k} \times \mathfrak{k} \to Y$ to define a wedge product

$$\wedge_\kappa : \Omega^1(M; \mathfrak{k}) \times \Omega^1(M; \mathfrak{k}) \to \Omega^2(M; Y)$$

by

$$(\alpha \wedge_\kappa \beta)(v, w) := \kappa(\alpha_p(v), \beta_p(w)) - \kappa(\beta_p(v), \alpha_p(w)), \quad v, w \in T_p(M).$$

The following theorem describes a situation where we have a global smooth group
cocycle associated to the cocycle obtained by composing a cocycle of product type with
the de Rham differential ${}_{3M,c}(Y) \to \Omega_c^2(M; Y)$. The reason behind the existence of the
global cocycle lies in the fact that all periods of $\omega_{M,\kappa}$ lie in the kernel $H^1_{dR,c}(M; Y)$ of
d (see [Ne02a, Section 8] for more details on the existence of global smooth cocycles).

Theorem V.9. *Let $G^+ := C_c^\infty(M, K)$. Then the map*

$$h: G^+ \times G^+ \to \Omega_c^2(M; Y), \quad h(f, g) := \delta^l(f) \wedge_\kappa \delta^r(g)$$

*defines a smooth $\Omega_c^2(M; Y)$-valued group 2-cocycle on G^+, so that we obtain a central
Lie group extension $\hat{G}^+ := G^+ \times_h \Omega_c^2(M; Y)$. The corresponding Lie algebra cocycle
Dh from (5.1) is given by*

$$Dh(\xi, \eta) = 2d\xi \wedge_\kappa d\eta \quad \text{for } \xi, \eta \in C_c^\infty(M; \mathfrak{k}).$$

*The map $\gamma: {}_{3M,c}(Y) \to \Omega_c^2(M; Y), [\beta] \mapsto 2d\beta$ satisfies $\gamma \circ \omega_{M,\kappa} = Dh$ and induces
a Lie algebra homomorphism*

$$\gamma_{\mathfrak{g}}: \hat{\mathfrak{g}} = \mathfrak{g} \oplus_{\omega_{M,\kappa}} {}_{3M,c}(Y) \to \hat{\mathfrak{g}}^+ := \mathfrak{g} \oplus_{Dh} \Omega_c^2(M; Y), \quad (X, [\beta]) \mapsto (X, 2d\beta).$$

*This homomorphism is G^+-equivariant with respect to the action on $\hat{\mathfrak{g}}^+$ induced by
the adjoint action of $\hat{G}^+$, which is given by*

$$\mathrm{Ad}_{\hat{\mathfrak{g}}^+}(g).(\xi, z) = \left(\mathrm{Ad}(g).\xi, z - d(\kappa(\delta^l(g), \xi))\right).$$

Proof. This follows with the same arguments as in the proof of [MN02, Th. III.9]. For non-compact manifolds we have to use Lemma II.2 for the smoothness of the maps $\delta^l, \delta^r : C_c^\infty(M, K) \to \Omega_c^1(M; \mathfrak{k})$.   □

Period maps for $C^\infty(M, S; K)$

Now we turn to the period groups $\Pi_{(M,S)}$ for the Lie algebra cocycles $\omega_{(M,S)}$ associated to the Lie algebras $C^\infty(M, S; \mathfrak{k})$, where M is compact and $S \subseteq M$ a closed subset.

Lemma V.10. *For each $\alpha \in C^\infty((I, \partial I), (M, S))$ let*

$$\alpha_K : C^\infty(M, S; K) \to C^\infty(I, \partial I; K)$$

denote the corresponding group homomorphism. Then

$$\mathrm{per}_{\omega_{(I,\partial I)}} \circ \pi_2(\alpha_K) = I_\alpha \circ \mathrm{per}_{\omega_{(M,S)}}.$$

Proof. First we recall from Lemma A.16 that the map α_K is a Lie group homomorphism. Let $G := C^\infty(M, S; K)_e$ and $\Omega_{(M,S)} \in \Omega^2(G, \mathfrak{z}_{(M,S)}(Y))$ denote the left invariant 2-form corresponding to $\omega_{(M,S)}$. Then $I_\alpha \circ \Omega_{(M,S)}$ is a Y-valued left invariant 2-form on G whose value in $\mathbf{1}$ is $I_\alpha \circ \omega_{(M,S)}$. Further $\alpha_K^* \Omega_{(I,\partial I)}$ is a left invariant 2-form on G whose value in $\mathbf{1}$ is given by

$$(\xi, \eta) \mapsto \omega_{(I,\partial I)}(\xi \circ \alpha, \eta \circ \alpha) = [\kappa(\xi \circ \alpha, d(\eta \circ \alpha))]$$

$$= [\kappa(\alpha^*\xi, \alpha^*(d\eta))] = \int_I \kappa(\alpha^*\xi, \alpha^*(d\eta)) = \int_\alpha \kappa(\xi, d\eta) = I_\alpha(\omega_{(M,S)}(\xi, \eta)).$$

This implies

$$\alpha_K^* \Omega_{(I,\partial I)} = I_\alpha \circ \Omega_{(M,S)}$$

for each $\alpha \in C^\infty((I, \partial I), (M, S))$, and hence the assertion.   □

Lemma V.11. *If we identify $\mathfrak{z}_{\mathbb{S}^1}(Y), \mathfrak{z}_{(I,\partial I)}(Y)$, and $\mathfrak{z}_{\mathbb{R},c}(Y)$ with Y via the integration maps from Lemma II.8, then*

$$\Pi_{\mathbb{S}^1} = \Pi_{(I,\partial I)} = \Pi_{\mathbb{R}}.$$

Proof. According to Corollary A.15, the natural inclusion

$$C^\infty((I, \partial I); K) \hookrightarrow C_*^\infty(\mathbb{S}^1; K)$$

induced from the canonical map $\alpha \in C^\infty((I, \partial I), (\mathbb{S}^1, *))$ is a weak homotopy equivalence. Therefore $\pi_2(\alpha_K)$ is an isomorphism, and Lemma V.10, applied to $(M, S) = (\mathbb{S}^1, 1)$, implies that

$$\Pi_{(I,\partial I)} = I_\alpha \circ \Pi_{\mathbb{S}^1} \cong \Pi_{\mathbb{S}^1}$$

because the map $I_\alpha : \mathfrak{z}_{\mathbb{S}^1}(Y) \to Y$ is the integration isomorphism which we ignore by identifying $\Pi_{\mathbb{S}^1}$ and $\Pi_{(I,\partial I)}$ as subsets of Y.

To obtain $\Pi_{\mathbb{R}} = \Pi_{(I,\partial I)}$, we first use Theorem A.13 and a diffeomorphism $\alpha \colon \mathbb{R} \to I \setminus \partial I$ to see that the natural embedding

$$\phi_K \colon C_c^\infty(\mathbb{R}; K) \to C_c^\infty(I \setminus \partial I; K) \hookrightarrow C^\infty(I, \partial I; K)$$

is a weak homotopy equivalence. Moreover, $\mathbf{L}(\phi_K)^* \omega_{(I,\partial I)} = \omega_{\mathbb{R}}$, so that $\phi_K^* \Omega_{(I,\partial I)} = \Omega_{\mathbb{R}}$, and by integration over $\mathbb{R}$ we obtain $\Pi_{(I,\partial I)} = \Pi_{\mathbb{R}}$. $\square$

Proposition V.12. *For each κ the period group $\Pi_{(M,S)}$ is contained in $H_{dR}^1(M, S; Y)$, and we have*

$$H_{dR}^1(M, S; \mathbb{Z}) \cdot \Pi_{\mathbb{S}^1} \subseteq \Pi_{(M,S)} \subseteq H_{dR}^1(M, S; \Pi_{\mathbb{S}^1}).$$

If $\Pi_{\mathbb{S}^1}$ is discrete, then

$$\Pi_{(M,S)} = H_{dR}^1(M, S; \Pi_{\mathbb{S}^1}) = H_{dR}^1(M, S; \mathbb{Z}) \cdot \Pi_{\mathbb{S}^1}.$$

Proof. In the situation of Lemma V.10, the homomorphism $\pi_2(\alpha_K)$ only depends on the homotopy class of α (Lemma A.16). Therefore Lemma V.10 implies that the restriction of I_α to $\Pi_{(M,S)}$ depends only on the homotopy class of α, hence $\Pi_{(M,S)} \subseteq H_{dR}^1(M, S; Y)$ by Lemma II.10. From Lemmas V.10 and V.11 we further get

$$\Pi_{(M,S)} \subseteq H_{dR}^1(M, S; \Pi_{(I,\partial I)}) = H_{dR}^1(M, S; \Pi_{\mathbb{S}^1}).$$

To prove the inclusion

$$H_{dR}^1(M, S; \mathbb{Z}) \cdot \Pi_{\mathbb{S}^1} \subseteq \Pi_{(M,S)},$$

let $[\zeta] \in H_{dR}^1(M, S; \mathbb{Z})$. Then Lemma II.3 implies the existence of $f \in C^\infty(M, S; \mathbb{T})$ with $\delta(f) = \zeta$. Let $0 \in \mathbb{T} \cong \mathbb{R}/\mathbb{Z}$ denote the identity element in $\mathbb{T}$. The map f induces a smooth group homomorphism

$$f_K \colon C^\infty(I, \partial I; K) \to C^\infty(M, S; K), \quad \phi \mapsto \phi \circ f$$

(Lemma A.16). We now get from Lemma V.10 for each $\alpha \in C^\infty((I, \partial I), (M, S))$ the relation

$$I_\alpha \circ \mathrm{per}_{\omega_{(M,S)}} \circ \pi_2(f_K) = \mathrm{per}_{\omega_{(I,\partial I)}} \circ \pi_2(\alpha_K) \circ \pi_2(f_K) = \mathrm{per}_{\omega_{(I,\partial I)}} \circ \pi_2((f \circ \alpha)_K),$$

where $f \circ \alpha$ is viewed as a map in $C^\infty((I, \partial I), (\mathbb{T}, \{0\}))$. This map factors through a smooth map $I/\partial I \cong \mathbb{T} \to \mathbb{T}$, and $\pi_2((f \circ \alpha)_K)$ is the multiplication with the winding number $\deg(f \circ \alpha)$ of this map ([MN02, Lemma I.10]). For each

$$[\sigma] \in \pi_2(C^\infty(\mathbb{T}, \{0\}; K)) \cong \pi_2(C_*^\infty(\mathbb{S}^1; K))$$

we then have

$$I_\alpha(\mathrm{per}_{\omega_{(M,S)}}(\pi_2(f_K)[\sigma])) = \deg(f \circ \alpha)\,\mathrm{per}_{\omega_{(I,\partial I)}}([\sigma]) = I_\alpha(\zeta) \cdot \mathrm{per}_{\omega_{(I,\partial I)}}([\sigma]).$$

Since the I_α separate points on $H_{dR}^1(M, S; Y)$, it follows that

$$\mathrm{per}_{\omega_{(M,S)}}(\pi_2(f_K)[\sigma]) = [\zeta] \cdot \mathrm{per}_{\omega_{(I,\partial I)}}([\sigma]),$$

and hence that

$$H^1_{dR}(M, S; \mathbb{Z}) \cdot \Pi_{\mathbb{S}^1} = H^1_{dR}(M, S; \mathbb{Z}) \cdot \Pi_{(I,\partial I)} \subseteq \Pi_{(M,S)}.$$

If $\Pi_{\mathbb{S}^1}$ is discrete, then we apply Proposition IV.23 to obtain the asserted equalities.

$\square$

Corollary V.13. *If $\Pi_{\mathbb{S}^1}$ is discrete, then $\Pi_{(M,S)}$ is discrete for each pair (M, S).*

Proof. Proposition V.12 implies that $\Pi_{(M,S)} \subseteq H^1_{dR}(M, S; \Pi_{\mathbb{S}^1})$, and the latter group is discrete by Theorem II.7.

$\square$

Remark V.14. In view of the preceding corollary, everything reduces to the study of the period map

$$\mathrm{per}_{\omega_{\mathbb{S}^1}} : \pi_3(K) \cong \pi_2(C^\infty(\mathbb{S}^1; K)) \to Y.$$

It is not necessary to know $\pi_2(G)$ explicitly.

$\square$

Proposition V.15. *Suppose that $Y = \mathbb{R}$ and $\Gamma = \mathbb{Z}$, so that $T_\Gamma = \mathbb{T}$. We further assume that $\mathfrak{k}$ is compact and simple and that κ in normalized in such a way that $\kappa(i\check{\alpha}, i\check{\alpha}) = -2$, where $\check{\alpha} \in \mathfrak{k}_{\mathbb{C}}$ is a coroot corresponding to a long root. For $G = C^\infty(M, S; K)_e$ we then have*

$$\Pi_{(M,S)} = H^1_{dR}(M, S; \mathbb{Z}).$$

Proof. We first recall from the calculations in Appendix IIa to Section II in [Ne01a] that under the present assumptions we have $\Pi_{(I,\partial I)} = \Pi_{\mathbb{S}^1} = \mathbb{Z}$ (see also [MN02, Th. II.9]). Therefore Proposition V.12 directly leads to

$$\begin{aligned}
H^1_{dR}(M, S; \mathbb{Z}) \cdot \Pi_{\mathbb{S}^1} &= \mathbb{Z} \cdot H^1_{dR}(M, S; \mathbb{Z}) \\
&= H^1_{dR}(M, S; \mathbb{Z}) \subseteq \Pi_{(M,S)} \\
&\subseteq H^1_{dR}(M, S; \Pi_{\mathbb{S}^1}) = H^1_{dR}(M, S; \mathbb{Z}).
\end{aligned}$$

$\square$

Applying Proposition V.15 to the group $C^\infty(M, S; K)$ from Example II.12, we obtain a cocycle on the Lie algebra of a Fréchet–Lie group for which the period group $\Pi_{(M,S)}$ is discrete but not finitely generated.

Period maps for $C_c^\infty(M; K)$

Let M be a connected non-compact manifold and Y a s.c.l.c. space. For a proper smooth map $\alpha : \mathbb{R} \to M$ and $\zeta \in Z^1_{dR,c}(M; Y)$ the integral

$$I_\alpha(\zeta) := \int_\alpha \zeta := \int_{\mathbb{R}} \alpha^* \zeta$$

is defined because $\alpha^*\zeta$ has compact support. We thus obtain a linear map

$$I_\alpha : Z^1_{dR,c}(M; Y) \to Y$$

which is easily seen to be continuous.

Lemma V.16. *For each $\alpha \in C_p^\infty(\mathbb{R}, M)$ let*

$$\alpha_K : C_c^\infty(M; K) \to C_c^\infty(\mathbb{R}; K), \quad f \mapsto f \circ \alpha$$

denote the corresponding Lie group homomorphism. Then

$$\mathrm{per}_{\omega_\mathbb{R}} \circ \pi_2(\alpha_K) = I_\alpha \circ \mathrm{per}_{\omega_M} . \tag{5.4}$$

Proof. From Lemma A.12 we recall that α_K is a Lie group homomorphism. The remaining argument can be copied from Lemma V.10. □

Proposition V.17. *For each non-compact manifold M and each κ we have*

$$\Pi_M = H^1_{dR,c}(M; \Pi_\mathbb{R}).$$

Proof. In the situation of Lemma V.16, the homomorphism $\pi_2(\alpha_K)$ only depends on the homotopy class of α (Lemma A.16). Therefore Lemma V.10 implies that the restriction of I_α to Π_M depends only on the homotopy class of α, hence $\Pi_M \subseteq H^1_{dR,c}(M; Y)$ by Lemma II.10. From Lemma V.16 we further get $\Pi_M \subseteq H^1_{dR,c}(M; \Pi_\mathbb{R})$.

To prove the converse inclusion $H^1_{dR,c}(M; \Pi_\mathbb{R}) \subseteq \Pi_M$, we first recall from Corollary IV.22 that

$$H^1_{dR,c}(M, \Pi_\mathbb{R}) = H^1_{dR,c}(M; \mathbb{Z}) \cdot \Pi_\mathbb{R}.$$

It therefore suffices to prove $H^1_{dR,c}(M; \mathbb{Z}) \cdot \Pi_\mathbb{R} \subseteq \Pi_M$. Let $[\zeta] \in H^1_{dR}(M; \mathbb{Z})$. Then Proposition IV.20 implies the existence of $f \in C_c^\infty(M, \mathbb{T})$ with $\delta(f) = \zeta$. Let $0 = \mathbb{Z} \in \mathbb{T} \cong \mathbb{R}/\mathbb{Z}$ denote the identity element in $\mathbb{T}$. The map f induces a smooth group homomorphism

$$f_K : C_c^\infty(\mathbb{T}; K) \to C_c^\infty(M; K), \quad f \mapsto f \circ \phi$$

(Lemma A.12). In view of Lemma V.16, we have for each $\alpha \in C_p^\infty(\mathbb{R}, M)$

$$I_\alpha \circ \mathrm{per}_{\omega_M} \circ \pi_2(f_K) = \mathrm{per}_{\omega_\mathbb{R}} \circ \pi_2(\alpha_K) \circ \pi_2(f_K) = \mathrm{per}_{\omega_\mathbb{R}} \circ \pi_2((f \circ \alpha)_K),$$

where $f \circ \alpha$ is viewed as a map in $C_c^\infty(\mathbb{R}, \mathbb{T})$. Viewing $\mathbb{R}$ as $\mathbb{T} \setminus \{0\}$, this map extends to a smooth map $\mathbb{T} \to \mathbb{T}$, and $\pi_2((f \circ \alpha)_K)$ is the multiplication with the winding number

$$\deg(f \circ \alpha) = \int_\alpha \zeta$$

of this map ([MN02, Lemma I.10]). For each $[\sigma] \in \pi_2(C_c^\infty(\mathbb{R}; K))$ we then have

$$I_\alpha(\mathrm{per}_{\omega_M}(\pi_2(f_K)[\sigma])) = \deg(f \circ \alpha)\,\mathrm{per}_{\omega_\mathbb{R}}([\sigma]) = I_\alpha(\zeta)\,\mathrm{per}_{\omega_\mathbb{R}}([\sigma]).$$

Since the I_α separate points on $H^1_{\mathrm{dR,c}}(M; Y)$ (here we need that M is non-compact), it follows that

$$\mathrm{per}_{\omega_M}(\pi_2(f_K)[\sigma]) = [\zeta] \cdot \mathrm{per}_{\omega_{\mathbb{R}}}([\sigma])$$

and hence that $H^1_{\mathrm{dR,c}}(M; \mathbb{Z}) \cdot \Pi_{\mathbb{R}} \subseteq \Pi_M$. □

Corollary V.18. *If $\Pi_{\mathbb{R}}$ is discrete, then Π_M is discrete for each non-compact connected manifold M.* □

For the following proposition we recall the space $V(\mathfrak{k})$ from Definition V.7.

Proposition V.19. *If $\dim K < \infty$, and $\kappa : \mathfrak{k} \times \mathfrak{k} \to V(\mathfrak{k})$ is the universal symmetric invariant bilinear map, then there exists for $Z := V(\mathfrak{k})/\Pi_{M,\kappa}$ a central Lie group extension*

$$Z \hookrightarrow \hat{G} \twoheadrightarrow G = C^\infty_c(M, K)_e.$$

Proof. In view of [MN02, Th. II.9], the period group $\Pi_{\mathbb{S}^1,\kappa} = \Pi_{\mathbb{R},\kappa}$ is discrete (cf. Lemma V.11), and Corollary V.18 now shows that Π_M is discrete. Therefore Theorem V.5 applies. □

Remark V.20. The main idea behind our identification of the period group for current groups is as follows. Let M be a compact manifold, $x_M \in M$, and

$$G := C^\infty_*(M; K) := \{f \in C^\infty(M; K) : f(x_M) = e\}.$$

The evaluation map

$$\mathrm{ev} : G \times M \to K, \quad (f, p) \mapsto f(p)$$

induces maps

$$\phi_{k,l} : \pi_k(G) \times \pi_l(M) \to \pi_{k+l}(K)$$

as follows. We view $\pi_k(M)$ as the set of arc-components in the space

$$C((I^n, \partial I^n), (M, x_M))$$

of continuous maps of pairs, where I is the unit interval. Then $\phi_{k,l}([f], [h])$ is the class defined by the map

$$I^{k+l} \to K, \quad (x, y) \mapsto f(x)(h(y)),$$

vanishing on the boundary

$$\partial I^{k+l} = (\partial I^k \times I^l) \cup (I^k \times \partial I^l).$$

In particular we obtain a map

$$\phi_{2,1} : \pi_2(G) \times \pi_1(M) \to \pi_3(K),$$

and our analysis of the period map is based on the commutative diagram

$$
\begin{array}{ccc}
\pi_2(G) \quad \times \pi_1(M) & \to & \pi_3(K) \\
\Big\downarrow \mathrm{per}_{\omega_M} \qquad \Big\downarrow \mathrm{id} & & \Big\downarrow \mathrm{per}_{\mathbb{S}^1} \\
H^1_{\mathrm{dR}}(M;Y) \times \pi_1(M) & \to & H^1_{\mathrm{dR}}(\mathbb{S}^1;Y) \cong Y
\end{array}
$$

The effectiveness of this picture comes from the fact that the natural pairing

$$
H^1_{\mathrm{dR}}(M;Y) \times \pi_1(M) \to Y
$$

defined by integration over loops is non-degenerate in the sense that the integrals separate points in $H^1_{\mathrm{dR}}(M;Y)$.

The arguments for non-compact manifolds essentially follow the same line, where we have to take smooth proper curves instead of loops.   □

VI Universal central extensions of current groups

For the special class of finite-dimensional semisimple Lie groups K, each Lie algebra cocycle $\omega \in Z^2_c(C^\infty_c(M,\mathfrak{k}), \mathfrak{z})$ is equivalent to a cocycle of product type ([Ma02]). This observation permits us to construct a universal central extension of the Lie algebra $\mathfrak{g} := C^\infty_c(M;\mathfrak{k})$. In the present section we show that this construction can be globalized in the sense that we construct a universal central extension of the connected Lie group $C^\infty_c(M;K)_e$.

First cyclic homology of function spaces

Definition VI.1. Let E, F and G be locally convex spaces over $\mathbb{K} \in \{\mathbb{R}, \mathbb{C}\}$. Then the *projective topology* on the tensor product $E \otimes F$ is defined by the seminorms

$$
(p \otimes q)(x) = \inf \left\{ \sum_{j=1}^n p(y_j)q(z_j) : x = \sum_j y_j \otimes z_j \right\},
$$

where p, resp., q is a continuous seminorm on E, resp., F (cf. [Tr67, Prop. 43.4]). We write $E \otimes_\pi F$ for the locally convex space obtained by endowing $E \otimes F$ with the locally convex topology defined by this family of seminorms. It is called the *projective tensor product of E and F*. It has the universal property that the continuous bilinear maps $E \times F \to G$ are in one-to-one correspondence with the continuous linear maps $E \otimes_\pi F \to G$ (here we need that G is locally convex). We write $E \hat\otimes_\pi F$ for the completion of the projective tensor product of E and F.   □

Definition VI.2. Let A be a unital locally convex topological algebra over $\mathbb{K} \in \{\mathbb{R}, \mathbb{C}\}$.

(a) We recall that the first *Hochschild homology space* $HH_1(A)$ is defined as

$$
HH_1(A) := Z_1(A)/\overline{B_1(A)},
$$

where

$$Z_1(A) := \ker b_A \subseteq A \otimes A, \quad b_A(a \otimes b) = [a, b] = ab - ba$$

and

$$B_1(A) := \mathrm{span}\{xy \otimes z - x \otimes yz + zx \otimes y \colon x, y, z \in A\}.$$

Here we endow $A \otimes A$ with the projective tensor product topology.

Suppose that A is commutative. Then $Z_1(A) = A \otimes A$. Let M be a *continuous A-module*, i.e., M is a locally convex space with an A-module structure given by a continuous bilinear map $A \times M \to M$. For a linear map $D \colon A \to M$ the bilinear map

$$A \otimes A \to M, \quad x \otimes y \mapsto x.Dy$$

annihilates $B_1(A)$ if and only if D is a derivation. Hence $HH_1(A)$ has the universal property of the universal differential module $\Omega^1(A)$ with respect to the differential

$$d \colon A \to HH_1(A), \quad a \mapsto [\mathbf{1} \otimes a].$$

This means that for each continuous derivation $D \colon A \to M$ there exists a unique continuous linear map $\phi \colon HH_1(A) \to M$ with $D = \phi \circ d$ (cf. [Ma02]). Therefore $HH_1(A)$ is isomorphic to the topological module $\Omega^1(A)$ of Kähler differentials on A ([Lo98, Prop. 1.1.10]).

(b) The first *cyclic homology space* of A can be obtained as the quotient

$$HC_1(A) := Z_1^\lambda(A)/\overline{B_1^\lambda(A)},$$

where

$$Z_1^\lambda := \ker b_A \subseteq \Lambda^2(A), \quad b_A(a \wedge b) := [a, b],$$

and

$$B_1^\lambda(A) := \mathrm{span}\{xy \wedge z - x \wedge yz + zx \wedge y \colon x, y, z \in A\}$$

(cf. [Lo98, Th. 2.15]).

If A is commutative, then $a \otimes b + b \otimes a - 1 \otimes ab \in B_1(A)$ implies that the universal differential $d \colon A \to HH_1(A)$ satisfies

$$\mathrm{im}(d) = [\mathbf{1} \otimes A] \cong \mathbf{1} \otimes A + B_1(A) = \{a \otimes b + b \otimes a \colon a, b \in A\} + B_1(A).$$

Hence

$$HH_1(A)/\overline{\mathrm{im}\,d} \cong \Lambda^2(A)/\overline{B_1^\lambda(A)} \cong HC_1(A)$$

(cf. [Lo98, Prop. 2.1.14]).   □

Let M be a finite-dimensional manifold and $A := C_c^\infty(M; \mathbb{K})$. According to [Gl01c], the multiplication on $C_c^\infty(M; \mathbb{K})$ is a continuous bilinear map, so that A is a locally convex topological algebra. This is not obvious because the topology on $C_c^\infty(M; \mathbb{K})$ is the locally convex direct limit topology which differs from the direct limit topology with respect to the subspaces $C_{X_n}^\infty(M; \mathbb{K})$, where $(X_n)_{n\in\mathbb{N}}$ is an exhaustive sequence of compact submanifolds with boundary in M. Hence there is no a priori reason for a bilinear map on $C_c^\infty(M; \mathbb{K})$ to be continuous if all the restrictions to the subspaces $C_{X_n}^\infty(M; \mathbb{K})$ are continuous.

Let $A_+ := \mathbb{K}\mathbf{1} + A \subseteq C^\infty(M; \mathbb{K})$. In this section we will show that, as locally convex spaces, we have

$$HH_1(A) := HH_1(A_+) \cong \Omega_c^1(M; \mathbb{K})$$

and

$$HC_1(A) \cong \Omega_c^1(M; \mathbb{K})/dA = {}_{3M.c}(\mathbb{K}).$$

Theorem VI.3 (Glöckner's Theorem). $\Omega_c^1(M; \mathbb{K})$ is a continuous module of $C_c^\infty(M; \mathbb{K})$.

Proof. This follows from [Gl01d, Th. 5.1] because the module structure is induced by the bundle map given in a point $p \in M$ by the scalar multiplication $\mathbb{K} \times T_p(M)^* \to T_p(M)^*$. $\qquad\square$

Theorem VI.4. $HH_1(C_c^\infty(M; \mathbb{K})) \cong \Omega_c^1(M; \mathbb{K})$.

Proof (cf. [Ma02, Th. 11]). We will show that the continuous derivation $d\colon A = C_c^\infty(M\ \mathbb{K}) \to \Omega_c^1(M; \mathbb{K})$ has the universal property of the universal differential module of A. From this the assertion follows, as $HH_1(A)$ can be viewed as the universal differential module of A (Definition VI.2).

We consider the map

$$\tau\colon C^\infty(M \times M; \mathbb{K}) \to \Omega^1(M; \mathbb{K}), \qquad \tau(F)(x)(v) := dF(x, x)(0, v).$$

Via the natural embedding

$$A_+ \otimes A_+ \to C^\infty(M \times M, \mathbb{K}), \qquad (f, g) \mapsto ((x, y) \mapsto f(x)g(y)),$$

we view $A_+ \otimes A_+$ (the algebraic tensor product) as a subalgebra of $C^\infty(M \times M, \mathbb{K})$. This embedding is topological on the subspaces of the form

$$C_X^\infty(M; \mathbb{K}) \otimes_\pi C_X^\infty(M; \mathbb{K})$$

for compact subsets $X \subseteq M$ ([Gr55, Ch. 2, p.81]). Let

$$I := \{F \in A_+ \otimes A_+ \colon (\forall x \in M) F(x, x) = 0\}.$$

This is an ideal of $A_+ \otimes A_+$ which can also be viewed as the kernel of the multiplication map $\mu\colon A_+ \otimes A_+ \to A_+$. Note that $\tau(f \otimes g) = f \cdot dg \in \Omega_c^1(M; \mathbb{K})$ for $f, g \in A_+$. (1) Let $(\phi_j)_{j\in J}$ be a locally finite partition of unity in A for which $\operatorname{supp}(\phi_j)$ is contained in a coordinate neighborhood $U_j \subseteq M$ with U_j diffeomorphic to $\mathbb{R}^d$, $d := \dim M$.

With this partition of unity we write each $\alpha \in \Omega_c^1(M; \mathbb{K})$ as

$$\alpha = \sum_j \phi_j \alpha,$$

where the sum is finite because only finitely many of the supports of the functions ϕ_j intersect the support of α. As $U_j \cong \mathbb{R}^d$ and $\mathrm{supp}(\phi_j)$ is a compact subset of U_j, there exist functions $\overline{y}_1^j, \ldots, \overline{y}_d^j \in A$ such that on $\mathrm{supp}(\phi_j)$ the differentials $d\overline{y}_i^j$, $i = 1, \ldots, d$, are linearly independent. Then we write

$$\phi_j \alpha = \sum_{i=1}^d \alpha_i^j \, d\overline{y}_i^j$$

with $\alpha_i^j \in A$.

(2) $\tau(A \otimes A) = \tau(A_+ \otimes A_+) = \Omega_c^1(M; \mathbb{K})$: This follows from

$$\alpha = \sum_j \sum_i \alpha_i^j \, d\overline{y}_i^j = \sum_{j,i} \tau(\alpha_i^j \otimes \overline{y}_i^j).$$

(3) As $\mu(A_+ \otimes 1) = A_+$ and $\tau(A_+ \otimes 1) = 0$, we have $\tau(I) = \tau(A_+ \otimes A_+) = \Omega_c^1(M; \mathbb{K})$ by (2). Let $N := \ker(\tau|_I)$. We claim that $N = \overline{I^2}$. The inclusion $I^2 \subseteq N$ follows directly from

$$\tau(FG) = F\tau(G) + \tau(F)G, \tag{6.1}$$

which also shows that N is an ideal of $A_+ \otimes A_+$. As τ is continuous and I is closed, we also obtain $\overline{I^2} \subseteq N$. Now let $F \in N$. Since F can be written as a finite sum

$$F = \sum_{i,j} (\phi_i \otimes \phi_j) F,$$

where each summand is contained in the ideal N, it suffices to assume that $\mathrm{supp}(F) \subseteq U_i \times U_j \cong \mathbb{R}^{2d}$ for some pair $(i, j) \in J^2$. Then we have

$$F(x, y) = \sum_{l=1}^d (x_l - y_l) F_l(x, y)$$

with

$$F_l(x, y) := \frac{1}{2} \int_0^1 \frac{\partial F}{\partial x_l}(tx + (1-t)y, y) - \frac{\partial F}{\partial y_l}(x, tx + (1-t)y) \, dt,$$

and it is easy to see that the supports of the functions F_l are compact. From

$$\tau(F)(x) = -\sum_{l=1}^d F_l(x, x) dx_l$$

we derive that the functions F_l vanish on the diagonal in $\mathbb{R}^d \times \mathbb{R}^d$, so that Lemma 5 in [Ma02] implies that $F_l \in C_c^\infty(M \times M, \mathbb{K})$ is contained in the closure $\overline{I}$ of the ideal $I \subseteq A_+ \otimes A_+$. Let $C \subseteq \mathbb{R}^d$ be a compact subset such that $C^0 \times C^0$ contains the support of all the functions F_l. We replace the coordinate functions x_j on $\mathbb{R}^d$ by functions $\overline{x}_j \in C_c^\infty(\mathbb{R}^d; \mathbb{K})$ with $\text{supp}(\overline{x}_j) \subseteq C$ and obtain

$$F(x, y) = \sum_{l=1}^{d} (\overline{x}_l - \overline{y}_l) F_l(x, y) \in I \cdot \overline{I} \subseteq \overline{I^2},$$

where the closure is taken in

$$C_{C \times C}^\infty(\mathbb{R}^{2d}, \mathbb{K}) \cong C_C^\infty(\mathbb{R}^d, C_C^\infty(\mathbb{R}^d, \mathbb{K})) \cong C_C^\infty(\mathbb{R}^d; \mathbb{K}) \hat{\otimes}_\pi C_C^\infty(\mathbb{R}^d, \mathbb{K})$$

(cf. [Gr55, Ch. 2, p.81]).

(4) The derivation $d\colon A \to \Omega_c^1(M; \mathbb{K})$ has the universal property of the universal topological differential module $\Omega^1(A)$: Let E be a topological A-module and $d_E\colon A \to E$ a continuous derivation. We will complete the proof by showing that there exists a continuous linear map $\Phi\colon \Omega_c^1(M; \mathbb{K}) \to E$ with $\Phi(f dg) = f d_E(g)$.

We have seen above that $\ker(\tau \mid_I) = \overline{I^2} \cap N = \overline{I^2}$ with respect to the relative topology, so that $\tau \mid_I$ leads to a continuous bijective linear map $I/\overline{I^2} \cong \Omega^1(A) \to \Omega_c^1(M; \mathbb{K})$. Therefore the natural map

$$A_+ \otimes A_+ \supseteq I \to E, \quad f \otimes g \mapsto f d_E(g)$$

yields a linear map

$$\Phi\colon \Omega_c^1(M; \mathbb{K}) \to E \quad \text{with } \Phi(f dg) = \Phi(\tau(f \otimes g)) = f d_E(g).$$

Hence it only remains to show that Φ is continuous when viewed as a linear map on $\Omega_c^1(M; \mathbb{K})$. As the topology on $\Omega_c^1(M; \mathbb{K})$ is the locally convex direct limit topology with respect to the subspaces $\Omega_X^1(M; \mathbb{K})$, $X \subseteq M$ compact, it suffices to verify that the restrictions $\Phi \mid_{\Omega_X^1(M; \mathbb{K})}$ are continuous.

The set $J_X := \{j \in J\colon \text{supp}(\phi_j) \cap X \neq \emptyset\}$ is finite, and for each $\alpha \in \Omega_X^1(M; \mathbb{K})$ we have

$$\alpha = \sum_{j \in J_X} \phi_j \alpha = \sum_{j \in J_X} \sum_i \alpha_i^j d\overline{y}_i^j.$$

Now

$$\Phi(\alpha) = \sum_{j \in J_X} \Phi(\phi_j \alpha) = \sum_{j \in J_X} \sum_i \alpha_i^j d_E(\overline{y}_i^j)$$

because the sum is finite. The functions $\overline{y}_i^j$ do not depend on α, and the multiplication with ϕ_j is a continuous endomorphism of $\Omega_c^1(M; \mathbb{K})$. Therefore the maps

$$\Omega_c^1(M; \mathbb{K}) \to A, \quad \alpha \mapsto \alpha_i^j$$

are continuous. Now the continuity of the module structure on E implies that Φ is continuous. $\square$

Corollary VI.5. *For $A = C_c^\infty(M; \mathbb{K})$ and $\mathbb{K} \in \{\mathbb{R}, \mathbb{C}\}$ we have*

$$HC_1(A) \cong HH_1(A)/dA \cong \Omega_c^1(M; \mathbb{K})/dC_c^\infty(M; \mathbb{K}).$$ $\square$

Universal central extensions

In this subsection we turn to the question whether for a finite-dimensional semisimple Lie group K the central extension of $C_c^\infty(M, K)_e$ from Proposition V.19 is universal. This question will be answered affirmatively if $\mathfrak{k}$ is finite-dimensional and semisimple. First we recall some concepts and a result from [Ne01c] on weakly universal central extensions of Lie groups and Lie algebras.

Definition VI.6 (cf. [Ne01c]). Let $\mathfrak{g}$ be a topological Lie algebra over $\mathbb{K} \in \{\mathbb{R}, \mathbb{C}\}$ and $\mathfrak{a}$ be a topological vector space considered as a trivial $\mathfrak{g}$-module. We call a central extension $q \colon \hat{\mathfrak{g}} = \mathfrak{g} \oplus_\omega \mathfrak{z} \to \mathfrak{g}$ with $\mathfrak{z} = \ker q$ (or simply the Lie algebra $\hat{\mathfrak{g}}$) *weakly universal for $\mathfrak{a}$* if the corresponding map $\delta_\mathfrak{a} \colon \mathrm{Lin}(\mathfrak{z}, \mathfrak{a}) \to H_c^2(\mathfrak{g}, \mathfrak{a}), \gamma \mapsto [\gamma \circ \omega]$ is bijective.

We call $q \colon \hat{\mathfrak{g}} \to \mathfrak{g}$ *universal for $\mathfrak{a}$* if for every central extension $q_1 \colon \hat{\mathfrak{g}}_1 \to \mathfrak{g}$ of $\mathfrak{g}$ by $\mathfrak{a}$ with a continuous linear section there exists a unique homomorphism $\phi \colon \hat{\mathfrak{g}} \to \hat{\mathfrak{g}}_1$ with $q_1 \circ \phi = q$. Note that this universal property immediately implies that two central extensions $\hat{\mathfrak{g}}_1$ and $\hat{\mathfrak{g}}_2$ of $\mathfrak{g}$ by $\mathfrak{a}_1$ and $\mathfrak{a}_2$ such that both $\hat{\mathfrak{g}}_1$ and $\hat{\mathfrak{g}}_2$ are universal for $\mathfrak{a}_1$ and $\mathfrak{a}_2$ are isomorphic. A central extension is said to be *(weakly) universal* if it is (weakly) universal for all locally convex spaces $\mathfrak{a}$. $\square$

Definition VI.7. We call a central extension $\hat{G} = G \times_f Z$ of the connected Lie group G by the abelian Lie group Z given by $f \in Z_s^2(G, Z)$ *weakly universal for the abelian Lie group A* if the map

$$\delta_A \colon \mathrm{Hom}(Z, A) \to H_s^2(G, A), \quad \gamma \mapsto [\gamma \circ f]$$

is bijective. It is called *universal for the abelian Lie group A* if for every central extension

$$q_1 \colon G \times_\phi A \to G, \quad \phi \in Z_s^2(G, A),$$

there exists a unique Lie group homomorphism $\psi \colon G \times_f Z \to G \times_\phi A$ with $q_1 \circ \psi = q$ (cf. Definition V.1). A central extensional is said to be *(weakly) universal* if it is (weakly) universal for all Lie groups A with $A_e \cong \mathfrak{a}/\pi_1(A)$ and $\mathfrak{a}$ s.c.l.c. $\square$

Definition VI.8. If $\mathfrak{g}$ is a locally convex Lie algebra, then we write $H_1(\mathfrak{g})$ for the completion of the quotient space $\mathfrak{g}/\overline{[\mathfrak{g}, \mathfrak{g}]}$. If $\mathfrak{g}$ is a Fréchet space, then $\mathfrak{g}/\overline{[\mathfrak{g}, \mathfrak{g}]}$ is also Fréchet, and no completion is necessary.

If G is a connected Lie group with Lie algebra $\mathfrak{g}$ and $\widetilde{G}$ its universal covering group, then we have a natural homomorphism $d_G \colon \widetilde{G} \to H_1(\mathfrak{g})$. Its kernel is denoted by $(\tilde{G}, \tilde{G})$. If G is finite-dimensional, then $(\tilde{G}, \tilde{G})$ is the commutator group of $\tilde{G}$.   □

Theorem VI.9. (Recognition Theorem) *Assume that $q: \hat{G} \to G$ is a central Z-extension of Lie groups over $\mathbb{K} \in \{\mathbb{R}, \mathbb{C}\}$ for which*

(1) *the corresponding Lie algebra extension $\hat{\mathfrak{g}} \to \mathfrak{g}$ is weakly $\mathbb{K}$-universal,*

(2) *$\hat{G}$ is simply connected, and*

(3) *$\pi_1(G) \subseteq (\tilde{G}, \tilde{G})$.*

If $\hat{\mathfrak{g}}$ is weakly universal for a s.c.l.c. space $\mathfrak{a}$, then $\hat{G}$ is weakly universal for each abelian Lie group A with $A_e \cong \mathfrak{a}/\pi_1(A)$.

Proof. The original statement of this theorem in [Ne01c, Th. IV.13] is formulated only for Fréchet–Lie groups, but one easily verifies that the proof yields the more general result stated above.   □

Theorem VI.10. *Let K be a finite-dimensional semisimple Lie group and $G := C_c^\infty(M, K)_e$. Further let $\mathfrak{z} := \mathfrak{z}_{M,c}(V(\mathfrak{k}))$ and $\omega = \omega_{M,\kappa} \in Z_c^2(\mathfrak{g}, \mathfrak{z})$ be a cocycle of product type given by $\omega(\eta, \xi) = [\kappa(\eta, d\xi)]$. Then the corresponding central Lie algebra extension $\hat{\mathfrak{g}} := \mathfrak{g} \oplus_\omega \mathfrak{z}$ is universal, and there exists a corresponding central Lie group extension $Z \hookrightarrow \hat{G} \twoheadrightarrow G$ with $Z \cong \pi_1(G) \times (\mathfrak{z}/\Pi_M)$ which is universal for all Lie groups A with $A_e \cong \mathfrak{a}/\Gamma$, where $\mathfrak{a}$ is a s.c.l.c. space and $\Gamma \subseteq \mathfrak{a}$ a discrete subgroup.*

Proof. First we show that $\hat{\mathfrak{g}}$ is perfect. In fact, for $x, y \in \mathfrak{k}$ and $f, g \in C^\infty(M; \mathbb{K})$ we have in $\hat{\mathfrak{g}}$ the relation

$$[f \otimes x, g \otimes y] - [g \otimes x, f \otimes y] = \big(fg \otimes [x, y] - gf \otimes [x, y], 2[fdg] \cdot \kappa(x, y)\big)$$
$$= \big(0, 2[fdg] \cdot \kappa(x, y)\big).$$

Since $V(\mathfrak{k})$ is spanned by $\mathrm{im}(\kappa)$, the fact that $\mathfrak{z}_{M,c}(\mathbb{K})$ is spanned by elements of the form $[f \cdot dg]$ implies that $\hat{\mathfrak{g}}$ is perfect.

Since $\hat{\mathfrak{g}}$ is perfect, for each locally convex space $\mathfrak{a}$ the natural map

$$\delta: \mathrm{Lin}(\mathfrak{z}, \mathfrak{a}) \to H_c^2(\mathfrak{g}, \mathfrak{a}), \quad \gamma \mapsto [\gamma \circ \omega]$$

is injective ([Ne01c, Rem. I.6]). It has been shown in [Ma02, Thm. 16] that δ is also surjective, so that $\hat{\mathfrak{g}}$ is weakly universal for all locally convex spaces $\mathfrak{a}$. Since $\hat{\mathfrak{g}}$ is perfect, it even is a universal central extension of $\mathfrak{g}$ ([Ne01c, Lemma I.12]).

Furthermore, the period map $\mathrm{per}_\omega: \pi_2(G) \to \mathfrak{z}$ has discrete image Π_ω (Proposition V.19). In view of Theorem V.8, Theorem V.5 now implies the existence of a central Lie group extension $Z \hookrightarrow \hat{G} \twoheadrightarrow G$ with $Z \cong (\mathfrak{z}/\Pi_\omega) \times \pi_1(G)$ corresponding to the Lie algebra extension $\mathfrak{z} \hookrightarrow \hat{\mathfrak{g}} \to \mathfrak{g}$ and such that the connecting homomorphism $\pi_1(G) \to \pi_0(Z)$ is an isomorphism.

To prove the universality of $\hat{G}$, we use the Recognition Theorem VI.9. For that we have to verify that

(1) $\hat{\mathfrak{g}}$ is weakly universal,

(2) $\pi_1(\hat{G}) = 1$,

(3) $\pi_1(G) \subseteq (\widetilde{G}, \widetilde{G})$.

Condition (1) has been verified above. Further (3) follows from the perfectness of $\mathfrak{g}$, which implies $(\widetilde{G}, \widetilde{G}) = \widetilde{G}$. It therefore remains to verify (2). For that we consider a part of the long exact homotopy sequence of the Z-principal bundle $q \colon \hat{G} \to G$:

$$\pi_2(G) \xrightarrow{\;\delta\;} \pi_1(Z) \to \pi_1(\hat{G}) \to \pi_1(G) \to \pi_0(Z). \tag{6.2}$$

According to [Ne02a, Prop. 5.11], we have $\delta = -\mathrm{per}_\omega$, so that $\pi_1(Z) = \Pi_\omega$ (as subsets of $\mathfrak{z}$) implies that δ is surjective. Moreover, the natural homomorphism $\pi_1(G) \to \pi_0(Z)$ is an isomorphism by the construction of $\hat{G}$, so that the exactness of (6.2) implies that $\hat{G}$ is simply connected. $\qquad\square$

Remark VI.11. If K is finite-dimensional and reductive, then $\widetilde{K} \cong \mathfrak{z}(\mathfrak{k}) \times (\widetilde{K}, \widetilde{K})$. Therefore $\pi_1(K)$ is contained in $(\widetilde{K}, \widetilde{K})$ if and only if $K \cong \mathfrak{z}(\mathfrak{k}) \times (K, K)$. In this case we have

$$C^\infty(M, K) \cong C^\infty(M, \mathfrak{z}(\mathfrak{k})) \times C^\infty(M, (K, K))$$

and hence we have for $G = C^\infty(M, K)_e$ the direct product decomposition

$$G = G_D \times G_Z \quad \text{with} \quad G_D := C^\infty(M, (K, K))_e \quad \text{and} \quad G_Z := C^\infty(M, \mathfrak{z}(\mathfrak{k})).$$

In this case the Lie algebra $\mathfrak{g} = C^\infty(M; \mathfrak{k})$ has the direct decomposition $\mathfrak{g} = \mathfrak{g}' \oplus \mathfrak{z}(\mathfrak{g})$ with $\mathfrak{g}' = C^\infty(M; \mathfrak{k}')$ and $\mathfrak{z}(\mathfrak{g}) = C^\infty(M; \mathfrak{z}(\mathfrak{k}))$, where $\mathfrak{k}'$, resp., $\mathfrak{g}'$ denote the commutator algebra. It is easy to see that every Lie algebra cocycle $\omega \in Z_c^2(\mathfrak{g}; Y)$ vanishes on $\mathfrak{g}' \times \mathfrak{z}(\mathfrak{g}) \subseteq \mathfrak{g} \times \mathfrak{g}$ because $\mathfrak{g}'$ is perfect. From that one further derives that a weakly universal central extension of $\mathfrak{g}$ can be obtained with

$$\mathfrak{z} := \mathfrak{z}_M(V(\mathfrak{k}')) \oplus \Lambda^2(\mathfrak{z}(\mathfrak{g})),$$

where for a locally convex space E the space $\Lambda^2(E)$ is defined as the quotient of $E \otimes_\pi E$ modulo the closure of the subspace spanned by the elements $e \otimes e, e \in E$. To describe the corresponding cocycle, we write $\xi \in \mathfrak{g}$ as $\xi = (\xi', \xi_{\mathfrak{z}})$ with $\xi' \in \mathfrak{g}'$ and $\xi_{\mathfrak{z}} \in \mathfrak{z}(\mathfrak{g})$. Then a weakly universal cocycle is given by

$$\omega(\xi, \eta) = ([\kappa_{\mathfrak{k}'}(\xi', d\eta')], \xi_{\mathfrak{z}} \wedge \eta_{\mathfrak{z}}).$$

Let $\hat{G}_D$ be the universal central extension of G_D from Theorem VI.10 and define $\hat{G} := \hat{G}_D \times \hat{G}_Z$, where $\hat{G}_Z$ is the 2-step nilpotent Lie algebra

$$\mathfrak{z}(\mathfrak{g}) \times_{\omega_Z} \Lambda^2(\mathfrak{z}(\mathfrak{g})) \quad \text{with} \quad \omega_Z(\xi, \eta) = \xi \wedge \eta,$$

viewed as a Lie group with the multiplication $x * y := x + y + \frac{1}{2}[x, y]$. Using Theorem VI.9, we see that $\hat{G}_Z$ is a weakly universal central extension of $G_Z \cong \mathfrak{g}_Z$. Theorems VI.9 and VI.10 now imply that $\hat{G}$ is a weakly universal central extension of G. $\qquad\square$

Appendix A. Homotopy groups of smooth current groups

In this section we show that the homotopy groups of the Lie groups of smooth maps $C_c^\infty(M; K)$, resp., $C^\infty(M, S; K)$ introduced in Section I coincide with the homotopy groups of the corresponding groups of continuous maps $C_0(M; K)$, resp., $C_0(M \setminus S; K)$. The latter groups are usually better accessible by means of topological methods.

More specifically, for the group $C_c^\infty(M; K)$ of compactly supported smooth functions on a manifold M with values in a Lie group K the main result will be that the inclusion $C_c^\infty(M; K) \hookrightarrow C_0(M; K)$ is a weak homotopy equivalence. For the group $C^\infty(M, S; K)$ of smooth maps on a compact manifold M vanishing with all derivatives on a closed subset S we show that the inclusion $C^\infty(M, S; K) \hookrightarrow C_0(M \setminus S; K)$ is a weak homotopy equivalence.

In the present paper the results of this section are mainly needed to get information on the second homotopy group which is important for period maps associated to Lie algebra cocycles (cf. Section V). Moreover, the results of this appendix are of independent interest in many other contexts, where they provide valuable information on the topology of current groups.

Groups of compactly supported functions

Lemma A.1. *For each compact subset E of $C_c^\infty(M; K)$ there exists a compact subset $X \subseteq M$ with $E \subseteq C_X^\infty(M; K)$.*

Proof. Let $\mathfrak{k} := \mathbf{L}(K)$ be the Lie algebra of K, $U \subseteq \mathfrak{k}$ be an open 0-neighborhood, and $\phi: U \to \phi(U)$ a chart with $\phi(0) = e$. Then there exists an open 0-neighborhood $U_0 \subseteq U$ such that we obtain a local chart for $G := C_c^\infty(M; K)$ by $\phi_G(f) := \phi \circ f$ (Definition I.2(b)). Let $V := \{f \in C_c^\infty(M; \mathfrak{k}): f(M) \subseteq U_0\}$ and observe that

$$\phi_G: V \to \phi_G(V) = \{f \in C_c^\infty(M; K): f(M) \subseteq \phi(U_0)\}.$$

Then for each $f \in G$ the set $f\phi_G(V)$ is an open neighborhood, and the map

$$\phi_f: V \to f\phi_G(V), \quad \xi \mapsto f\phi_G(\xi)$$

is a diffeomorphism. Let $W \subseteq V$ be a closed 0-neighborhood with $\phi_G(W)\phi_G(W) \subseteq \phi_G(V)$. Since $\overline{\phi_G(W)}$ is the intersection of all sets $\phi_G(W)N$, where N is an identity neighborhood in $C_c^\infty(M; K)$, $\overline{\phi_G(W)} \subseteq \phi_G(V)$, so that the closedness of W implies that $\phi_G(W)$ is closed.

Since the compact set E is covered by the open sets $f\phi_G(W^0)$, $f \in E$, there exist $f_1, \ldots, f_n \in E$ with

$$E \subseteq f_1\phi_G(W^0) \cup \ldots \cup f_n\phi_G(W^0).$$

The closedness of $\phi_G(W)$ implies that each set $E \cap f_j \phi_G(W)$ is compact, so that for each j the closed set

$$\phi_{f_j}^{-1}(E \cap f_j \phi_G(W)) = W \cap \phi_{f_j}^{-1}(E) \subseteq C_c^\infty(M; \mathfrak{k}) = \varinjlim C_X^\infty(M; \mathfrak{k})$$

is compact, so that there exists a compact subset $X_j \subseteq M$ with $\phi_{f_j}^{-1}(E \cap f_j \phi_G(W)) \subseteq C_{X_j}^\infty(M; \mathfrak{k})$ ([He89, Prop. 1.5.3]). Let

$$X := X_1 \cup \ldots \cup X_n \cup \operatorname{supp}(f_1) \cup \ldots \cup \operatorname{supp}(f_n).$$

Then X is compact and $E \subseteq C_X^\infty(M; K)$.   □

Lemma A.2. *Let E be a compact space and $f: E \to C_c^\infty(M; K)$ a continuous map. Then there exists a compact subset $X \subseteq M$ and a continuous map $f_X: E \to C_X^\infty(M; K)$ such that $f = \eta_X \circ f_X$ holds for the inclusion map $\eta_X: C_X^\infty(M; K) \to C_c^\infty(M; K)$.*

Proof. Since $C_c^\infty(M; K)$ is Hausdorff, the set $f(E)$ is compact. In view of Lemma A.1, there exists a compact subset $X \subseteq M$ with $f(E) \subseteq C_X^\infty(M; K)$. Let $f_X: E \to C_X^\infty(M; K)$ denote the corestriction of f to $C_X^\infty(M; K)$. Since η_X is a topological embedding (Remark I.3), the map f_X is continuous. It obviously satisfies $f = \eta_X \circ f_X$.   □

Proposition A.3. *Let $X_n \subseteq M$ be compact with $X_n \subseteq X_{n+1}^0$ and $M = \bigcup_n X_n$. Then the map*

$$\varinjlim C_{X_n}^\infty(M; K) \to C_c^\infty(M; K)$$

is a weak homotopy equivalence. In particular $\pi_m(C_c^\infty(M; K)) \cong \varinjlim \pi_m(C_{X_n}^\infty(M; K))$ for each $m \in \mathbb{N}_0$.

Proof. Lemma A.2 first implies that each continuous map $f: \mathbb{S}^m \to C_c^\infty(M; K)$ factors through some inclusion $C_{X_n}^\infty(M; K) \to C_c^\infty(M; K)$. If two such maps f_1, f_2 are homotopic, then each homotopy $h: \mathbb{S}^m \times [0, 1] \to C_c^\infty(M; K)$ also factors through some group $C_{X_k}^\infty(M; K)$. This implies that the natural map

$$\varinjlim \pi_m(C_{X_n}^\infty(M; K)) \cong \pi_m(\varinjlim C_{X_n}^\infty(M; K)) \to \pi_m(C_c^\infty(M; K))$$

is bijective, i.e., that the continuous map $\varinjlim C_{X_n}^\infty(M; K) \to C_c^\infty(M; K)$ is a weak homotopy equivalence.   □

Remark A.4. A similar argument as the one leading to Proposition A.3 shows that the map

$$\varinjlim C_{X_n}(M; K) \to C_c(M; K)$$

is a weak homotopy equivalence.   □

If M and N are topological spaces, we write $[M, N]$ for the set of homotopy classes of continuous maps $f : M \to N$. If, in addition, $x_M \in M$ and $x_N \in N$ are base points, then $C_*(M, N) := \{f \in C(M, N) : f(x_M) = x_N\}$ denotes the set of base point preserving continuous maps and $[M, N]_*$ denotes the corresponding set of homotopy classes. We recall that if M is locally compact, then homotopy classes correspond to arc components in the compact open topology.

Eventually we want to show that the map

$$C_c^\infty(M; K) \to C_c(M; K)$$

is a weak homotopy equivalence, so that the homotopy groups of $C_c^\infty(M; K)$ are the limits of the corresponding homotopy groups of $C_X(M; K)$. These groups are more approachable since they are isomorphic to $C_*(X/\partial X; K)$, where $X/\partial X$ is a compact space, with the image of ∂X as the base point.

If M is a compact manifold with boundary, then the homotopy groups $\pi_m(C_*(M/\partial M; K))$ might be well accessible. Note that if ∂M is empty, then $C_*(M/\partial M; K)$ should be read as the group $C(M; K)$.

Lemma A.5. *Let $X_1, X_2 \subseteq M$ be compact subsets with $X_1 \subseteq X_2^0$ and $f \in C_{X_1}(M; K)$. Then every neighborhood of f contains a map f' in $C_{X_2}^\infty(M; K)$. The image of the homomorphism*

$$\eta : \pi_0(C_{X_2}^\infty(M; K)) \to \pi_0(C_{X_2}(M; K))$$

contains the image of $\pi_0(C_{X_1}(M; K))$. Moreover, if f is contained in $C_{X_1}(M; K)_e$, then we may choose $f' \in C_{X_2}^\infty(M; K)_e$.

Proof. The first assertion follows from [Ne02a, Th. A.3.7]. Since the groups $C_X(M; K)$ and $C_X^\infty(M; K)$ are Lie groups, their connected components are open, so that every connected component of $C_{X_2}(M; K)$ meeting $C_{X_1}(M; K)$ contains a smooth element.

If the map $f \in C_{X_1}(M; K)$ is sufficiently close to e in the sense that $f(M) \subseteq V$ for some chart e-neighborhood $V \subseteq K$ diffeomorphic to an open convex set, we find $f_1 \in C_{X_2}^\infty(M; K)$ with $f_1(M) \subseteq V$. Now any two smooth maps $f_1, f_2 \in C_{X_2}^\infty(M; K)$ with $f_j(M) \subseteq V$ are smoothly homotopic, hence contained in the same connected component of $C_{X_2}^\infty(M; K)$.

If $f \in C_{X_1}(M; K)$ is contained in the identity component, then there exists a continuous curve $\gamma : [0, 1] \to C_{X_1}(M; K)$ with $\gamma(0) = e$ and $\gamma(1) = f$. For a sufficiently fine subdivision $0 = t_0 < t_1 < \ldots < t_N = 1$ we now find smooth maps $f_j \in C_{X_2}^\infty(M; K)$ close to $\gamma(t_j)$ in the sense that $(f_j^{-1} \cdot \gamma(t_i))(M) \subseteq V$, where for $j < N$ the maps f_j and f_{j+1} are smoothly homotopic. Hence f_N is contained in the identity component of $C_{X_2}^\infty(M; K)$. $\square$

Lemma A.6. *The map $\iota : C_c^\infty(M; K) \to C_c(M; K)$ induces an isomorphism*

$$\pi_0(\iota) : \pi_0(C_c^\infty(M; K)) \to \pi_0(C_c(M; K)).$$

Proof. The surjectivity of $\pi_0(\iota)$ follows directly from Lemma A.5. If $f \in C_c^\infty(M; K)$ satisfies $[f] \in \ker \pi_0(\iota)$, then there exists a compact subset $X \subseteq M$ and a continuous map $\gamma : [0, 1] \to C_X(M; K)$ with $\gamma(0) = e$ and $\gamma(1) = f$ (Lemma A.2). Let $Y \subseteq M$ be a compact subset with $X \subseteq Y^0$. Then Lemma A.5 implies that we can approximate f by smooth functions f' in the identity component of $C_Y^\infty(M; K)$. It follows in particular that f is contained in the identity component of $C_Y^\infty(M; K)$, hence also in the identity component of $C_c^\infty(M; K)$. This shows that $\pi_0(\iota)$ is injective.   □

In M we fix a base point x_M and in any group we consider the unit element e as the base point. We write $C_*^\infty(M; K) \subseteq C^\infty(M; K)$ for the subgroup of base point preserving maps and observe that

$$C^\infty(M; K) \cong C_*^\infty(M; K) \rtimes K$$

as Lie groups, where we identify K with the subgroup of constant maps. This relation already leads to

$$\pi_k(C^\infty(M; K)) \cong \pi_k(C_*^\infty(M; K)) \times \pi_k(K), \quad k \in \mathbb{N}_0. \tag{A.1}$$

In particular we have

$$\pi_0(C^\infty(M; K)) \cong \pi_0(C_*^\infty(M; K))$$

if K is connected.

On the other hand, we have for each topological group G and each $k \in \mathbb{N}$ the relation

$$\pi_k(G) \cong \pi_0(C_*(\mathbb{S}^k, G)) = \pi_0(C_*(\mathbb{S}^k, G_e)) = \pi_0(C(\mathbb{S}^k, G_e)), \tag{A.2}$$

where G_e denotes the arc-component of the identity in G.

The following theorem is one of the two main results of this section. It provides a valuable tool to determine the homotopy groups of groups of smooth maps in terms of the corresponding groups of continuous maps.

Theorem A.7. *If M is a connected σ-compact finite-dimensional manifold and K a Lie group, then the inclusion $C_c^\infty(M; K) \to C_c(M; K)$ is a weak homotopy equivalence. If M is compact and $x_M \in M$ is a base point, then the inclusion*

$$C_*^\infty(M; K) \to C_*(M; K) := \{f \in C(M; K): f(x_M) = e\} \tag{A.3}$$

is a weak homotopy equivalence.

Proof. We have to show that the inclusion induces for each $k \in \mathbb{N}_0$ an isomorphism

$$\pi_k(C_c^\infty(M; K)) \to \pi_k(C_c(M; K)).$$

For $k = 0$ this is Lemma A.6. If M is compact, then

$$\pi_0(C_c^\infty(M; K)) = \pi_0(C^\infty(M; K)) \cong \pi_0(C_*^\infty(M; K)) \times \pi_0(K)$$

and
$$\pi_0(C_c(M; K)) = \pi_0(C(M; K)) \cong \pi_0(C_*(M; K)) \times \pi_0(K),$$

so that (A.3) follows from Lemma A.6. We only observe that if f_t is a homotopy between f_0 and f_1 in $C_c^\infty(M; K)$ and $x_M \in M$ is a base point, then $f_t(x) f_t(x_M)^{-1}$ is a homotopy between f_0 and f_1 in $C_*^\infty(M; K)$.

Next we assume that $k \geq 1$ and observe that the inclusions

$$C_*(\mathbb{S}^k, C_c^\infty(M; K)) = C_*(\mathbb{S}^k, C_c^\infty(M; K)_e)$$
$$\hookrightarrow C(\mathbb{S}^k, C_c^\infty(M; K)_e)$$
$$\hookrightarrow C(\mathbb{S}^k, C_c(M; K)_e)$$
$$\hookrightarrow C(\mathbb{S}^k, C_c(M; K))$$
$$\cong C_c(\mathbb{S}^k \times M; K)$$

are continuous homomorphisms of Lie groups, where

$$C(\mathbb{S}^k, C_c(M; K)_e) \hookrightarrow C(\mathbb{S}^k, C_c(M; K))$$

is an open embedding. For the group of connected components, we obtain for $k \geq 1$ with (A.2) the homomorphisms

$$\pi_k(C_c^\infty(M; K)) \cong \pi_0\big(C_*(\mathbb{S}^k, C_c^\infty(M; K))\big) \cong \pi_0\big(C(\mathbb{S}^k, C_c^\infty(M; K)_e)\big)$$
$$\to \pi_0\big(C(\mathbb{S}^k, C_c(M; K)_e)\big) \cong \pi_k\big(C_c(M; K)\big).$$

If $f : \mathbb{S}^k \times M \to K$ is a continuous map with compact support corresponding to an element of $C_*(\mathbb{S}^k; C_c(M; K)_e)$, then Lemma A.5 first implies that every neighborhood of f contains a smooth map with compact support. So every connected component of $C_c(\mathbb{S}^k \times M; K)$ contains an element of $C(\mathbb{S}^k, C_c^\infty(M; K))_e$ by the openness argument from above. This means that the homomorphism $\pi_k(C_c^\infty(M; K)) \to \pi_k(C_c(M; K))$ is surjective. To see that it is injective, suppose that $\sigma \in C(\mathbb{S}^k, C_c^\infty(M; K)_e)$ satisfies $\sigma \in C(\mathbb{S}^k, C_c(M; K)_e)_e \cong C_c(\mathbb{S}^k \times M; K)_e$. From Lemma A.6 we obtain

$$C_c^\infty(\mathbb{S}^k \times M; K) \cap C_c(\mathbb{S}^k \times M; K)_e \subseteq C_c^\infty(\mathbb{S}^k \times M; K)_e,$$

so that approximating σ by elements in $C_c^\infty(\mathbb{S}^k \times M; K)$ (Lemma A.5), we see that we may even approximate it by elements in $C_c^\infty(\mathbb{S}^k \times M; K)_e$, which implies that σ lies in the identity component of $C(\mathbb{S}^k, C_c^\infty(M; K)_e)$. This proves that the homomorphisms $\pi_k(C_c^\infty(M; K)) \to \pi_k(C_c(M; K))$, $k \in \mathbb{N}_0$, are isomorphisms.   □

Theorem A.7 can also be extended to non-connected manifolds M as follows. Let $M = \bigcup_{j \in J} M_j$ be the decomposition of M into connected components M_j. Here one can use

$$C_c(M; K) = \bigoplus_{j \in J} C_c(M_j; K),$$

and for each compact subset $X \subseteq M$ we have the finite sum decomposition

$$C_X(M; K) = \bigoplus_{X \cap M_j \neq \emptyset} C_{X \cap M_j}(M_j; K).$$

If M has only finitely many connected components, then there is no problem, but if M has infinitely many connected components, then one has to take the direct sum topology on $C_c(M; \mathfrak{k})$ into account and the corresponding Lie group topology on $C_c(M; K)$.

Lemma A.8 and Proposition A.9 provide additional information on the homotopy type of the topological current groups.

Lemma A.8. *If M is a locally compact space, then the inclusion $\eta \colon C_c(M; K) \to C_0(M; K)$ induces an isomorphism $\pi_0(C_c(M; K)) \to \pi_0(C_0(M; K))$.*

Proof. Let $f \in C_0(M; K)$. Then there exists a compact subset $X \subseteq M$ such that $f(M \setminus X)$ is contained in an identity neighborhood of K which is diffeomorphic to a convex 0-neighborhood U in $\mathfrak{k}$, where 0 corresponds to $e \in K$. Using a continuous function $h \in C_c(M; \mathbb{R})$ which is 1 on X and satisfies $h(M) \subseteq [0, 1]$, we define a function $\widetilde{f} \in C_c(M; K)$ by $\widetilde{f} = f$ on X and $\widetilde{f} = hf$ on $M \setminus X$, where we consider $f|_{M \setminus X}$ as a function with values in U. Then

$$F \colon M \times [0, 1] \to K, \quad F(x, t) := \begin{cases} f(x) & \text{for } x \in X \\ (t + (1 - t)h(x))f(x) & \text{for } x \in M \setminus X \end{cases}$$

is a homotopy between f and $\widetilde{f}$, and we see that $\pi_0(\eta)$ is surjective.

A similar argument shows that for $f, g \in C_c(M; K)$ any path joining f and g in $C_0(M; K)$ can be deformed to a path lying completely inside of $C_X(M; K)$ for a compact subset X of M. Therefore $\pi_0(\eta)$ is injective. $\square$

Proposition A.9. *If M is a locally compact space, then the inclusion $\eta \colon C_c(M; K) \to C_0(M; K)$ is a weak homotopy equivalence.*

Proof. Let $M_\infty = M \cup \{\infty\}$ denote the one-point compactification of M. For every compact space X we have an embedding of topological groups

$$C(X, C_0(M; K)) \cong C(X, C_*(M_\infty; K)) \hookrightarrow C(X, C(M_\infty; K)) \cong C(X \times M_\infty; K),$$

which easily leads to the isomorphism

$$C(X, C_0(M; K)) \cong C_0(X \times M; K).$$

In view of Lemma A.8, there exists for each $f \in C_0(X \times M; K)$ some compact subset $Y \subseteq M$ and a continuous map $f_Y \in C(X, C_Y(M; K)) \subseteq C(X \times Y; K)$ homotopic to f. The same argument applies to $[0, 1] \times X$ instead of X, so that we see that the inclusion $C_c(M; K) \to C_0(M; K)$ induces a bijection $[X, C_c(M; K)] \to [X, C_0(M; K)]$ on the level of homotopy classes.

Applying this to $X := \mathbb{S}^k, k \in \mathbb{N}$, we obtain with Lemma A.8 that the natural map

$$\pi_k(C_c(M; K)) \cong [\mathbb{S}^k, C_c(M; K)]_* \cong [\mathbb{S}^k, C_c(M; K)_e] \to [\mathbb{S}^k, C_0(M; K)_e]$$
$$\cong [\mathbb{S}^k, C_0(M; K)]_* \cong \pi_k(C_0(M; K))$$

is bijective, hence an isomorphism of groups. $\square$

Theorem A.10. *For each σ-compact connected finite-dimensional manifold M and each Lie group K the inclusion map*

$$C_c^\infty(M; K) \to C_0(M; K) \cong C_*(M_\infty; K)$$

is a weak homotopy equivalence.

Proof. We only have to combine Proposition A.9 with Theorem A.7. $\square$

Example A.11. For $M = \mathbb{R}^n$ we obtain with Theorem A.10 for each $k \in \mathbb{N}_0$:

$$\pi_k(C_c^\infty(\mathbb{R}^n; K)) \cong \pi_k(C_*(\mathbb{R}^n_\infty; K)) \cong \pi_k(C_*(\mathbb{S}^n; K)) \cong \pi_{k+n}(K). \square$$

Lemma A.12. *Let $\phi: N \to M$ be a smooth proper map.*

(i)   *The map*

$$\phi_K: C_c^\infty(M; K) \to C_c^\infty(N; K), \quad f \mapsto f \circ \phi$$

   is a morphism of Lie groups.

(ii)   *Let $\phi_\infty: M_\infty \to N_\infty$ denote the continuous extension of ϕ to the one-point compactifications. Then for each $k \in \mathbb{N}_0$ the map*

$$\pi_k(\phi_K): \pi_k(C_c^\infty(M; K)) \to \pi_k(C_c^\infty(N; K))$$

   only depends on the homotopy class of ϕ_∞ in the set $[M_\infty, N_\infty]_*$ of pointed homotopy classes.

Proof. (i) It is clear that ϕ_K maps $C_c^\infty(N; K)$ into $C_c^\infty(M; K)$ and that it is a group homomorphism. It therefore suffices to show smoothness in some identity neighborhood.

   Let $U \subseteq K$ be an open identity neighborhood and $\psi: U \to W$ a chart of K where $W \subseteq \mathfrak{k}$ is an open subset and $\psi(e) = 0$. Then there exists an open 0-neighborhood $V \subseteq W$ such that

$$C_c^\infty(N, W) := \{f \in C_c^\infty(N; K): f(N) \subseteq \psi^{-1}(V)\}$$

is an open subset of $C_c^\infty(N; K)$ ([Gl01b]). Now it suffices to see that the map

$$C_c^\infty(M, V) \to C_c^\infty(N, V), \quad f \mapsto f \circ \phi$$

is smooth. As this map is the restriction of a linear map, we only have to show that it is continuous.

For each compact subset $X \subseteq M$ we have

$$C_X^\infty(M; K) \circ \phi \subseteq C_{\phi^{-1}(X)}^\infty(M; K),$$

so that the assertion follows from the observation that for each $n \in \mathbb{N}$ the map $d^n(f \circ \phi)$ depends continuously on f, when considered as an element of $C(T^n(N), \mathfrak{k})_c$ (cf. Definition I.2).

(ii) Let $\eta_M: C_c^\infty(M; K) \to C_*(M_\infty; K)$ denote the natural inclusion. Then $\eta_N \circ \phi_K = \widetilde{\phi}_K \circ \eta_M$ holds with

$$\widetilde{\phi}_K: C_*(M_\infty; K) \to C_*(N_\infty; K), \quad f \mapsto f \circ \phi.$$

We know from Theorem A.10 that the maps η_M and η_N are weak homotopy equivalences. Therefore it suffices to show that the maps $\pi_k(\widetilde{\phi}_K)$ only depend on the homotopy class of ϕ. If $\phi, \psi: M \to N$ are proper and smooth such that ϕ_∞ and ψ_∞ are homotopic, then it is easy to see that the maps $\widetilde{\phi}_K$ and $\widetilde{\psi}_K$ are homotopic, hence induce the same homomorphisms on homotopy groups. $\qquad \square$

Homotopy groups of groups defined by vanishing conditions

In this subsection we discuss the other major class of groups of smooth maps $C^\infty(M, S; K)$. Theorem A.13 is a variant of Theorem A.7 for this context.

Theorem A.13. *Let M be a compact manifold, $S \subseteq M$ a closed subset, and let $C^\infty(M, S; K)$ be the subgroup of $C^\infty(M; K)$ consisting of all smooth maps vanishing together with all their partial derivatives on S. Then the inclusion*

$$\eta: C_c^\infty(M \setminus S; K) \to C^\infty(M, S; K)$$

is a weak homotopy equivalence.

Proof. As M is compact, the group $C^\infty(M, S; K)$, when considered as a group of maps $M \setminus S \to K$, is contained in $C_0(M \setminus S; K)$. The inclusion $C_c^\infty(M \setminus S; K) \to C_0(M \setminus S; K)$ is a weak homotopy equivalence by Theorem A.10, so that all the maps $\pi_k(\eta), k \in \mathbb{N}_0$, are injective. It therefore remains to show that they are also surjective.

So let

$$\sigma \in C_*(\mathbb{S}^k, C^\infty(M, S; K)) \subseteq C_*(\mathbb{S}^k, C_0(M \setminus S; K)) \subseteq C_0(\mathbb{S}^k \times (M \setminus S); K).$$

Then there exists a compact subset $X \subseteq M \setminus S$ such that $\sigma(\mathbb{S}^k \times (M \setminus X^0))$ is contained in an identity neighborhood of K which is diffeomorphic to a convex 0-neighborhood U in $\mathfrak{k}$, where 0 corresponds to $e \in K$. Let $\phi: U \to \phi(U) \subseteq K$ denote the corresponding chart and $h \in C_c^\infty(M \setminus S; \mathbb{R})$ with $h(X) = \{1\}$ and $h(M) \subseteq [0, 1]$. We now define

$$\widetilde{\sigma}: \mathbb{S}^k \times M \to K, \quad \widetilde{\sigma}(t, x) := \begin{cases} \sigma(t, x) & \text{for } x \in X \\ \phi\big(h(x)\phi^{-1}(\sigma(t, x))\big) & \text{for } x \notin X. \end{cases}$$

As $\sigma(\mathbb{S}^k \times (M \setminus X^0))$ is a compact subset of $\phi(U)$, it easily follows that $\tilde{\sigma}$ is continuous and that $t \mapsto \tilde{\sigma}(t, \cdot)$ yields a continuous map $\mathbb{S}^k \to C_c^\infty(M \setminus S; K)$. In fact, the support of each map $\tilde{\sigma}(t, \cdot)$ is contained in the support of h. Moreover,

$$F : [0, 1] \times \mathbb{S}^k \times M \to K,$$

$$F(s, t, x) := \begin{cases} \sigma(t, x) & \text{for } x \in X \\ \phi\big([sh(x) + (1 - s)] \cdot \phi^{-1}(\sigma(t, x))\big) & \text{for } x \notin X \end{cases}$$

is a homotopy between σ and $\tilde{\sigma}$ preserving base points. This implies that the map $\pi_k(\eta)$ is surjective. $\qquad\square$

Observe that Theorem A.13 does not imply that $C_c^\infty(M \setminus S; K)$ is dense in $C^\infty(M, S; K)$. This will be shown in Theorem A.18 below.

Corollary A.14. *Let M be a compact manifold and $\emptyset \neq S \subseteq M$ a closed subset. Then the inclusion*

$$\zeta : C^\infty(M, S; K) \to C_0(M \setminus S; K) \cong C_*(M/S; K)$$

is a weak homotopy equivalence.

Proof. According to Theorem A.10, the inclusion $C_c^\infty(M \setminus S; K) \to C_0(M \setminus S; K)$ is a weak homotopy equivalence, and this map is the composition of ζ and the inclusion map η from Theorem A.10. This implies that ζ also is a weak homotopy equivalence. $\qquad\square$

Corollary A.15. *For a compact manifold M and $k \in \mathbb{N}_0$ we have*

$$\pi_k(C^\infty(M, S; K)) \cong \pi_k(C_*(M/S; K))$$

and in particular

$$\pi_k(C^\infty(I, \partial I; K)) \cong \pi_k(C_*(\mathbb{S}^1; K)) \cong \pi_{k+1}(K).$$

Proof. For $M = I$ and $S = \partial I$ we have $M/S \cong \mathbb{S}^1$ and therefore

$$\pi_k(C^\infty(I, \partial I; K)) \cong \pi_k(C_*(\mathbb{S}^1; K)) \cong \pi_{k+1}(K). \qquad\square$$

Lemma A.16. *For each $\alpha \in C^\infty((M', S'), (M, S))$ let*

$$\alpha_K : C^\infty(M, S; K) \to C^\infty(M', S'; K), \quad f \mapsto f \circ \alpha.$$

Then α_K is a homomorphism of Lie groups and the homomorphisms $\pi_k(\alpha_K)$ only depend on the homotopy class of α in the space $C((M', S'), (M, S))$.

Proof. First we observe that the chain rule for Taylor expansions implies that α_K does indeed map $C^\infty(M, S; K)$ into $C^\infty(M', S'; K)$. That α_K is a homomorphism of Lie groups follows by similar arguments as in the proof of Lemma A.12 (i).

Viewing α as a continuous map $(M', S') \to (M, S)$ of space pairs, we see that it induces a continuous map

$$\alpha^*: C_*(M/S; K) \to C_*(M'/S'; K), \quad f \mapsto f \circ \alpha.$$

Since the inclusion $C^\infty(M, S; K) \to C_*(M/S; K)$ is a weak homotopy equivalence (Corollary A.14), the maps $\pi_k(\alpha_K)$ are conjugate to the maps $\pi_k(\alpha^*)$. It is easy to see that $\pi_k(\alpha^*)$ only depends on the homotopy class of α because for each continuous map $\sigma: \mathbb{S}^k \to C_*(M'/S'; K)$ the map $\alpha^* \circ \sigma: \mathbb{S}^k \to C_*(M/S; K)$ depends continuously on α. $\qquad\square$

Lemma A.17. *For each locally convex space Y the space $C^\infty(M, S; Y)$ is a closed subspace of $C^\infty(M; Y)$ invariant under multiplication with elements of $C^\infty(M; \mathbb{R})$.*

Proof. This follows directly from the Leibniz formula for the higher partial derivatives of a product of two functions. $\qquad\square$

Theorem A.18 (Approximation Theorem). *If M is compact, then $C_c^\infty(M \setminus S; K)$ is dense in the Lie group $C^\infty(M, S; K)$.*

Proof. First we reduce the problem to the assertion that for the Lie algebra $\mathfrak{k}$ of K the subspace $C_c^\infty(M \setminus S; \mathfrak{k})$ is dense in $C^\infty(M, S; \mathfrak{k})$.

Let $U \subseteq K$ be an open identity neighborhhod and $\phi: V \to U$ a chart of K with $V \subseteq \mathfrak{k}$ an open convex subset and $\phi(0) = e$. Then $\{f \in C^\infty(M, S; K): f(M) \subseteq U\}$ is an open subset of $C^\infty(M, S; K)$ because it is already open in the compact open topology. We choose an open convex 0-neighborhood $V_1 \subseteq V$ with $\phi(V_1)^{-1}\phi(V_1) \subseteq \phi(V)$.

Let $f \in C^\infty(M, S; K)$. As f vanishes on S, the set $f^{-1}(\phi(V_1))$ is an open subset of M containing S. Therefore its complement X is a compact subset of $M \setminus S$. Arguing as in the proof of Lemma A.8, we find a function $\tilde{f} \in C_c^\infty(M \setminus S; K)$ with $\tilde{f}|_X = f|_X$ and $\tilde{f}(M \setminus X) \subseteq \phi(V_1)$. Now it suffices to show that $h := f^{-1}\tilde{f}$, whose values are contained in $\phi(V_1)^{-1}\phi(V_1) \subseteq \phi(V)$, is contained in the closure of $C_c^\infty(M \setminus S; K)$. As $\phi^{-1} \circ h: M \to \mathfrak{k}$ is a well-defined smooth map, we see that it suffices to prove the theorem for $\mathfrak{k}$ instead of K. In this setting we have to show that if $V \subseteq \mathfrak{k}$ is an open convex 0-neighborhood with $f(M) \subseteq V$, then f can be approximated by functions in $C_c^\infty(M; \mathfrak{k})$ whose values lie in V.

Let $f \in C^\infty(M, S; \mathfrak{k})$. Using Lemma A.17 and a smooth partition of unity on M, we may assume that the support of f lies in a coordinate neighborhood which we may identify with $\mathbb{R}^n$. We are therefore led to the following situation. We consider a smooth function $f \in C_c^\infty(\mathbb{R}^n; \mathfrak{k})$ all of whose derivatives vanish on the closed subset $S \subseteq \mathbb{R}^n$, and we are looking for a sequence of functions with compact support in $\mathbb{R}^n \setminus S$ converging to f in $C^\infty(\mathbb{R}^n; \mathfrak{k})$ whose supports are uniformly contained in a compact set. The existence of such a sequence is proved in Proposition A.22 below. $\qquad\square$

An Approximation Lemma

Let $\emptyset \neq S \subseteq \mathbb{R}^d$ be a closed subset, Y a Banach space, and $f \in C_X^\infty(\mathbb{R}^d; Y)$ for a compact subset $X \subseteq \mathbb{R}^d$ such that f and all its partial derivatives vanish on $S \cap X$. We want to see that f is contained in the closure of the subspace $C_c^\infty(\mathbb{R}^d \setminus S; Y) \cap C_X^\infty(\mathbb{R}^d; Y)$. In the following $d(S, x)$ denotes the euclidean distance of the set S and x. We write $\| \cdot \|$ for the euclidean norm on $\mathbb{R}^d$.

Lemma A.19. *For each $k \in \mathbb{N}$ and each $f \in C_c^\infty(\mathbb{R}^d, S; Y)$ there exists a constant $C_k > 0$ with*

$$\| f(x) \| \leq C_k d(S, x)^k.$$

Proof. We prove the assertion by induction over k. For $k = 0$ it follows from the compactness of the support of f.

Now we assume that the assertion holds for $k \in \mathbb{N}_0$. Let $h \in C_c^\infty(\mathbb{R}^d, S; Y)$. Then the induction hypothesis applies to $dh \in C_c^\infty(\mathbb{R}^d, S; \mathrm{Lin}(\mathbb{R}^d; Y))$, and we obtain a constant D_k with $\| dh(x) \| \leq D_k d(S, x)^k$ for all $x \in \mathbb{R}^d$. For $x \in \mathbb{R}^d$ we find an $x_0 \in S$ with $\| x - x_0 \| \leq 2d(S, x)$. Then

$$h(x) = h(x_0) + \int_0^1 dh(x_0 + t(x - x_0))(x - x_0)\, dt = \int_0^1 dh(x_0 + t(x - x_0))(x - x_0)\, dt$$

leads to

$$\| h(x) \| \leq \| x - x_0 \| \sup_{0 \leq t \leq 1} \| dh(x_0 + t(x - x_0)) \|$$

$$\leq 2d(S, x) D_k \sup_{0 \leq t \leq 1} d(S, x_0 + t(x - x_0))^k$$

$$\leq 2 D_k d(S, x) 2^k d(S, x)^k = 2^{k+1} D_k d(S, x)^{k+1}.$$

This completes the induction, and hence the proof of the lemma. $\qquad\square$

Now let δ be a smooth function supported in the closed unit ball $B_1(0)$ in $\mathbb{R}^d$ with $\int_{\mathbb{R}^d} \delta(x)\, dx = 1$ and $\mathrm{im}(\delta) \subseteq [0, 1]$. We define

$$\delta_n(x) := n^d \delta(nx)$$

and observe that these functions form a smooth Dirac sequence. For each multiindex $J = (j_1, \ldots, j_d) \in \mathbb{N}_0^d$ we have

$$\| \partial^J \delta_n \|_\infty = n^{d+|J|} \| \partial^J \delta \|_\infty.$$

Let $S_n := \left\{ x \in \mathbb{R}^d : d(S, x) \leq \frac{2}{n} \right\}$ and

$$\chi_{S_n}(x) := \begin{cases} 1 & \text{for } x \in S_n \\ 0 & \text{for } x \notin S_n \end{cases}$$

the characteristic function of S_n. Then we define

$$\phi_n(x) := 1 - (\delta_n * \chi_{S_n})(x) = 1 - \int_{x-S_n} \delta_n(y)\, dy \in [0, 1].$$

Then each function ϕ_n is smooth with $\phi_n(x) = 1$ for $d(S, x) \geq \frac{3}{n}$ and $\phi_n(x) = 0$ for $d(S, x) \leq \frac{1}{n}$.

Lemma A.20. *For each multiindex J there exists a constant D_J such that*

$$\|\partial^J \phi_n(x)\| \leq D_J d(S, x)^{-|J|}, \quad x \in \mathbb{R}^d, n \in \mathbb{N}.$$

Proof. For $|J| = 0$ the assertion follows from $\operatorname{im}(\phi_n) \subseteq [0, 1]$.

Suppose that $|J| > 0$ and that $d(S, x) \in \left[\frac{1}{n}, \frac{3}{n}\right]$. Otherwise $\partial^J \phi_n(x)$ vanishes anyway. Then we have

$$\|\partial^J \phi_n(x)\| = \|((\partial^J \delta_n) * \chi_{S_n})(x)\| \leq \operatorname{vol}(B_{\frac{1}{n}}(0))\|\partial^J \delta_n\|_\infty$$

$$\leq Cn^{-d} n^{d+|J|}\|\partial^J \delta\|_\infty = Cn^{|J|}\|\partial^J \delta\|_\infty \leq C3^{|J|} d(S, x)^{-|J|}\|\partial^J \delta\|_\infty.$$

$\square$

Lemma A.21. *For all multiindices J with $|J| > 0$ we have uniformly $\partial^J \phi_n \cdot f \to 0$.*

Proof. Combining Lemma A.19 and A.20, we get for each $k \in \mathbb{N}$ a constant C_k with

$$\|(\partial^J \phi_n(x)) f(x)\| \leq C_k d(S, x)^{-|J|} d(S, x)^{|J|+k} = C_k d(S, x)^k.$$

As $\partial^J \phi_n(x) = 0$ for $d(S, x) \geq \frac{3}{n}$ (here we need $|J| > 0$), this leads to

$$\|(\partial^J \phi_n(x)) f(x)\| \leq C_k 3^k n^{-k}$$

for all $x \in \mathbb{R}^d$, and this implies the assertion. $\square$

Proposition A.22. *For each locally convex space Y and $f \in C_c^\infty(\mathbb{R}^d, S; Y)$ we have $\phi_n f \to f$ in $C^\infty(\mathbb{R}^d; Y)$.*

Proof. As every locally convex space can be embedded into a product of Banach spaces, it suffices to assume that Y is a Banach space. Since the supports of the functions $\phi_n f$ and f are contained in one compact subset of $\mathbb{R}^d$, we have to show $\|\partial^J (\phi_n f - f)\|_\infty \to 0$ for all multiindices J.

For $|J| = 0$ this follows easily from the support properties of ϕ_n and $\|f(x)\| \leq Cd(S, x)$.

Next we note that for each multiindex J the function $\partial^J f$ also has the property that all its partial derivatives vanish on S. Therefore Lemma A.21 implies that $\partial^{J'} \phi_n \cdot \partial^J f \to 0$ uniformly whenever $|J'| > 0$. In view of the Leibniz rule, the problem reduces to showing that $\phi_n \partial^J f$ converges uniformly to $\partial^J f$, but this follows from the case $|J| = 0$, applied to $\partial^J f$ instead of f. $\square$

Appendix B. Locally convex direct limit spaces

In this section we discuss the discreteness of certain subgroups of direct limits of locally convex spaces. In this paper we only use Lemma B.4. Nevertheless Proposition B.3 provides a much more direct way to prove the discreteness of the groups $H^k(M; Y, \Gamma)$ if Y is finite-dimensional and $\Gamma \subseteq Y$ is a discrete subgroup (cf. Corollary IV.21).

Lemma B.1. *If X is a locally convex space, $Y \subseteq X$ a closed subspace and $F \subseteq X$ a finite-dimensional subspace complementing Y, then $X \cong Y \oplus F$ as topological vector spaces.*

Proof. The quotient map $q \colon X \to X/Y$ induces an isomorphism $q\,|_F \colon F \to X/Y$. Hence q has a continuous linear section $\sigma \colon X/Y \to X$ whose range is F, and therefore the addition map $a \colon Y \times F \to X$ is a topological isomorphism because $a^{-1}(x) = \bigl(x - \sigma(q(x)), \sigma(q(x))\bigr)$ is continuous. $\qquad\square$

Lemma B.2. *Let X be a locally convex space which is the locally convex direct limit of the subspaces X_n, $n \in \mathbb{N}$, where each X_n is a closed subspace of X_{n+1}. Further let $F \subseteq X$ be a subspace such that for each $n \in \mathbb{N}$ the intersection $F_n := F \cap X_n$ is finite-dimensional. Then the following assertions hold:*

(i)   *There exists a continuous linear projection $p \colon X \to F$ with $p(X_n) = F_n$ for each $n \in \mathbb{N}$. In particular we have $X \cong \ker p \oplus F$.*

(ii) *F is closed.*

(iii) *F is the topological direct limit of the subspaces F_n, $n \in \mathbb{N}$, which means that F carries the finest locally convex topology.*

Proof. (i) We argue by induction. As F_1 is finite-dimensional, the Hahn–Banach Theorem yields a continuous extension $p_1 \colon X_1 \to F_1$ of the identity map id_{F_1}. Then p_1 can be viewed as a continuous projection of X_1 to F_1.

Now let $n \in \mathbb{N}$ and assume that $p_n \colon X_n \to F_n$ is a continuous projection. Then we choose a complement E_{n+1} of F_n in F_{n+1}. As X_n is a closed subspace of the locally convex space $X_n + F_{n+1} = X_n \oplus E_{n+1}$, it follows from Lemma B.1 that $X_n + F_{n+1} \cong X_n \oplus E_{n+1}$ as topological vector spaces. The linear map $q_n := p_n \oplus \mathrm{id}_{E_{n+1}}$ is a continuous projection of $X_n + F_{n+1}$ onto F_{n+1}. We use the Hahn–Banach Theorem again to extend q_n to a continuous linear map $p_{n+1} \colon X_{n+1} \to F_{n+1}$ which then also is a continuous projection. We thus obtain a sequence $(p_n)_{n \in \mathbb{N}}$ of continuous linear maps $p_n \colon X_n \to F$ with $p_{n+1}\,|_{X_n} = p_n$. Now the universal property of X yields the existence of a continuous linear map $p \colon X \to F$ with $p\,|_{X_n} = p_n$ for each $n \in \mathbb{N}$. As $p\,|_F = \mathrm{id}_F$, we are done.

(ii) follows from (i).

(iii) Let Z be a locally convex space and $f \colon F \to Z$ be a linear map. We claim that f is continuous. To this end, we consider the map $h := f \circ p \colon X \to Z$. Then $h\,|_{X_n} = (f\,|_{F_n}) \circ p_n$, and p_n is continuous, as well as the map $f\,|_{F_n}$ on the finite-dimensional vector space F_n. Therefore all the restrictions $h|_{X_n}$ are continuous, and we conclude that h is continuous, which in turn implies that f is continuous. The fact that all linear maps from F to locally convex spaces are continuous shows that F carries the finest locally convex topology. Furthermore, F is countably dimensional because all the spaces F_n are finite-dimensional. Using [KK63], we now conclude that the topology on F coincides with the finite open topology, i.e., the direct limit topology with respect to the directed system of all finite-dimensional subspaces. As the sequence $(F_n)_{n\in\mathbb{N}}$ is cofinal, this topology coincides with the direct limit topology with respect to the sequence $(F_n)_{n\in\mathbb{N}}$.   □

Proposition B.3. *Let X be a locally convex space which is the locally convex direct limit of the subspaces $X_n, n \in \mathbb{N}$, with $X_n \subseteq X_{n+1}$, where X_n is closed in X_{n+1}. Let further $\Gamma \subseteq X$ be a subgroup such that for each $n \in \mathbb{N}$ the group $\Gamma \cap X_n$ is discrete and finitely generated. Then Γ is a discrete subgroup of X.*

Proof. For each $n \in \mathbb{N}$ we consider the finite-dimensional subspace $F_n := \operatorname{span} \Gamma_n$ for the discrete finitely generated subgroup $\Gamma_n := \Gamma \cap X_n$ of X_n. Let $F := \bigcup_n F_n = \operatorname{span} \Gamma$. We claim that $F_n = F \cap X_n$ holds for each $n \in \mathbb{N}$. Fix $n, m \in \mathbb{N}$ with $n < m$. As Γ_n is discrete in the finite-dimensional space F_n, there exists a basis B_n of F_n with $\Gamma_n = \operatorname{span}_{\mathbb{Z}} B_n$. Further $\Gamma_n = \Gamma \cap X_n = \Gamma_m \cap X_n$ is a pure subgroup of Γ_m, so that Γ_m/Γ_n is a free abelian group. Hence we find a subset $C_m \subseteq \Gamma_m$ such that the image of C_m is a basis in $(F_m + X_n)/X_n \cong F_m/F_m \cap X_n$ generating the subgroup $(\Gamma_m + X_n)/X_n \cong \Gamma_m/\Gamma_n$. Now $B_m := B_n \cup C_m$ is a basis of F_m with $\Gamma_m = \operatorname{span}_{\mathbb{Z}} B_m$. In particular, it follows that $F_m \cap X_n = \operatorname{span}_{\mathbb{R}} B_n = F_n$. As m was arbitrary, we conclude that $F \cap X_n = F_n$.

Next Lemma B.2 applies to the subspace $F \subseteq X$ and shows that F is closed and carries the finite open topology. Let $O := (F \setminus \Gamma) \cup \{0\}$. For each $n \in \mathbb{N}$ we then have $O \cap F_n = (F_n \setminus \Gamma_n) \cup \{0\}$, which is an open set because Γ_n is discrete in F_n. Therefore O is an open subset of F (Lemma B.2(iii)), and since F carries the subspace topology of X, there exists an open subset $O_X \subseteq X$ with $O_X \cap F = O$. Now O_X is an open 0-neighborhood in X with $O_X \cap \Gamma = \{0\}$. This shows that Γ is discrete.   □

Lemma B.4. *Let $X = \varinjlim X_j$ be a locally convex direct limit of the spaces X_j.*

(i)   *If $F \subseteq X$ is a closed subspace, then $X/F \cong \varinjlim X_j/(F \cap X_j)$.*

(ii) *A subspace $F \subseteq X$ is closed if and only if all intersections $F \cap X_j$ are closed.*

Proof. (i) (cf. [Kö79, p.42]) Since F is closed, all the spaces $F_j := F \cap X_j$ are closed. Let $Z := \varinjlim X_j/F_j$ denote the locally convex direct limit of the spaces X_j/F_j. Then we have natural continuous maps $\phi_j : X_j/F_j \to X/F$ which define a continuous linear map $\phi : Z \to X/F$. On the other hand the continuous linear maps $X_j \to Z$ combine to a continuous linear map $X \to Z$ which then factors through a continuous linear map $\psi : X/F \to Z$. Now $\phi \circ \psi = \operatorname{id}_{X/F}$ and $\psi \circ \phi = \operatorname{id}_Z$ imply (i).

(ii) If F is closed, then the subspaces $F \cap X_j$ are trivially closed in X_j. If, conversely, this condition is satisfied, then we can form the locally convex direct limit space $Z := \varinjlim X_j/(F \cap X_j)$. The natural maps $X_j \to Z$ are continuous, hence combine to a continuous map $X \to Z$ whose kernel F is a closed subspace.   □

Problem B.1. Does Proposition B.3 also hold without the assumption that the groups $\Gamma \cap X_n$ are finitely generated? If this is true, then the proof of the discreteness of the groups $H^1_{\mathrm{dR},c}(M; \Gamma)$ in Section IV would be much easier because we would not need the complicated approximation procedure from Section III.   □

References

[Br93]   Bredon, G. E., *Topology and Geometry*, Grad. Texts in Math. 139, Springer-Verlag, Berlin 1993.

[Br97]   —, *Sheaf Theory*, Grad. Texts in Math. 170, Springer-Verlag, Berlin 1997.

[BJ73]   Bröcker, Th., and K. Jänich, *Einführung in die Differentialtopologie*, Springer-Verlag, Berlin 1973.

[tD00]   tom Dieck, T., *Topologie*, 2nd Ed., de Gruyter, Berlin–New York, 2000.

[Fe88]   Feigin, B. L., On the cohomology of the Lie algebra of vector fields and the current algebra, *Selecta Math. Soviet.* 7 (1) (1988), 49–62.

[Fu70]   Fuchs, L., *Infinite Abelian Groups, I*, Academic Press, New York 1970.

[Gl01a]   Glöckner, H., Infinite-dimensional Lie groups without completeness restrictions, in *Geometry and Analysis on Finite- and Infinite-Dimensional Lie Groups* (A. Strasburger et al, eds.), Banach Center Publications 55, Warszawa 2002, 43–59.

[Gl01b]   —, Lie group structures on quotient groups and universal complexifications for infinite-dimensional Lie groups, *J. Funct. Anal.* 195 (2001), 347–409.

[Gl01c]   —, Algebras whose groups of units are Lie groups, *Studia Math.* 153 (2002), 147–177.

[Gl01d]   —, *Differentiable mappings between spaces of sections*, submitted, Nov. 6, 2001.

[Gl02]   —, Patched locally convex spaces, almost local mappings, and diffeomorphism groups of non-compact manifolds, Manuscript, TU Darmstadt, 26.6.02.

[Gr55]   Grothendieck, A., *Produits tensoriels topologiques et espaces nucléaires*, Mem. Amer. Math. Soc. 16 (1955), Providence, R.I., 1955.

[He89]   Hervé, M., *Analyticity in Infinite Dimensional Spaces*, de Gruyter Stud. Math. 10, Walter de Gruyter, Berlin–New York 1989.

[KK63]   Kakutani, S., and V. Klee, The finite topology on a linear space, *Arch. Math.* 14 (1963), 55–58.

[Kö69]   Köthe, G., *Topological Vector Spaces I*, Grundlehren Math. Wiss. 159, Springer-Verlag, 1969.

[Kö79]   —, *Topological Vector Spaces II*, Grundlehren Math. Wiss. 237, Springer-Verlag, 1979.

[KM97]   Kriegl, A., and P. Michor, *The Convenient Setting of Global Analysis*, Math. Surveys and Monographs 53, Amer. Math. Soc., Providence, R.I., 1997.

[Lo98]   Loday, J.-L., *Cyclic Homology*, Grundlehren Math. Wiss. 301, Springer-Verlag, Berlin 1998.

[LMNS96]   Losev, A., G. Moore, N. Nekrasov, and S. Shatashvili, Four-dimensional avatars of two-dimensional RCFT, in *Strings 95* (Los Angeles, CA, 1995), World Scientific Publishing, River Edge, N.J., 1996, 336–362.

[LMNS98]   —, Central extensions of gauge groups revisited, *Selecta Math.* (N.S.) 4 (1998), 117–123.

[Ma02]   Maier, P., Central extensions of topological current algebras, in *Geometry and Analysis on Finite- and Infinite-Dimensional Lie Groups* (A. Strasburger et al., eds.), Banach Center Publications 55, Warszawa 2002, 61–76.

[MN02]   Maier, P., and K. - H. Neeb, *Central extensions of current groups*, *Math. Ann.* 362 (2003), 367–415.

[Mic87]   Mickelsson, J., Kac-Moody groups, topology of the Dirac determinant bundle, and fermionization, *Commun. Math. Phys.* 110 (1987), 173–183.

[Mic89]   —, *Current Algebras and Groups*, Plenum Press, New York 1989.

[Mil83]   Milnor, J., Remarks on infinite-dimensional Lie groups, in *Relativity, groups and topology II* (B. DeWitt ed.), Les Houches 1983, North-Holland, Amsterdam 1984, 1007–1057.

[Ne01a]   Neeb, K.-H., *Borel-Weil Theory for Loop Groups*, in *Infinite Dimensional Kähler Manifolds* (A. Huckleberry, T. Wurzbacher, eds.), DMV-Seminar 31, Birkhäuser Verlag, 2001, 179–222.

[Ne01b]   —, *Representations of infinite dimensional groups*, in *Infinite Dimensional Kähler Manifolds* (A. Huckleberry, T. Wurzbacher, eds.), DMV-Seminar 31, Birkhäuser Verlag, 2001, 131–178.

[Ne01c]   —, Universal central extensions of Lie groups, *Acta Appl. Math.* 73 (2002), 175–219.

[Ne02a]   —, Central extensions of infinite-dimensional Lie groups, *Ann. Inst. Fourier* 52 (2002), 1365–1442.

[Ne02b]   —, Lie theoretic K-groups and Steinberg–Lie groups, in preparation.

[Ne03]   —, Locally convex root graded Lie algebras, *Travaux Math.* 14 (2003), 25–120.

[NW03]   Neeb, K.-H., and F. Wagemann, Lie groups structures on the group of smooth maps from a non-compact manifold into a Lie group, in preparation.

[Pa66]   Palais, R. S., Homotopy theory of infinite dimensional manifolds, *Topology* 5 (1965), 1–16.

[PS86]   Pressley, A., and G. Segal, *Loop Groups*, Oxford University Press, Oxford 1986.

[Sch75]   Schubert, H., *Topologie*, Teubner Verlag, Stuttgart 1975.

[Se76]   Seligman, G. B., *Rational Methods in Lie Algebras*, Lecture Notes in Pure Appl. Math. 17, Marcel Dekker, New York 1976.

[SS78]   Steen, L. A., and J. A. Seebach, *Counterexamples in Topology*, Dover Publ. Inc., New York 1978.

[Tr67]   Treves, F., *Topological Vector Spaces, Distributions, and Kernels*, Academic Press, New York 1967.

[Wa72]   Warner, G., *Harmonic Analysis on Semisimple Lie Groups I*, Springer-Verlag, Berlin–Heidelberg–New York, 1972.

[Wu01]   Wurzbacher, T., Fermionic second quantization and the geometry of the restricted Grassmannian, in *Infinite Dimensional Kähler Manifolds* (A. Huckleberry, T. Wurzbacher, eds.), DMV-Seminar 31, Birkhäuser Verlag, 2001.

Traces and characteristic classes on loop spaces

Sylvie Paycha and Steven Rosenberg*

Laboratoire de Mathématiques, Université Blaise Pascal (Clermont II)
Complexe Universitaire des Cézeaux
63177 Aubière Cedex, France
email: `sylvie.paycha@math.univ-bpclermont.fr`

Department of Mathematics and Statistics, Boston University
111 Cummington St., Boston, MA 02215, U.S.A.
email: `sr@math.bu.edu`

Abstract. We construct Chern–Weil classes on infinite dimensional vector bundles with structure group $Cl_0^*(M, E)$, the group of zeroth order invertible classical pseudo-differential operators acting on a finite rank vector bundle E over a closed manifold M. $Cl_0^*(M, E)$ is the structure group of geometric bundles naturally associated to loop spaces of Riemannian manifolds. Mimicking the finite dimensional Chern–Weil construction, we replace the ordinary trace on matrices by different linear functionals on the Lie algebra of $Cl_0^*(M, E)$. We use (i) traces built from the leading symbol, and (ii) a linear map which considers all terms in the asymptotic expansion of a heat kernel regularized trace. For a specific bundle on loop spaces, the first approach yields non-vanishing Chern classes in all degrees. The second approach produces connection independent cohomology classes under stringent conditions. For the tangent bundle to a loop group, the first method gives a vanishing first Chern class, while the second method recovers the first Chern class investigated by Freed, and explains why this class is not connection independent.

2000 Mathematics Subject Classification: 53C05; 58J40.

1 Introduction

Infinite dimensional vector bundles with connections are frequently encountered in mathematical physics; basic examples include the tangent bundle of loop spaces [6] and infinite rank vector bundles associated to families of Dirac operators [4]. In this paper, we construct Chern forms and Chern classes for a class of vector bundles including the tangent bundle TLM to a loop space LM, and produce examples of non-vanishing classes. In light of Freed's curious example [6] of a connection dependent first Chern

---
*Partially supported by the NSF and the CNRS.

form on loop groups, an impossibility in finite dimensions, it seems worthwhile to examine extensions of Chern–Weil theory to infinite dimensions.

In contrast to finite dimensions, on infinite dimensional bundles one first has to choose the topology on the fiber and determine a structure group. One obvious choice, modeling the fiber on a Hilbert space H and the structure group on GL(H), leads to a trivial theory by Kuiper's theorem [10]. Since a direct topological approach to characteristic classes seems difficult, we follow the geometric approach of Chern–Weil theory, which both historically preceded topological approaches and is perhaps more elementary. In our approach, we assume that our infinite dimensional bundle $\mathcal{E}$ has (i) fibers modeled on the space of sections of a finite rank bundle E over a closed manifold M, in either a Sobolev or C^∞ topology, and (ii) a connection whose connection one-form takes values in $Cl_{\leq 0} = Cl_{\leq 0}(M, E)$, the space of classical pseudo-differential operators (ΨDOs) of nonpositive order acting on the fibers. As in finite dimensions, $Cl_{\leq 0}$ should be the Lie algebra of the structure group. In our case, the structure group is therefore Cl_0^*, the space of zeroth order invertible (and hence elliptic) ΨDOs on sections of E. This framework includes the case of TLM [13].

Chern–Weil theory produces characteristic classes from invariant polynomials on the Lie algebra $Cl_{\leq 0}$. Avoiding the difficult question of determining all such invariants, we focus on those polynomials which produce the Chern classes in finite dimensions, namely $\mathrm{Tr}(\Lambda^k A)$, the trace of exterior powers of a matrix. However, powers of the curvature need not be trace class for our structure group. One main topic of the paper is the investigation in §3 of alternative traces on $Cl_{\leq 0}$. One of these traces, the leading symbol trace, produces nonvanishing Chern classes.

In general, the leading symbol trace picks out the leading term in the asymptotic expansion $\mathrm{Tr}(\Omega e^{-\varepsilon Q})$, where Ω is the curvature of the connection and Q is a generalized Laplacian on the fibre, while the weighted traces of e.g. [17] pick out the finite term. As a second main topic, in §4 we show that certain asymptotic coefficients are closed, and that the corresponding cohomology classes are independent of the connection under more stringent conditions. Thus, in contrast to finite dimensions, it is qualitatively harder to show that characteristic forms are connection independent. In fact, Freed's example occurs as an asymptotic coefficient which is closed but does not satisfy the stringent conditions. Thus both the leading symbol traces of §3 and the results of §4 give extensions of Chern–Weil theory that improve the weighted trace approach of [5, 14, 17].

In more detail, in §2 we review classical Chern–Weil theory, with an emphasis on traces as morphisms $\lambda : \mathrm{Ad}\, P \to \mathbb{C}$ from the adjoint bundle of a principal bundle P to the trivial $\mathbb{C}$ bundle. Here the structure group and the base may be infinite dimensional, and we are thinking of P as the principal bundle associated to $\mathcal{E}$. These morphisms λ produce characteristic forms and classes as in finite dimensions (Theorem 2.2).

In §3, we introduce two types of traces in infinite dimensions, each of which can be interpreted as generalizations of the ordinary trace on matrices. The first example is the Wodzicki residue, the unique trace on the space of classical ΨDOs. However, we show in §3.1 that the associated Chern forms vanish on loop groups, confirming results in [14]. We produce more interesting examples by noting that the Lie algebra of the structure group $\mathcal{C}l_0^*$ admits a family of "symbol traces" of the form $A \mapsto \Lambda(\sigma_0^A)$, where σ_0^A is the leading symbol of A and Λ is a distribution on the cosphere bundle of M. In the main section §3.2, we show that the associated Chern classes are non-zero in general for the structure group of loop spaces (Theorem 3.3). We also present evidence that, despite appearances, these classes are not given by integration over the fiber of Chern classes of a finite dimensional bundle.

In §3.3, we relate the symbol traces to regularization techniques familiar in statistical mechanics. In particular, in certain cases the symbol trace $\Lambda^Q(\sigma_0^A)$ equals the leading term in the asymptotic expansion $\mathrm{tr}(Ae^{-\varepsilon Q})$, where Λ^Q is a distribution associated to a generalized Laplacian Q (Propostion 3.4). This applies to loop spaces: $Q = D^*D$ at the loop γ, with D covariant differentiation in the direction $\dot\gamma$ along γ. We also build characteristic classes from symbol traces on the smaller algebra $\mathcal{C}l_{\leq p}$ of ΨDOs of order at most p for $p < 0$, and discuss their dependence on the choice of connection (Theorem 3.6). This is relevant to the loop group case, as the Levi-Civita connection one-form takes values in such an algebra.

In §4, we extend the discussion of §3.3 to relate symbol traces to other terms in the asymptotics of $\mathrm{tr}(Ae^{-\varepsilon Q})$, and in particular to the finite part. This finite part regularization of $\mathrm{tr}(A)$ is well known in the physics literature, but the corresponding Chern–Weil construction (i.e. replacing A by powers of the curvature form Ω) does not produce closed forms in general because of the Q dependence.

To analyze this difficulty, we consider the entire asymptotic series $\mathrm{tr}_\varepsilon^Q(\Omega^k) := \mathrm{tr}(\Omega^k e^{-\varepsilon Q})$ as a $2k$-form with values in the sum of (i) a formal Laurent series in $\varepsilon^{\frac{1}{q}}$ (for some $q \in \mathbb{N}$) and (ii) $\log \varepsilon$ times a power series in ε. We modify the given connection ∇ on $\mathcal{E}$ to a connection ∇_ε^Q with connection one-form taking values in the power series ring $\mathcal{C}l_{\leq 0}[[\varepsilon]]$. We use ∇_ε^Q to determine which coefficients of $\mathrm{tr}_\varepsilon^Q(\Omega^k)$ are closed (Theorem 4.4), and when their cohomology classes are independent of the connection (Theorem 4.6). Roughly speaking, the number of coefficients which are closed grows linearly in $-d := -\mathrm{ord}([\nabla, Q])$. Thus, the more (covariantly) constant Q is, the greater the number of closed forms. For example, when $d < \mathrm{ord}(Q)$ as for loop groups, the coefficient of the most divergent term is closed; for $k = 1$, this coefficient is precisely the first Chern form considered by Freed. The number of coefficients whose cohomology class is constant for a family of connections ∇_t also depends linearly on $-d$, provided the order of $\frac{d}{dt}\nabla_t$ is sufficiently smaller than d. From these theorems, we can see precisely why Freed's first Chern class is connection dependent.

The second author would like to thank Université Blaise Pascal for its hospitality during the preparation of this article. Conversations with Simon Scott on this subject are also gratefully acknowledged. We also thank the editor and a referee for helpful comments, and in particular for the referee's simplified proof of (4.4).

2 Chern–Weil calculus

In this section, we review finite dimensional Chern–Weil calculus as in e.g. [3] and check its extension to the infinite dimensional setting. We emphasize the role of linear functionals on the Lie algebra of the structure group, as the choice for functionals is the main topic of §3.

Let B be a finite dimensional manifold, G a Lie group and $P \to B$ a smooth principal G-bundle. A smooth representation $\rho : G \to \mathrm{GL}(W)$ of G on a finite dimensional vector space W induces an associated smooth vector bundle $\mathcal{W} := P \times_\rho W \to B$. In particular, the adjoint representation $\rho : G \to \mathrm{Aut}(\mathrm{Lie}(G))$ determines a bundle $\mathrm{Ad}\, P$.

This framework extends to Kriegl and Michor's "convenient setting" for global analysis [11], which includes principal bundles for regular Fréchet Lie groups G over Fréchet manifolds. We will work with the space $\mathcal{C}l_0^*$ of invertible zeroth order $\Psi\mathrm{DOs}$ acting on smooth sections of a vector bundle over a closed manifold M. The Fréchet topology on $\mathcal{C}l_0^*$ is induced from the standard Fréchet topology on the coefficients of the homogeneous symbols σ_i of a $\Psi\mathrm{DO}$ T and the C^k topology on the smoothing part $T - \sum_i \sigma_i$. (The σ_i and the smoothing part depend on the choice of a partition of unity on M and a cutoff function in the cotangent variables, which we make once and for all.) This puts a regular Fréchet Lie group structure on $\mathcal{C}l_0^*$.

We briefly recall the geometric constructions we need in the Banach and Fréchet setting, referring the reader to [11] for details. The finite dimensional constructions must be modified, as a representation $G \to \mathrm{GL}(W)$ fails to be continuous in any reasonable sense once G and W are infinite dimensional. Indeed, $\mathrm{GL}(W)$ cannot be equipped with an appropriate Lie group structure in general; that is, if G is Banach and W is either Banach or Fréchet, one does not expect $\mathrm{GL}(W)$ to be a Lie group for the topology in which the representation is expected to be continuous. Even worse, if W is Frechet, $\mathrm{GL}(W)$ is never even a topological group unless W is a Banach space, in which case it is a Banach Lie group in the operator norm topology [15].

To circumvent these difficulties, one works with the group action associated to a representation [11, §49.1]. In more detail, let G be a regular Fréchet, resp. Banach Lie group with Lie algebra A, and let $P \to B$ be a smooth principal bundle equipped with a connection given by a Lie algebra valued connection one-form $\omega \in \Omega^1(P, A)$. Let W be a Fréchet, resp. Banach vector space (and therefore a regular space in the sense of [11]). A representation ρ' of G on W which induces a jointly smooth map $\rho : G \times W \to W$, $\rho(g, w) = \rho'(g)(w)$, determines an associated vector bundle $\mathcal{W} := P \times_\rho G \to B$ [11, §37.12]. Note that the associated bundle is constructed just as in finite dimensions, but the smoothness requirement of the representation has been restated. The space $\Omega(B, \mathcal{W})$ of $\mathcal{W}$-valued forms on B can be identified via a canonical isomorphism with the space $(\Omega(P) \otimes W)_b$ of basic forms on P with values in the trivial bundle $P \times W$ [11, §37.31]. Recall that a form is basic if it is G-invariant and horizontal. Moreover, the connection one-form ω on P with curvature form $\Omega^P \in \Omega^2(P, A)$ induces a covariant derivative ∇ on smooth sections of $\mathcal{W}$ [11,

§37.26], and its curvature $\Omega^W \in \Omega^2(B, \mathrm{Hom}(\mathcal{W}))$ is related to Ω^P via the canonical isomorphism above [11, §37.32].

Let $W = A$ be the Fréchet, resp. Hilbert Lie algebra of G. Recall that the adjoint representation $\mathrm{Ad} : G \to \mathrm{Aut}(A)$ is the differential of conjugation in G: $\mathrm{Ad}_g\, a := (D_e C_g)a$, where $C_g : G \to G$ is $C_g(h) = ghg^{-1}$. The differential of Ad, $\mathrm{ad} = D\,\mathrm{Ad} : A \to \mathrm{End}(A)$, is given by $\mathrm{ad}_b(a) = [b, a]$. It is immediate that the adjoint representation satisfies the joint smoothness condition above. In particular, a connection one-form $\theta \in \Omega^1(P, A)$ yields a connection ∇^{ad} on $\mathrm{Ad}\, P$, with $\nabla^{\mathrm{ad}} = d + [\theta, \cdot]$. (For this reason, our $\mathrm{Ad}\, P$ is often denoted $\mathrm{ad}\, P$.)

A linear form on A:

$$\lambda : A \to \mathbb{C}$$

such that $\mathrm{Ad}^* \lambda := \lambda \circ \mathrm{Ad} = \lambda$ induces a bundle morphism

$$\lambda : \mathrm{Ad}\, P \to B \times \mathbb{C}$$

defined as follows. Given a local trivialization (U, Φ), where $U \subset B$ is open and $\Phi : \mathrm{Ad}\, P|_U \to U \times A$ is an isomorphism, and a local section $\sigma \in \Gamma(\mathrm{Ad}\, P|_U)$, we set

$$\lambda(\sigma) := \lambda(\Phi(\sigma)).$$

This definition is independent of the local trivialization. Indeed, given another local trivialization (V, Ψ), at $b \in U \cap V$ we have

$$\lambda(\Phi(\sigma)) = \lambda(\mathrm{Ad}_g \Psi(\sigma)) = \lambda(\Psi(\sigma)), \quad \text{for some } g = g_b \in G.$$

The connection ∇^{ad} on $\mathrm{Ad}\, P$ induced by a connection θ on P induces in turn a connection ∇^* on the dual bundle $\mathrm{Ad}\, P^*$ (i.e. $(\mathrm{Ad}\, P)^*$), which is locally described by $\nabla^{\mathrm{ad}^*} = d + \mathrm{ad}_\theta^*$. Since $\mathrm{Ad}^* \lambda = \lambda$ implies $\mathrm{ad}^*\lambda = 0$, we have $\nabla^{\mathrm{ad}^*}\lambda = d\lambda = 0$, since λ is locally constant. Summarizing, we have:

Lemma 2.1. *Let* $\lambda : \mathrm{Ad}\, P \to B \times \mathbb{C}$ *be the linear morphism induced by a linear form* $\lambda : A \to \mathbb{C}$ *with* $\mathrm{Ad}^* \lambda = \lambda$. *Let* $\nabla^{\mathrm{ad}} = d + [\theta, \cdot] = d + \mathrm{ad}_\theta$ *be a connection on* $\mathrm{Ad}\, P$ *induced by a connection* θ *on* P. *Then*

$$d \circ \lambda = \lambda \circ \nabla^{\mathrm{ad}}. \tag{2.1}$$

Proof. Since $d\lambda = 0$, we have $d \circ \lambda = \lambda \circ d$ locally. However, $\mathrm{ad}^*\lambda = 0$ implies $\lambda \circ d = \lambda \circ (d + \mathrm{ad}_\theta) = \lambda \circ \nabla^{\mathrm{ad}}$, so $d \circ \lambda = \lambda \circ \nabla^{\mathrm{ad}}$ globally. □

Abusing notation, we will sometimes denote $\nabla^{\mathrm{ad}}\alpha$ by $[\nabla, \alpha]$, for α an $\mathrm{Ad}\, P$-valued form, in analogy to the local description $\nabla^{\mathrm{ad}} = d + [\theta, \cdot]$, with the understanding that $[\nabla, \alpha]$ is a superbracket with respect to the $\mathbb{Z}_2$-grading on differential forms.

The lemma leads to the main result of this section. To set the notation, let $\mathcal{E} \to B$ be a vector bundle with structure group a Fréchet or Banach Lie group G and with fiber modeled on a vector space V. The associated principal G-bundle $P^{\mathcal{E}}$ is given by gluing copies of G over B via the transition maps of $\mathcal{E}$. Strictly speaking, if $\{U_\alpha\}$ is an open cover of B which trivializes $\mathcal{E}$ and with transition maps $g_{\alpha\beta}(x) \in G$, $x \in U_\alpha \cap U_\beta$, $g_{\alpha\beta}(x) : V \to V$, then $P^\mathcal{E} = \coprod_\alpha (U_\alpha \times G)/(x, g) \sim (x, g_{\alpha\beta}(x)g)$ with the quotient topology. As an example applicable to loop spaces, let M, X be smooth finite dimensional manifolds, $E \to X$ a finite rank vector bundle, $B := C^\infty(M, X)$, and let $\mathcal{E} \to B$ have fiber $\mathcal{E}_b = C^\infty(M, b^*E)$. Let $\{V_\beta\}$ be the path components of B, and pick $b_\beta \in V_\beta$. Then the structure group of $\mathcal{E}|_{V_\beta}$ is $G = G_\beta = C^\infty(M, \mathrm{Aut}(b_\beta^*E))$, and the associated G-bundle has fiber $P^\mathcal{E}|_b = C^\infty(M, \mathrm{Aut}(b^*E)) = C^\infty(M, b^*P^E)$, where P^E is the frame bundle of E. In particular, if $E = TX$ is the tangent bundle to a paralellizable n-manifold such as a Lie group, then $P^\mathcal{E}|_b \simeq C^\infty(M, \mathrm{GL}_n(\mathbb{C}))$.

A G-connection on $\mathcal{E}$ induces a connection one-form on $P^\mathcal{E}$, just as a connection induces a connection one-form on the G-frame bundle in finite dimensions. In particular, a connection ∇ on $\mathcal{E}$ induces a connection ∇^{ad} on $\mathrm{Ad}\, P^\mathcal{E}$, and the curvature Ω of ∇ lies in $\Omega^2(B, \mathrm{Ad}\, P^\mathcal{E})$.

Theorem 2.2. *Let $P = P^\mathcal{E}$ be the principal bundle associated to a vector bundle with connection $(\mathcal{E}, \nabla) \to B$ with structure group G. Let Ω be the curvature of ∇. Let λ be as in Lemma 2.1. Then for any analytic function $f : \mathbb{C} \to \mathbb{C}$, the form $\lambda(f(\Omega))$ is closed, and its de Rham cohomology class in $H^*(B; \mathbb{C})$ is independent of the choice of ∇.*

As usual, we mean that the degree k piece of $\lambda(f(\Omega))$ is a closed $2k$-form, for all $k \in \mathbb{N}$.

Proof. The usual finite dimensional proof (see e.g. [3]) runs through, with ordinary traces replaced by λ.

In more detail, $\lambda(f(\Omega))$ is closed because $\lambda(\Omega^k)$ is closed for any $k \in \mathbb{N}$, which we check in a local trivialization of $\mathrm{Ad}\, P$. We have

$$d\,\lambda(\Omega^k) = \lambda(\nabla^{\mathrm{ad}}\Omega^k) = \lambda\left(\sum_{j=1}^k \Omega^{j-1}(\nabla^{\mathrm{ad}}\Omega)\Omega^{k-j}\right) = 0,$$

where we have used the Bianchi identity $\nabla^{\mathrm{ad}}\Omega = 0$ in the last identity.

To check that the corresponding de Rham class is independent of the choice of connection, we consider a differentiable one-parameter family of connections $\{\nabla_t, t \in \mathbb{R}\}$ on $\mathcal{E}$. More precisely, connections are elements of the smooth one-forms $\Omega^1(\mathrm{Ad}\, P)$, i.e. smooth bundle maps $\alpha : TM \to \mathrm{Ad}\, P$ in the Fréchet topologies. Differentiable families of connections are defined similarly. ∇_t induces a family of connections ∇_t^{ad} on $\mathrm{Ad}\, P$. Then

$$\frac{d}{dt}\lambda(\Omega_t^k) = \lambda\left(\sum_{j=1}^k \Omega_t^{k-j}\left(\dot{\nabla}_t\nabla_t + \nabla_t\dot{\nabla}_t\right)\Omega_t^{j-1}\right) = \lambda\left(\sum_{j=1}^k \Omega_t^{k-j}(\nabla_t^{\mathrm{ad}}\dot{\nabla}_t)\Omega_t^{j-1}\right)$$

$$= \lambda\left(\nabla_t^{\mathrm{ad}}\sum_{j=1}^k \Omega_t^{k-j}\dot{\nabla}_t\Omega_t^{j-1}\right) = d\lambda\left(\sum_{j=1}^k \Omega_t^{k-j}\dot{\nabla}_t\Omega_t^{j-1}\right). \tag{2.2}$$

In the first equality, we use $\nabla_t^2 = \Omega_t$, in the second we have extended the bracket connection to forms, and in the third we have used the Bianchi identity. (2.2) shows that the dependence on the connection is measured by an exact form and hence vanishes in cohomology. □

This yields the usual Chern–Weil classes:

Corollary 2.3. *Let $G \subset GL(n, \mathbb{C})$ be a finite dimensional Lie group, and let $\mathcal{E} \to B$ be a vector bundle with structure group G. Let ∇ be a connection on $\mathcal{E}$ with curvature Ω. For any analytic function f, the forms $\mathrm{tr}(f(\Omega)) \in \Omega^*(B, \mathbb{C})$ are closed and their de Rham cohomology classes are independent of the choice of ∇.*

This follows from Theorem 2.2 by passing from $\mathcal{E}$ to $P^{\mathcal{E}}$ and using $\lambda = \mathrm{tr}$, the ordinary trace on matrices.

Remark. For $GL(n, \mathbb{C})$ and $U(n)$, all characteristic classes are generated by $\mathrm{tr}(\Omega^k)$, $k \in \mathbb{N}$. However, we do not capture the Euler class for $SO(n, \mathbb{R})$ by this procedure, as this class is generated by the non-linear, but Ad-invariant function $\sqrt{\det}$. We can treat this case either by using the identity $\det(1 + A) = \sum_k \mathrm{tr}(\Lambda^k A)$, or by noting that the proof of Theorem 2.2 does not use the linearity of λ.

Notation. Throughout the paper, "ΨDOs" means classical pseudo-differential operators, and $Cl(M, E)$ denotes the space of all classical ΨDOs acting on smooth sections of the finite dimensional Hermitian bundle E over a closed Riemannian manifold M. $Cl_k(M, E)$ denotes the subspace of ΨDOs of order $k \in \mathbb{R}$. $Cl_{\leq k}(M, E)$, resp. $Cl_{<k}(M, E)$ denotes the space of ΨDOs of order at most k, resp. less than k. $Cl_k^*(M, E)$ denotes the set of invertible operators in $Cl_k(M, E)$. $\mathcal{E}ll^+(M, E)$ denotes the space of positive order, elliptic operators in $Cl(M, E)$ with positive definite leading symbol.

A bundle $\mathcal{E}$ with fiber modeled on $C^\infty(M, E)$ or on $H^s(M, E)$ is a ΨDO *bundle* if the transition maps lie in the regular Fréchet Lie group $Cl_0^*(M, E)$. Here $C^\infty(M, E)$, $H^s(M, E)$ are the spaces of smooth and s-Sobolev class sections of E, respectively. Ad $P^{\mathcal{E}}$ is a bundle of algebras locally modeled on $Cl_{\leq 0}(M, E)$, and will be denoted $Cl_{\leq 0}(\mathcal{E})$. Note that Ad $P^{\mathcal{E}}$ equals the bundle $\mathcal{E} \times_{Cl_0^*(M,E)} Cl_{\leq 0}(M, E)$ associated to the adjoint representation. In §4, we also consider the larger bundle $Cl(\mathcal{E}) = \mathcal{E} \times_{Cl_0^*(M,E)} Cl(M, E)$, associated to the adjoint representation of $Cl_0^*(M, E)$ on the algebra $Cl(M, E)$; here $Cl(M, E)$ is given the inductive limit topology of the usual Fréchet topology on $Cl_{\leq k}(M, E)$.

A connection on a ΨDO bundle $\mathcal{E}$ is a ΨDO *connection* if its connection one-form takes values in $Cl(M, E)$ in any local trivialization.

Remark. If θ is the locally defined connection one-form of a ΨDO connection on a bundle $\mathcal{E}$ modeled on $C^\infty(M, E)$, then under a gauge change g, θ transforms to

$g^{-1}\theta g + g^{-1}dg$. Since $g^{-1}dg$ is zeroth order if g is nonconstant, the connection one-form is usually of non-negative order. (For left-invariant connections on loop groups, $g^{-1}dg$ vanishes, and θ can be of any order.) When θ is of non-positive order, it is a bounded operator and hence extends to a connection on the extension of $\mathcal{E}$ to an $H^s(M, E)$-bundle. We call a connection with connection one-form taking values in $\mathcal{C}l_{\leq 0}(M, E)$ a $\mathcal{C}l_{\leq 0}$-*connection*. The curvature of such a connection is a $\mathcal{C}l_{\leq 0}$-valued two-form on the base.

3 Examples of traces and corresponding Chern classes in infinite dimensions

In this section we examine two examples of Theorem 2.2. The trace is furnished by the Wodzicki residue in the first example, and by various traces applied to the leading order symbol of a zeroth order ΨDO in the second. We also consider various traces applied to the leading order symbol of ΨDOs of negative order, for which an extension of Theorem 2.2 is needed.

In each case, we begin with a Fréchet or Hilbert vector bundle $\mathcal{E}$ over a base space B, with fiber modeled on $C^\infty(M, E)$ or $H^s(M, E)$, with structure group given by $\mathcal{C}l_0^* = \mathcal{C}l_0^*(M, E)$. We will consider a connection on B with values in the corresponding Lie algebra $\mathcal{C}l_{\leq 0} = \mathcal{C}l_{\leq 0}(M, E)$. In the language of §2, we pass from $\mathcal{E}$ to the corresponding principal bundle $P = P^{\mathcal{E}}$ with fiber modeled on $\mathcal{C}l_0^*$. Then Ad P has fiber modeled on $\mathcal{C}l_{\leq 0}$, and we can apply the Chern–Weil machinery of §2, using either the Wodzicki residue or the leading symbol traces for the functional λ.

Note that in this section, we are treating the structure group $\mathcal{C}l_0^*$ as a generalization of $GL(n, \mathbb{C})$. As in finite dimensions, we focus only on invariant polynomials on the Lie algebras given by traces. We do not discuss the interesting question of whether all such polynomials on $\mathcal{C}l_{\leq 0}$ are generated by these traces.

Exactly how these examples generalize the finite dimensional situation is open to interpretation. When the manifold is reduced to a point, the leading symbol of an endomorphism in the fiber, a "zeroth order ΨDO," is just the endomorphism itself, and the only trace, up to normalization, is the ordinary trace on a vector space. In contrast, in finite dimensions the Wodzicki residue vanishes. So in this interpretation, the Wodzicki residue is a purely infinite dimensional phenomenon, while the symbol trace generalizes the finite dimensional theory.

On the other hand, both the Wodzicki residue and the symbol trace appear in the most divergent term in asymptotic expansions: the Wodzicki residue of an operator A is the residue of the pole of the zeta function regularization $\text{Tr}(AQ^{-s})$ at $s = 0$ (for any positive elliptic operator Q), and the symbol trace is related to the coefficient of the most divergent term in the heat operator regularization $\text{Tr}(Ae^{-\varepsilon Q})$ as $\varepsilon \to 0$. (The last statement is proved in Proposition 3.4.) Since the two corresponding "regularizations" in finite dimensions using a positive definite matrix Q simply reduce to $\text{Tr}(A)$, we can alternatively view both examples as proper generalizations of the finite dimensional Chern–Weil theory.

Similarly, on the smaller algebra $Cl_{\leq p}(M, E)$ for $p < 0$, there are leading symbol traces that are also related to the coefficient of the leading term (usually the "most divergent" term) in the heat operator regularization. Because this is a proper subalgebra of the Lie algebra $Cl_{\leq 0}(M, E)$ of the structure group, we cannot expect a full Chern–Weil theory for $Cl_{\leq p}(M, E)$-connections. Nevertheless, in §3.3, we produce closed characteristic forms for these connections and show that the characteristic classes obtained this way are independent of the choice of connection, provided the two connections differ by a $Cl_{\leq p}(M, E)$-valued one-form. In §4, we improve this result and the results in [17] by formally keeping track of all terms in the relevant asymptotic expansions.

In summary, there seems to be no canonical generalization or regularization of finite dimensional Chern–Weil theory free from drawbacks: using the operator trace on trace class operators is too restrictive for zeroth order operators, the better adapted leading symbol traces vanish on trace class operators and operators of negative order, and the weighted traces of §4 are not true traces. Moreover, there is no canonical interpretation of whether a specific method is indeed a proper generalization, as the Wodzicki residue can be interpreted either as an extension of the finite dimensional theory or as a purely infinite dimensional pheonomenon. The particular choice of regularization depends on a combination of physical motivation, computability and nontriviality results.

3.1 The Wodzicki residue

Recall that the Wodzicki residue $\mathrm{res_w}\, A$ of a ΨDO A acting on sections of a bundle E over a closed manifold M is defined to be the residue of the pole term of $\mathrm{Tr}(AQ^{-s})$ at $s = 0$, for an elliptic operator Q with certain technical conditions. Alternatively, $\mathrm{res_w}\, A$ is proportional to the coefficient of $\log \varepsilon$ in the asymptotic expansion of $\mathrm{Tr}(Ae^{-\varepsilon Q})$ as $\varepsilon \to 0$, for $Q \in \mathcal{E}ll^+(M, E)$. The strengths of the Wodzicki residue are (i) its local nature:

$$\mathrm{res_w}\, A = \frac{1}{(2\pi)^n} \int_{S^*M} \mathrm{tr}\, \sigma_{-n}^A(x, \xi)\, d\xi\, dx,$$

where $n = \dim(M)$, S^*M is the unit cosphere bundle of M, and σ_{-n}^A is the $(-n)^{\mathrm{th}}$ homogeneous piece of the symbol of A; and (ii) the fact that it is the unique trace on $Cl = Cl(M, E)$, up to normalization. Its drawback is its vanishing on all differential and multiplication operators, all trace class operators (and so all ΨDOs of order less than $-n$) and all operators of non-integral order.

Given an infinite dimensional bundle $\mathcal{E}$ over a base B with fibers modeled either on $H^s(M, E)$ (with $s \gg 0$) or on $C^\infty(M, E)$, and a connection on $\mathcal{E}$ with curvature

$\Omega \in \Omega^2(B, \mathcal{C}l)$, we can form the kth Wodzicki–Chern form by setting

$$c_k^{\text{w}}(\Omega) = \text{res}_{\text{w}}\Omega^k \in \Omega^{2k}(B).$$

By Theorem 2.2, $c_k^{\text{w}}(\Omega)$ is closed and independent of the connection.

As an example, we show that the Wodzicki–Chern forms vanish for current groups $\mathcal{C} = H^{s+1}(M, G)$, the space of H^{s+1}-maps from a closed Riemannian manifold M to a Lie group G. (The same vanishing holds for Fréchet current groups.) The tangent space at any map f is the space of sections $H^s(M, f^*TG)$. Since TG is canonically trivial, so is $T\mathcal{C}$. For the trivial connection, we certainly have the vanishing of the Wodzicki–Chern forms. It follows that the Wodzicki–Chern classes vanish for any ΨDO connection.

We now check that the Levi-Civita connection on a current group is a ΨDO connection for a semi-simple Lie group G of compact type. (These assumptions ensure that the Killing form is nondegenerate and that the adjoint representation is antisymmetric for this form.) $\mathcal{C}$ is a Hilbert Lie group with Lie algebra $H^s(M, A)$, the space of H^s sections of the trivial bundle $M \times A$, where $A = \text{Lie}(G)$. Thus the tangent bundle $T\mathcal{C}$ is a ΨDO bundle with fibers modeled on $H^s(M, A)$. For Δ the Laplacian on functions on M, we set $Q_0 := \Delta \otimes 1_A$, a second order elliptic operator acting densely on $H^s(M, A)$. Q_0 is non-negative for the scalar product $\langle \cdot, \cdot \rangle_0 := \int_M \text{dvol}(x)(\cdot, \cdot)$, where $(\cdot, \cdot)$ is minus the Killing form. $T\mathcal{C}$ has a left-invariant weight $Q_\gamma = L_\gamma Q_0 L_\gamma^{-1}$ (i.e. a family of elliptic operators on the fibers), where L_γ is left translation by $\gamma \in \mathcal{C}$.

$\mathcal{C}$ has a left-invariant Sobolev s-metric defined by

$$\langle \cdot, \cdot \rangle^s := \langle Q_0^{\frac{s}{2}} \cdot, Q_0^{\frac{s}{2}} \cdot \rangle_0,$$

where Q_0 is really $Q_0 + P$, for P the orthogonal projection of Q_0 onto its kernel. The corresponding left-invariant Levi-Civita connection has the global expression $\nabla^s = d + \theta^s$, with θ^s a left-invariant $\text{End}(T\mathcal{C})$-valued one-form on $\mathcal{C}$ induced by the $\text{End}(H^s(M, A))$-valued one-form on $H^s(M, A)$

$$\theta_0^s(U) = \frac{1}{2}\left(\text{ad}_U + Q_0^{-s}\text{ad}_U\, Q_0^s - Q_0^{-s}\text{ad}_{Q_0^s U}\right), \tag{3.1}$$

for $U \in H^s(M, A)$ [6, (1.9)]. By inspection, θ^s takes values in $\mathcal{C}l_{\leq 0}(M, M \times A)$.

The fact that Wodzicki–Chern classes vanish on current groups is not surprising, since the same argument works on any parallelizable manifold. It is more surprising that these classes vanish on any loop space, even when the target manifold is not parallelizable [14]. Thus the Wodzicki residue, the natural first choice for a trace functional, yields a Chern–Weil theory that is currently vacuous. As a result we look for other functionals with nontrivial Chern–Weil theory.

3.2 Leading symbol traces

The uniqueness of the trace on $\mathcal{C}l$ defined by the Wodzicki residue does not rule out the existence of other traces on subalgebras of $\mathcal{C}l$. Indeed, the ordinary operator trace on $Cl_{\leq -n}$ is an example. In this subsection, we will introduce a family of traces on $Cl_{\leq 0}$ and show that they produce non-vanishing Chern classes on the universal bundle associated to the gauge group for $\mathcal{E} = TLM$, the tangent bundle to the free loop space of a Riemannian manifold M. To our knowledge, this is the first example of non-vanishing Chern classes of infinite dimensional bundles above c_1.

We first produce a "trace" on $Cl_{\leq p}$ for fixed $p \leq 0$ with values in S^*M, and an associated family of true traces. A description of all traces on e.g. $Cl_{\leq p}$ for fixed $p \leq 0$ is an interesting question; we have preliminary results with J.-M. Lescure. Let $\mathcal{D}'(X)$ denote the space of complex valued distributions on a compact manifold X.

Lemma 3.1. *For $p \leq 0$, the map $\mathrm{Tr}_p : Cl_{\leq p}(M, E) \rightarrow C^\infty(S^*M)$ defined by $\mathrm{Tr}_p(A) = \mathrm{tr}_x(\sigma_p^A(x, \xi))$ has $\mathrm{Tr}_p(A + B) = \mathrm{Tr}_p(A) + \mathrm{Tr}_p(B)$, $\mathrm{Tr}_p(\lambda A) = \lambda \mathrm{Tr}_p(A)$, and $\mathrm{Tr}_p(AB) = \mathrm{Tr}_p(BA)$. For any $\Lambda \in \mathcal{D}'(S^*M)$, the map $\mathrm{Tr}_p^\Lambda : Cl_{\leq p} \rightarrow \mathbb{C}$ given by $\mathrm{Tr}_p^\Lambda(A) = \Lambda(\mathrm{Tr}_p(A))$ is a trace.*

Proof. Certainly taking the pth order symbol is linear. When $p = 0$, since the leading order symbol is multiplicative, we have

$$\mathrm{tr}_x \, \sigma_0^{AB} = \mathrm{tr}_x(\sigma_0^A \cdot \sigma_0^B) = \mathrm{tr}_x(\sigma_0^B \cdot \sigma_0^A) = \mathrm{tr}_x \, \sigma_0^{BA}.$$

When $p < 0$, for $A, B \in Cl_{\leq p}(M, E)$, the products AB and BA lie in $Cl_{\leq 2p}(M, E)$ so that we have

$$\mathrm{tr}_x \, \sigma_p^{AB} = 0 = \mathrm{tr}_x \, \sigma_p^{BA}.$$

The proof of the second statement is immediate. $\square$

In this subsection, we focus on the case $p = 0$, leaving the case $p < 0$ for the next subsection. For convenience we set $\mathrm{Tr} := \mathrm{Tr}_0$, $\mathrm{Tr}^\Lambda := \mathrm{Tr}_0^\Lambda$. When the distribution is given by $\Lambda(\phi) = \int_{S^*M} f(x, \xi)\phi(x, \xi)$ for all $\phi \in C^\infty(S^*M)$, we simply write Tr^f.

Remarks. (i) When $p < 0$, for $r \in [2p, p]$, $\mathrm{Tr}_r A = \mathrm{tr}_x(\sigma_r^A(x, \xi))$ is also a trace, as $\mathrm{Tr}_r(AB)$ trivially vanishes for $r > 2p$. The proof of the lemma covers the case $r = 2p$.

(ii) Let $Q \in \mathcal{E}ll^+(M, E)$ have scalar leading symbol $\sigma_L^Q(x, \xi) = f(x, \xi)\mathrm{Id}$. Define $\tilde{f} \in C^\infty(S^*M)$ by

$$\tilde{f} = \frac{(n-1)!\Gamma\left(\frac{n}{q}\right)\dim(E)}{q(2\pi)^n} f(x, \xi)^{-\frac{n}{q}},$$

where $n = \dim(M)$, $q = \mathrm{ord}(Q)$. Then $\mathrm{Tr}^{\tilde{f}}(A)$ is the leading term in the asymptotics of $\mathrm{Tr}(Ae^{-\varepsilon Q})$ if $\mathrm{ord}(A) = 0$ (see Proposition 3.4).

Recall that the ring of characteristic classes for e.g. $U(n)$ bundles is generated by the Chern classes $c_k = [\mathrm{Tr}(\Lambda^k \Omega)]$, or equivalently by the components $\nu_k = [\mathrm{Tr}(\Omega^k)]$ of the Chern character. Note we are momentarily distinguishing between $\mathrm{Tr}(\Lambda^k A)$

and $\text{Tr}(A^k)$ for a matrix A. We will concentrate on Chern forms, and abuse notation by writing $c_k = [\text{Tr}(\Omega^k)]$.

Definition. Let $\mathcal{E}$ be a ΨDO-bundle over B modeled on $H^s(M, E)$ or $C^\infty(M, E)$, and let ∇ be a ΨDO-connection on $\mathcal{E}$. The kth Chern class of ∇ with respect to $\Lambda \in \mathcal{D}'(S^*M)$ is defined to be the de Rham cohomology class

$$[c_k^\Lambda(\Omega)] = [\Lambda(\text{tr}_x \sigma_0^{\Omega^k}(x, \xi))] \in H^{2k}(B; \mathbb{C}). \tag{3.2}$$

As before, when $\Lambda(\phi) = \int_{S^*M} f(x, \xi) \phi(x, \xi)$ we set $c_k^f = c_k^\Lambda$.

Remarks. (1) As an example, if $f = 1 \in C^\infty(S^*M)$ is the constant map with value 1 on S^*M, then

$$c_k^f(\Omega) = \int_{S^*M} \text{tr}_x \, \sigma_0^{\Omega^k}(x, \xi).$$

At another extreme, if $\Lambda = \delta_{(x_0, \xi_0)}$ is a delta function, then

$$c_k^\Lambda(\Omega) = \text{tr}_{x_0} \, \sigma_0^{\Omega^k}(x_0, \xi_0).$$

(2) As in the previous remark, we can define Chern classes $c_{r,k}^\Lambda$ for connections with curvature forms taking values in $\mathcal{C}l_{\leq p}$ for any $r \in [2p, p]$. Note that for $r < p$ and e.g. $\Lambda(\phi) = \int_{S^*M} \phi(x, \xi)$, these classes are defined only after a choice of coordinates on M, E and a partition of unity on M, since integrals of non-leading order symbols depend on such choices.

The following result justifies this definition:

Theorem 3.2. *Let ∇ be a $\mathcal{C}l_{\leq 0}$-connection on a ΨDO bundle $\mathcal{E}$. Then the differential forms $c_k^\Lambda(\Omega)$ are closed, and their cohomology classes are independent of the choice of $\mathcal{C}l_{\leq 0}$-connection.*

Proof. By Lemma 3.1, Tr^Λ is a trace on $\mathcal{C}l_{\leq 0}(M, E)$, so we can apply Lemma 2.1 to the principal bundle $P = P^{\mathcal{E}}$ built from $\mathcal{E}$ to get the relation $d \circ \text{Tr}^\Lambda = \text{Tr}^\Lambda \circ \nabla^{\text{ad}}$. We then apply Theorem 2.2 to get the corresponding Chern classes $[c_k^\Lambda(\Omega)]$.   □

Remark. For the linear functionals Tr_p^Λ, the proof that the Chern forms are closed goes through. However, the proof of their independence of choice of connection breaks down, since the class of connections with curvature forms lying in $\mathcal{C}l_{\leq p}$ is not connected. In the next subsection, we nevertheless show that the independence holds on a restricted class of connections.

When the structure group reduces to a gauge group, we can construct an example of non-zero Chern classes $[c_k^f(\Omega)]$. Fix $n > k$ and consider the Grassmannian $\text{BU}(n) = \text{Gr}(n, \infty)$ with its universal vector bundle E_n. We consider the pullback bundle $E =$

$\pi^* E_n$ over $S^1 \times \mathrm{BU}(n)$, with π the projection onto $\mathrm{BU}(n)$. We now "loopify" to form $B = L(S^1 \times \mathrm{BU}(n))$, the free loop space of $S^1 \times \mathrm{BU}(n)$, with bundle $\mathcal{E}$ whose fiber over a loop γ is the space of smooth sections of $\gamma^* E$ over S^1. ($\mathcal{E}_\gamma$ is the space of loops in E lying over γ, suitably interpreted at self-intersection points of γ, so we will write $\mathcal{E} = L\pi^* E_n$.) Since $\gamma^* E$ is (non-canonically) isomorphic to the trivial bundle $S^1 \times \mathbb{C}^n$ over S^1, it is easily checked that the structure group for $\mathcal{E}$ is the gauge group $\mathcal{G}$ of this trivial bundle. Indeed, as in the example before Theorem 2.2 with $M = S^1$, $X = S^1 \times \mathrm{BU}(n)$, the structure group of $\mathcal{E}$ is $C^\infty(S^1, \mathrm{Aut}(\gamma^* E)) \simeq C^\infty(S^1, \mathrm{GL}_n(\mathbb{C}))$.

Take a hermitian connection ∇ on E_n (e.g. the universal connection $P\,dP$, where P_x is the projection of $\mathbb{C}^\infty$ onto the n-plane x) and its pullback connection $\pi^* \nabla$ on $S^1 \times \mathrm{BU}(n)$. As in the case of the tangent bundle to a loop space, we can take an L^2 or pointwise connection ∇^0 on $\mathcal{E}$ by setting

$$\nabla^0_X Y(\gamma)(\theta) = (\pi^* \nabla)_{X(\theta)} Y(\theta),$$

for X a vector field along γ (i.e. a tangent vector in B at γ) and Y a local section of $\mathcal{E}$. The curvature Ω^0 acts pointwise and hence is a multiplication operator: $(\Omega^0_{\gamma} u)(\theta) = (\pi^* \Omega)_{\gamma(\theta)} u(\theta)$, where Ω is the curvature of ∇. In particular, its symbol is independent of ξ.

Pick the distribution $\delta = (1, +)$ on $C^\infty(S^* S^1) = C^\infty(S^1 \times \{\pm \partial_\theta\})$: i.e.

$$\delta(f(\theta, \partial_\theta), g(\theta, -\partial_\theta)) = \frac{1}{2\pi} \int_{S^1} f(\theta, \partial_\theta)\, d\theta.$$

We claim that $[c_k^\delta(\Omega^0)]$ is nonzero in $H^{2k}(B; \mathbb{C})$. To see this, let $a = a_{2k} \in H_{2k}(\mathrm{BU}(n), \mathbb{C})$ be such that $\langle c_k(E_n), a \rangle = 1$. Define $c \in H^{2k}(B)$ to be $c = \beta_* a$, where $\beta : \mathrm{BU}(n) \to L(S^1 \times \mathrm{BU}(n))$ is given by $\beta(x)(\theta) = (\theta, x)$. Now

$$\langle [c_k^\delta(\Omega^0)], c \rangle = \langle [c_k^\delta(\Omega^0)], \beta_* a \rangle = \langle [\beta^* c_k^\delta(\Omega^0)], a \rangle. \tag{3.3}$$

For $\gamma \in L(S^1 \times \mathrm{BU}(n))$, we have

$$c_k^\delta(\Omega^0)(\gamma) = \frac{1}{2\pi} \int_{S^1} \mathrm{tr}\big(\sigma_0^{(\Omega^0)^k}(\gamma(\theta), \partial_{\gamma(\theta)})\big)\, d\theta = \frac{1}{2\pi} \int_{S^1} \mathrm{tr}\big(\pi^* \Omega^k_{\gamma(\theta)}\big)\, d\theta. \tag{3.4}$$

For a tangent vector $X \in T_x \mathrm{BU}(n)$, it is immediate that $\beta_*(X) \in T_{\beta(x)} B$ has $\beta_*(X)(\theta, x) = (0, X)$. Thus by (3.4),

$$\beta^* c_k^\delta(\Omega^0)(X_1, \ldots, X_{2k}) = \frac{1}{2\pi} \int_{S^1} \mathrm{tr}(\pi^* \Omega^k)((0, X_1), \ldots, (0, X_{2k}))$$
$$= \mathrm{tr}(\Omega^k)(X_1, \ldots, X_{2k}). \tag{3.5}$$

Combining (3.3) and (3.5), we get

$$\langle [c_k^\delta(\Omega^0)], c \rangle = \langle [\mathrm{tr}(\Omega^k)], a \rangle = 1.$$

In particular, the class $[c_k^\delta(\Omega^s)]$ must be non-zero.

Theorem 3.3. *The cohomology classes* $[c_k^\delta(\Omega)]$ *are non-zero in general. In particular, the corresponding classes for the universal bundle* $E\mathcal{G}$ *are nonzero in the cohomology of the classifying space* $B\mathcal{G}$, *where* $\mathcal{G}$ *is the gauge group of the trivial bundle* $S^1 \times \mathbb{C}^n$ *over* S^1.

We have shown the first statement. To explain the second statement, note that although the structure group of $L\pi^*\gamma_n$ is the gauge group of the trivial bundle $S^1 \times \mathbb{C}^n$ over S^1, the curvature of the connection will take values in $\mathcal{C}l_{\leq 0}(S^1, S^1 \times \mathbb{C}^n)$ of this bundle. As a result, the classifying space is really $B\mathcal{C}l_0^*$. It can be shown [18] that the principal symbol map is the time one map of a deformation retraction of $\mathcal{C}l_0^*$ onto the gauge group of the trivial bundle over S^*S^1, which is just two copies of $\mathcal{G}$. Thus $B\mathcal{C}l_0^*$ is homotopy equivalent to $B\mathcal{G} \coprod B\mathcal{G}$, and each $[c_k^{(1,\pm)}]$ is non-zero in one copy of $B\mathcal{G}$. The proof of the second statement depends on the existence of a universal connection on $E\mathcal{G}$ over $B\mathcal{G}$ [18].

In fact, $B\mathcal{G}$ equals $L_0\,\mathrm{BU}(n)$, the space of contractible loops on $\mathrm{BU}(n)$ [2]. It is known that $H^*(B\mathcal{G}, \mathbb{C})$ is a super-polynomial (i.e. super-commutative) algebra with one generator in each degree $k \in \{1, \dots, 2n\}$. In analogy with finite dimensions, we conjecture that $[c_k^\delta]$ is a nonzero multiple of the generator in degree $2k$. For $k = 1$, this is clear.

We now outline a conjectured construction of geometric representatives of the odd generators in $H^*(B\mathcal{G})$. The tangent bundle TLM of any loop space has a canonical vector field, namely $\dot{\gamma} \in T_\gamma LM$. Note that for any connection on a bundle over LM, for any $\Lambda \in \mathcal{D}'(S^*S^1)$ we have

$$di_{\dot\gamma}c_k^\Lambda(\Omega) = di_{\dot\gamma}c_k^\Lambda(\Omega) + i_{\dot\gamma}dc_k^\Lambda(\Omega) = L_{\dot\gamma}c_k^\Lambda(\Omega), \tag{3.6}$$

where i is interior product and L is Lie derivative. We state without (the elementary) proof that (3.6) implies

$$di_{\dot\gamma}c_k^\Lambda(\Omega) = c_k^{\partial\Lambda}(\Omega), \tag{3.7}$$

where $\partial\Lambda$ is the derivative of Λ as a distribution on S^*S^1. In particular, we see that $i_{\dot\gamma}c_k^{(1,\pm)}(\Omega)$ are closed forms. It remains to be seen if

$$[i_{\dot\gamma}c_k^{(1,\pm)}(\Omega)] \in H^{2k-1}\big(B\mathcal{G} \coprod B\mathcal{G}\big)$$

are non-zero.

We can also use (3.7) to understand the dependence of $[c_k^\Lambda(\Omega)]$ on Λ. Since Λ is a zero current on S^*S^1 and hence is trivially closed, and since exact zero currents $\partial\Lambda$ produce vanishing Chern classes by (3.7), we see that the space of classes

$$\{[c_k^\Lambda(\Omega)] : f \in \mathcal{D}'(S^*S^1)\} \in H^{2k}(B\mathcal{G})$$

is isomorphic to the zeroth cohomology group of complex currents on S^*S^1. (Here we are extending the usual confusion of functions f and one-forms $fd\theta$ on S^1 to a confusion of zero- and one-currents.) This cohomology group is isomorphic to $H_0(S^*S^1)$, and is spanned by $(1, \pm)$. (The reader may wish to check directly from

(3.7) that all δ-function currents on one copy of S^1 produce the same cohomology class. A more interesting exercise is to show that these delta functions produce the same cohomology class as one of $(1, \pm)$ using Fourier series.)

Remarks. (i) The general case, where the gauge group is associated to the bundle E over a closed manifold M, is more complicated. The cohomology of $B\mathcal{G}$ is known, and in general has odd dimensional cohomology [2]. We do not know at present which part of $H^*(B\mathcal{G})$ is spanned by $[c_k^\Lambda(\Omega^{E\mathcal{G}})]$, where we use the universal connection mentioned above. We also do not know how to produce geometric representatives of odd dimensional classes in $H^*(B\mathcal{G})$, nor do we know how the Chern classes depend on the distribution.

(ii) In the loop group case, Freed showed [6] that the curvature Ω^s of the H^s Levi-Civita connection is a ΨDO of order -1 for $s > 1/2$. For this connection, the Chern forms built from σ_0 trivially vanish. In the next subsection, we discuss the k-th Chern forms one can build using the symbol of order $-k$. For loop groups, the first Chern form requires the additional analysis in §4, while higher powers of the curvature are trace class operators, requiring no regularization.

(iii) For the distribution Λ given by integration over S^*M, the symbol trace looks like an integration over the fiber. Nevertheless, to the best of our knowledge, our Chern classes are not given by an integration of characteristic classes of an associated finite dimensional bundle.

In more detail, let $\mathcal{E}$ be a bundle over B with structure group Cl_0^* and with a ΨDO connection ∇. Let $\mathcal{G}$ be the gauge group of π^*E over S^*M. As in [18], $\mathcal{E}$ reduces to a $\mathcal{G}$-bundle $\mathcal{F}'$ with connection ∇', where the connection one-form of ∇' is the zeroth order symbol of the connection one-form of ∇. The curvature of ∇' equals the zeroth order symbol of the curvature of ∇. The fiber of $\mathcal{F}'$ is still $C^\infty(M, E)$, and we can form a $\mathcal{G}$-bundle $\mathcal{F}$ over B with fiber $C^\infty(S^*M, \pi^*E)$ using the same gluing maps as for $\mathcal{F}'$. (While $\mathcal{G}$ acts on fibers of $\mathcal{F}'$ as zeroth order ΨDOs, it acts on fibers of $\mathcal{F}$ as multiplication operators.) The connection one-form for ∇' still transforms correctly on $\mathcal{F}$, and so defines a connection on $\mathcal{F}$, also denoted ∇'. The curvatures of the ∇' connections are equal.

$\mathcal{F}$ induces a finite dimensional bundle F over $B \times S^*M$, with fiber $\pi^*E|_{(x,\xi)}$ over (b, x, ξ). However, ∇' induces a connection ∇^F on F only after we specify how to differentiate in S^*M directions. Assume that we can specify these differentiations so that the curvature Ω^F is flat in S^*M directions. Ω^F will still agree with $\Omega^{\mathcal{F}}$ in B directions, and

$$[c_k^\Lambda(\Omega^{\mathcal{E}})] = [c_k^\Lambda(\Omega^{\mathcal{F}'})] = [c_k^\Lambda(\Omega^{\mathcal{F}})] = \left[\int_{S^*M} \mathrm{tr}((\Omega^{\mathcal{F}})^k) \right]$$

$$= \int_{S^*M} [\mathrm{tr}((\Omega^F)^k)] = \int_{S^*M} c_k(F),$$

where $\int_{S^*M}$ outside the braces is the pushforward map from $H^*(B \times S^*M)$ to $H^*(B)$.

Thus under the assumption, the cohomology class $c_k^\Lambda(\mathcal{E})$ will indeed be the integration over the fiber of the Chern class of a finite dimensional bundle. However, the assumption is unreasonable: even if E is trivial, as for loop spaces, there is no canonical identification of the fibers of $\mathcal{E}$ with $H^s(M, E)$, so the trivial connection on E does not glue up to a connection on F which is trivial in S^*M directions.

3.3 Leading terms in heat-kernel asymptotic expansions

In this section, we consider traces Tr_p^Λ for $p \leq 0$. Theorem 2.2 no longer applies, as it did for $p = 0$, since Tr_p^Λ defines a trace only on the subalgebra $\mathcal{C}l_{\leq p}$ of the Lie algebra $\mathcal{C}l_{\leq 0}$. We cannot expect these traces to produce a full Chern–Weil theory on ΨDO bundles. However, they do yield characteristic classes which are independent of the choice of the connection in some restricted class of connections.

We first relate leading symbol traces to leading terms in heat-kernel asymptotic expansions. We then use Tr_p^Λ to prove that the leading term in the asymptotic expansion of $\mathrm{tr}(\Omega e^{-\varepsilon Q})$ is closed, where Q is a generalized Laplacian and Ω the curvature on a ΨDO bundle. In fact, we show that if Q has positive leading symbol $\sigma_L(Q)(x, \xi) = f(x, \xi)\mathrm{Id}$ and if Ω has integer order $a > -\dim(M)$, then this leading term is given by the leading symbol trace $\mathrm{Tr}_a^f(\Omega)$.

The following folklore result follows from the analysis developed in [7], while the analysis in the following proof is hidden in the local nature of the Wodzicki residue.

Proposition 3.4. *Let $A \in \mathcal{C}l_{\leq 0}(M, E)$ have integral order $a > -n = -\dim(M)$, and let Q be an elliptic ΨDO of order q with positive scalar leading symbol $\sigma_L(Q)(x, \xi) = f(x, \xi)\mathrm{Id}$. Let $c = c(n, a, q)$ be*

$$c = \frac{\Gamma(\frac{n+a}{q})\dim(E)(n-1)!}{q(2\pi)^n}.$$

Then as $\varepsilon \to 0$,

$$\mathrm{tr}(Ae^{-\varepsilon Q}) = c \int_{S^*M} \mathrm{tr}\,(\sigma_a(A)(x, \xi))\,(f(x, \xi))^{-\frac{n+a}{q}} \cdot \varepsilon^{-\frac{n+a}{q}} + \mathrm{o}\left(\varepsilon^{-\frac{n+a}{q}}\right).$$

In particular, if $\sigma_L(Q)(x, \xi) = \|\xi\|^k$ for some k, then

$$\mathrm{tr}(Ae^{-\varepsilon Q}) = c \int_{S^*M} \mathrm{tr}\,(\sigma_a(A)) \cdot \varepsilon^{-\frac{n+a}{q}} + \mathrm{o}\left(\varepsilon^{-\frac{n+a}{q}}\right).$$

Proof. We want to compute the coefficient $a_0(A, Q)$ in the known asymptotic expansion

$$\mathrm{tr}(Ae^{-\varepsilon Q}) = \sum_{j=0}^{a+n} a_j(A, Q)\varepsilon^{\frac{j-a-n}{q}} + b_0(A, Q)\log\varepsilon + O(1), \tag{3.8}$$

for general $A \in \mathcal{C}l(M, E)$ of order a, and with $a_j(A, Q), b_0(A, Q) \in \mathbb{C}$. The coefficient $b_0(A, Q)$ satisfies $b_0(A, Q) = -\frac{1}{q}\mathrm{res}_w(A)$.

A Mellin transform yields

$$a_0(A, Q) = \mathrm{Res}_{z=\frac{n+a}{q}} \Gamma\left(\frac{n+a}{q}\right) \mathrm{tr}(AQ^{-z}),$$

(the case $A = 1$ considered in [10, (12)] easily extends to a general ΨDO A). Thus, for A as in the hypothesis,

$$a_0(A, Q) = \frac{\Gamma(\frac{n+a}{q})}{q\Gamma(n+a)} a_0\left(A, Q^{\frac{1}{q}}\right).$$

Thus it suffices to prove the formula for Q_1 of order one. Since $\mathrm{ord}(AQ_1^{-(n+a)}) = -n$, (3.8) becomes

$$\mathrm{tr}\left(AQ_1^{-(n+a)} e^{-\varepsilon Q_1}\right) = -\mathrm{res}_w\left(AQ_1^{-(n+a)}\right) \log \varepsilon + O(1).$$

Differentiating this expansion $n + a$ times (recall that $n + a$ is a positive integer) with respect to ε, we get

$$\mathrm{tr}(Ae^{-\varepsilon Q_1}) \sim (n + a - 1)! \, \mathrm{res}_w\left(AQ_1^{-(n+a)}\right)\varepsilon^{-(n+a)}.$$

The local formula for the Wodzicki residue yields:

$$\mathrm{res}_w(AQ_1^{-(n+a)}) = \frac{1}{(2\pi)^n} \int_{S^*M} \mathrm{tr}\left(\sigma_{-n}(AQ_1^{-(n+a)})\right)$$

$$= \frac{1}{(2\pi)^n} \int_{S^*M} \mathrm{tr}\left(\sigma_a(A)\sigma_{-(n+a)}(Q_1^{-(n+a)})\right)$$

$$= \frac{\dim(E)}{(2\pi)^n} \int_{S^*M} \mathrm{tr}(\sigma_a(A)) f(x, \xi)^{-(n+a)}.$$

Hence our original Q has

$$a_0(A, Q) = \frac{\Gamma(\frac{n+a}{q})}{q\Gamma(n+a)} a_0(A, Q^{\frac{1}{q}}) = c \int_{S^*M} \mathrm{tr}\,(\sigma_a(A))\,(f(x, \xi))^{-(n+a)/q}. \qquad \square$$

Lemma 3.5. *Let* $\mathcal{E} \to B$ *be a* ΨDO *bundle with a* ΨDO *connection* ∇ *whose connection one-form* θ *takes values in* $\mathcal{C}l_{\leq 0}(M, E)$. *Let* $A \in \Omega^k(B, \mathcal{C}l(\mathcal{E}))$ *be a* $\mathcal{C}l_{\leq a}(\mathcal{E})$-*valued form whose order* a *is independent of* $b \in B$.

(i) *For any distribution* $\Lambda \in \mathcal{D}'(S^*M)$, $d\mathrm{Tr}_a^\Lambda(A) = \mathrm{Tr}_a^\Lambda([\nabla, A])$. *In particular, if* $[\nabla, A] = 0$, *then* $\mathrm{Tr}_a^\Lambda(A) \in \Omega^k(B, \mathbb{C})$ *is closed.*

(ii) *Let* $Q = \{Q_b\} \in \Gamma(\mathcal{C}l(\mathcal{E}))$ *be a smooth family of elliptic operators of constant order* q *and with positive scalar leading symbol independent of* $b \in B$. *Define* $a_0(A, Q), b_0(A, Q) \in \Omega^k(B, \mathbb{C})$ *as in* (3.8). *If* $[\nabla, A] = 0$, *then* $b_0(A, Q)$ *is closed, and* $a_0(A, Q)$ *is closed if* $a \in \mathbb{Z}, a > -\dim(M)$.

Note that the condition on the leading symbol is independent of trivialization of $\mathcal{E}$.

Proof. (i) We have

$$d\text{Tr}_a^\Lambda(A) = \Lambda\,[d\text{tr}_x(\sigma_a(A))] = \Lambda\,[\text{tr}_x\,(\sigma_a(dA))]$$
$$= \Lambda\,[\text{tr}_x\,(\sigma_a(dA + [\theta, A]))] = \Lambda\,[\text{tr}_x\,(\sigma_a([\nabla, A]))].$$

Here we use the fact that θ has non-positive order, so that

$$\text{tr}_x\,(\sigma_a([\theta, A])) = \text{tr}_x\,([\sigma_0(\theta), \sigma_a(A)]) = 0,$$

if $[\theta, A]$ has expected order a. Finally, $\sigma_a([\theta, A]) = 0$ trivially if the order of $[\theta, A]$ is less than a.

(ii) $a_0(A, Q)$ is the leading term in the asymptotic expansion (3.8) and hence proportional to a leading symbol trace by the above proposition. It is therefore closed by (i). Since $b_0(A, Q) = -\frac{1}{q}\text{res}_w(A)$, it is closed by §3.1. $\qquad\square$

As a consequence, we can build "Chern–Weil type" closed forms from leading symbol traces Tr_p^Λ.

Theorem 3.6. *Let $\mathcal{E} \to B$ be a ΨDO bundle with a ΨDO connection ∇ whose connection one-form θ takes values in $Cl_{\leq 0}(\mathcal{E})$, and whose curvature two-form (which takes values in $Cl_{\leq 0}(\mathcal{E})$) has constant order a. Let $Q \in \Gamma\,(Cl(\mathcal{E}))$ be a smooth family of elliptic operators of constant order q and with positive scalar leading symbol independent of $b \in B$. In the notation of (3.8), the following elements of $\Omega^{2k}(B, \mathbb{C})$ are closed:*

(i) $\text{Tr}_{ka}^\Lambda(\Omega^k)$, *for any $\Lambda \in \mathcal{D}'(S^*M)$;*

(ii) $a_0(\Omega^k, Q)$, *for $ka \in \mathbb{Z}, a > -\frac{n}{k}$, where $n = \dim(M)$;*

(iii) $b_0(\Omega^k, Q)$. *Moreover, the cohomology class of $b_0(\Omega^k, Q)$ is independent of the choice of connection ∇.*

Let $\{\nabla_t : t \in [0, 1]\}$ be a smooth family of $Cl_{\leq 0}(\mathcal{E})$ connections such that $\dot{\nabla}_t \in \Omega^1(B, Cl_{\leq a}(\mathcal{E}))$ and $\Omega_t \in \Omega^2(B, Cl_{\leq a}(\mathcal{E}))$. The following de Rham cohomology classes are independent of t:

(iv) $[\text{Tr}_{ka}^\Lambda(\Omega_t^k)]$;

(v) $[a_0(\Omega_t^k, Q)]$ *for $ka \in \mathbb{Z}, a > -\frac{n}{k}$, where $n = \dim(M)$.*

Note that when $a = 0$, this gives back the results of Theorem 3.2.

Proof. (i)–(iii) follow from Lemma 3.5. The fact that $[b_0(\Omega^k, Q)]$ is independent of the choice of connection follows from the results of §3.1, since $b_0(\Omega^k, Q)$ is proportional to $\text{res}_w(\Omega^k)$. For (iv), we repeat (2.2), with λ replaced by Tr_{ka}^Λ. Note that we have to use Lemma 3.5 to swap the (covariant) differentiation and Tr_{ka}^Λ in this argument. Finally, since $a_0(\Omega_t^k, Q)$ is a leading order symbol by Proposition 3.4, we get (v). $\qquad\square$

Remark. Theorem 3.6 does not apply to Freed's conditional first Chern form on loop groups. Even though, as we will see in §4, this Chern form corresponds to the finite part $a_0(\Omega, Q_0)$, the curvature of the Levi-Civita connection for the $H^{1/2}$ metric on LG has order $a = -1 = -\dim(S^1)$, the borderline case for Theorem 3.6. Showing that Freed's conditional first Chern form on loop groups is closed [6] requires the more refined analysis of §4.

4 Characteristic classes and formal power series

In this section, we use heat kernel regularized traces to produce an asymptotic series of characteristic forms, provided the regularizing family of operators $\{Q_b\}$ is "fairly covariantly constant." This improves the weighted trace approach of [17], and is based on regularization techniques common in quantum field theory.

We begin with some calculations leading to Lemma 4.1, which measures the effect of trying to push a connection ∇ on a bundle $\mathcal{E}$ past a heat operator or a weighted trace. For $\{A_0, A_1, \ldots, A_n\} \subset \mathcal{C}l(M, E)$ and $(\sigma_0, \ldots, \sigma_n) \in (\mathbb{R}^+)^{n+1}$, the operator $A_0 e^{-\sigma_0 Q} A_1 e^{-\sigma_1 Q} \ldots A_n e^{-\sigma_n Q}$ is smoothing and hence trace class. We define *trace forms*

$$\langle A_0, A_1, \ldots, A_n \rangle_{\varepsilon, n, Q} := \int_{\Delta_n} \mathrm{tr} \left(A_0 e^{-\varepsilon \sigma_0 Q} A_1 e^{-\varepsilon \sigma_1 Q} \ldots A_n e^{-\varepsilon \sigma_n Q} \right), \quad (4.1)$$

where Δ_n is the standard n-simplex, in agreement with [9] (although there the A_i are bounded). In particular, we call the Q-weighted trace of A (with ε-cut-off) the linear functional $\langle A_0 \rangle_{\varepsilon, 0, Q} = \mathrm{tr}_\varepsilon^Q(A_0)$.

The concept of trace form and hence of weighted trace extends to sections of a ΨDO bundle. Recall that a ΨDO bundle $\mathcal{E}$ with structure group $\mathcal{C}l_0^*(M, E)$ has an associated bundle of algebras $\mathcal{C}l_{\leq 0}(\mathcal{E}) = \mathrm{Ad}\, P^{\mathcal{E}}$ with fibers modeled on $\mathcal{C}l_{\leq 0}(M, E)$. A weight is a section $Q \in \Gamma(\mathcal{C}l(\mathcal{E}))$ with Q elliptic with positive definite leading symbol and of constant order. These conditions are independent of local chart, since the transition maps are ΨDOs. In particular, if $\{g_b\}$ is the transition map between two trivializations of $\mathcal{E}$ over b, then Q_b transforms into $g_b^{-1} Q_b g_b$; the same holds for sections of $\mathcal{C}l(\mathcal{E})$. For $A \in \Gamma(\mathcal{C}l(\mathcal{E}))$, $\mathrm{tr}_\varepsilon^Q(A)$ is well-defined, since

$$\mathrm{tr}_\varepsilon^{g^{-1}Qg}(g^{-1}Ag) = \mathrm{Tr}(g^{-1}Age^{-\varepsilon g^{-1}Qg}) = \mathrm{Tr}(g^{-1}Agg^{-1}e^{-\varepsilon Q}g) = \mathrm{tr}_\varepsilon^Q(A). \quad (4.2)$$

In the same way, for $\{A_0, \ldots, A_n\} \subset \Gamma(\mathcal{C}l(\mathcal{E}))$, the trace form $\langle A_0, \ldots, A_n \rangle_{\varepsilon, n, Q}$ is well-defined .

We set $\mathrm{tr}^Q(A)$ to be the finite part of $\mathrm{tr}_\varepsilon^Q(A)$ as $\varepsilon \to 0$. In other words, $\mathrm{tr}^Q(A)$ is the coefficient of ε^0 in the asymptotic expansion (3.8). This is equivalent to taking the zeta function regularization $\mathrm{Tr}(AQ^{-z})|_{z=0}$, provided Q is invertible and (3.8) contains no log terms.

If $Q = Q_0 + Q_1$, with Q_0 elliptic of order $q_0 > 0$ and Q_1 of order $q_1 < q_0$, the Volterra formula [1, 3, 8] (the first and third references treat the Banach algebra setting) states

$$e^{-\varepsilon(Q_0+Q_1)} = \sum_{k=0}^{\infty}(-\varepsilon)^k \int_{\Delta^k} e^{-\sigma_0 \varepsilon Q_0} Q_1 e^{-\sigma_1 \varepsilon Q_0} Q_1 \cdots Q_1 e^{-\sigma_k \varepsilon Q_0} d\sigma_0 d\sigma_1 \ldots d\sigma_k.$$

The convergence holds in the trace operator norm topology, and so

$$\mathrm{tr}\left(e^{-\varepsilon(Q_0+Q_1)}\right) = \sum_{k=0}^{\infty}(-\varepsilon)^k \langle 1, Q_1, \ldots, Q_1 \rangle_{\varepsilon,k,Q_0}.$$

For the moment, let $\mathcal{E}$ be a trivial vector bundle over B modeled on $C^\infty(M, E)$ or $H^s(M, E)$, with structure group $Cl_0^* = Cl_0^*(M, E)$, and with the trivial connection d. Let Q be a weight on $\mathcal{E}$. For $h \in T_{b_0} B$, writing $Q_b = Q_{b_0} + dQ(b_0) \cdot h + \mathrm{o}(h)$ and substituting $Q_0 = Q_{b_0}$, $Q_1 = dQ(b_0) \cdot h + \mathrm{o}(h)$ in (4.1) yields

$$e^{-\varepsilon Q_b} - e^{-\varepsilon Q_{b_0}} = -\varepsilon \int_0^1 e^{-\varepsilon t Q_{b_0}} (dQ(b_0) \cdot h) e^{(1-t)\varepsilon Q_{b_0}} dt + \mathrm{o}(h)$$

in the trace operator norm topology. From this we derive Duhamel's formula:

$$de^{-\varepsilon Q} = -\varepsilon \int_0^1 e^{-\varepsilon t Q} dQ e^{-(1-t)\varepsilon Q} dt = -\varepsilon \int_0^1 e^{-(1-t)\varepsilon Q} dQ e^{-\varepsilon t Q} dt. \quad (4.3)$$

Remark. In this derivation we implicitly restrict attention to a compact subset K of B, so that the $\mathrm{o}(h)$ term is uniform on K (see [3]). This applies throughout this section. In particular, we check that a form ω is closed on B by evaluating $d\omega$ over every closed cycle in B. Since the image of a cycle is compact, $d\omega$ is well defined, and formulas like (4.3) are valid. Moreover, we can use Duhamel's formula to justify differentiating asymptotic expansions of the form $\mathrm{Tr}(A_b e^{-tQ_b})$ term by term, provided the asymptotic expansions contain $\varepsilon^{\pm k/q}$ terms (possibly with zero coefficients) with the k ranging over a subset of $\mathbb{Z}$ independent of b. This is certainly the case if the order of A is constant in b.

For $A, Q = Q_{b_0}$ as above, we also have

$$[e^{-\varepsilon Q}, A] = -\varepsilon \int_0^1 e^{-(1-t)\varepsilon Q} [Q, A] e^{-\varepsilon t Q} dt. \quad (4.4)$$

Indeed, differentiating the map $t \mapsto [e^{-tQ}, A]$ (which is differentiable as a bounded linear map from $H^{a+q}(\mathcal{E})$ to $H^0(\mathcal{E})$ for $a = \mathrm{ord}(A)$, $q = \mathrm{ord}(Q)$), we get $\left(\frac{d}{dt} + Q\right)[e^{-tQ}, A] = [A, Q]e^{-tQ}$. Solving this equation by (the other) Duhamel's formula for first order inhomogeneous linear differential equations gives $[e^{-\varepsilon Q}, A] = \int_0^1 e^{-sQ}[Q, A]e^{-(\varepsilon-s)Q} dt$, and substituting $t = \varepsilon s$ yields (4.4). This identity, which holds *a priori* in the space of bounded linear maps from $H^{a+q}(\mathcal{E})$ to $H^0(\mathcal{E})$, persists as long as both sides of the equation make sense.

Replacing A by $[A, Q]$ and Q by σQ in (4.4) yields

$$e^{-(1-\sigma)\varepsilon Q}[Q, A]e^{-\varepsilon\sigma Q}$$

$$= e^{-\varepsilon Q}[Q, A] + \varepsilon\sigma \int_0^1 e^{(-(1-\sigma_1)\sigma-(1-\sigma))\varepsilon Q}[Q, [Q, A]]e^{-\varepsilon\sigma\sigma_1 Q}d\sigma_1$$

$$= e^{-\varepsilon Q}[Q, A] + \varepsilon\sigma \int_0^1 e^{-(1-\sigma_1\sigma)\varepsilon Q}[Q, [Q, A]]e^{-\varepsilon\sigma\sigma_1 Q}d\sigma_1$$

$$= e^{-\varepsilon Q}[Q, A] + \varepsilon \int_0^\sigma e^{-(1-\sigma_1)\varepsilon Q}[Q, [Q, A]]e^{-\varepsilon\sigma_1 Q}d\sigma_1,$$

and so

$$[e^{-\varepsilon Q}, A] = -\varepsilon e^{-\varepsilon Q}[Q, A] - \varepsilon^2 \int_{\Delta_2} e^{-(1-\sigma_1)\varepsilon Q}[Q, [Q, A]]e^{-\sigma_1\varepsilon Q}d\sigma_1 d\sigma_0. \quad (4.5)$$

For a ΨDO A, define $[A]_Q^j$, $j \in \mathbb{N} \cup \{0\}$, by

$$[A]_Q^0 = A, \quad [A]_Q^{j+1} = [Q, [A]_Q^j] = (\text{ad } Q)^{j+1}(A).$$

We now make the important assumption that Q have scalar symbol. Iterating (4.5) gives

$$[e^{-\varepsilon Q}, A] = -\sum_{j=1}^{N-1} \frac{\varepsilon^j}{j!}[A]_Q^j e^{-\varepsilon Q} + R_{A,N}(\varepsilon), \quad (4.6)$$

for $N \in \mathbb{N}$, with

$$R_{A,N}(\varepsilon) = \varepsilon^N \int_{\Delta_N} e^{-\varepsilon(1-\sigma_1)Q}[A]_Q^N e^{-\sigma_1\varepsilon Q}d\sigma_1 \ldots d\sigma_N.$$

One can check that for $k > 0$, $R_{A,N}(\varepsilon) = O(\varepsilon^k)$ for $N = N(k) \gg 0$ (cf. [12, Lemma 4.2]). In particular, we have

$$\text{tr}_\varepsilon^Q([A, B]) = \text{tr}(e^{-\varepsilon Q}[A, B]) = \text{tr}([e^{-\varepsilon Q}, A]B)$$

$$= \sum_{j=1}^{N-1} \frac{\varepsilon^j}{j!}\text{tr}_\varepsilon^Q(A[B]_Q^j) + O(\varepsilon^k) \quad (4.7)$$

for $N \gg 0$. Letting $Q = Q_b$ vary again, and using (4.3), (4.6), we obtain

$$d e^{-\varepsilon Q} = -\sum_{j=1}^N \frac{\varepsilon^j}{j!}[d Q]_Q^{j-1} e^{-\varepsilon Q} + \tilde{R}_{dQ,N+1},$$

where $\tilde{R}_{dQ,N+1}(\varepsilon) := -\varepsilon \int_0^1 \sigma^{N+1} e^{-\varepsilon Q} R_{dQ,N+1}(\varepsilon\sigma) d\sigma$. Taking traces yields

$$\text{tr}\big((de^{-\varepsilon Q})A\big) = -\sum_{j=1}^N \frac{\varepsilon^j}{j!} \text{tr}_\varepsilon^Q\big([d\,Q]_Q^{j-1}A\big) + O(\varepsilon^k). \tag{4.8}$$

We now pass to the general setting by dropping the assumption that $\mathcal{E}$ is trivial. We assume that $\mathcal{E}$ has a ΨDO connection ∇.

Lemma 4.1. (i) *For* $\varepsilon > 0$,

$$[\nabla, e^{-\varepsilon Q}] = -\sum_{j=1}^N \frac{\varepsilon^j}{j!} [[\nabla, Q]]_Q^{j-1} e^{-\varepsilon Q} + \tilde{R}_{[\nabla,Q],N+1}(\varepsilon)$$

where $\tilde{R}_{[\nabla,Q],N+1}(\varepsilon) = -\varepsilon \int_0^1 \sigma^{N+1} e^{-\varepsilon Q} R_{[\nabla,Q],N+1}(\varepsilon\sigma) d\sigma$.

(ii) *For* $\alpha \in \Omega^*(B, Cl(\mathcal{E}))$ *and* $k > 0$, *there exists* $N \gg 0$ *such that*

$$[\nabla, \text{tr}_\varepsilon^Q](\alpha) := \big(\nabla\,\text{tr}_\varepsilon^Q - \text{tr}_\varepsilon^Q \nabla\big)(\alpha) = -\sum_{j=1}^N \frac{\varepsilon^j}{j!} \text{tr}_\varepsilon^Q\big([[\nabla, Q]]_Q^{j-1}\alpha\big) + O(\varepsilon^k).$$

Proof. Locally, we have $\nabla = d + \theta$ where θ is a local $Cl(M, E)$-valued one-form on B. We can apply (4.6), (4.8) to obtain

$$[\nabla, e^{-\varepsilon Q}] = d\,e^{-\varepsilon Q} + [\theta, e^{-\varepsilon Q}]$$

$$= -\sum_{j=1}^N \frac{\varepsilon^j}{j!}[d\,Q]_Q^{j-1} e^{-\varepsilon Q} + e^{-\varepsilon Q}\sum_{j=1}^N \frac{\varepsilon^j}{j!}[[Q, \theta]]_Q^{j-1} + \tilde{R}_{[\nabla,Q],N+1}(\varepsilon)$$

$$= -e^{-\varepsilon Q}\sum_{j=1}^N \frac{(-\varepsilon)^j}{j!}[[\nabla, Q]]_Q^{j-1} + \tilde{R}_{[\nabla,Q],N+1}(\varepsilon),$$

where $\tilde{R}_{[\nabla,Q],N+1}(\varepsilon) := \tilde{R}_{dQ,N+1}(\varepsilon) + \tilde{R}_{[\theta,Q],N+1}(\varepsilon)$. For α a $Cl(M, E)$-valued form on B, we get by (4.7)

$$[\nabla, \text{tr}_\varepsilon^Q](\alpha) = d\,\text{tr}(e^{-\varepsilon Q}\alpha) - \text{tr}\big(e^{-\varepsilon Q}[\nabla, \alpha]\big)$$

$$= \text{tr}\big((de^{-\varepsilon Q})\alpha\big) - \text{tr}(e^{-\varepsilon Q}[\theta, \alpha])$$

$$= -\sum_{j=1}^N \frac{\varepsilon^j}{j!}\text{tr}_\varepsilon^Q\big([[\nabla, Q]]_Q^{j-1}\alpha\big) + O(\varepsilon^k)$$

provided N is chosen so large that $\text{tr}_\varepsilon^Q(\alpha \tilde{R}_{[\nabla,Q],N+1}(\varepsilon)) = O(\varepsilon)$. $\qquad\square$

We now extend a familiar construction for ordinary algebras [8, 16] to bundles of algebras by considering $Cl(\mathcal{E})[[\varepsilon]]$, the space of formal power series in the variable ε with coefficients in $Cl = Cl(\mathcal{E})$. Thus an element $A(\varepsilon)$ of $Cl[[\varepsilon]]$ has the form

$A(\varepsilon) = \sum_{j=0}^{\infty} A_j \varepsilon^j$, $A_j \in \Gamma(\mathcal{C}l)$. The reader unhappy with these formal sums can just work with finite sums with error estimates, as in Theorems 4.4, 4.6 below.

Recall that the algebra $Cl(M, E)$ of classical (polyhomogeneous) ΨDOs is given by finite sums $\sum_{i=1}^{n} A_i$, where each A_i is polyhomogeneous in the sense that the symbol of A_i has an asymptotic expansion $\sigma(A_i) \sim \sum_{j=0}^{\infty} \sigma_{o_i-j}$, $o_i = \mathrm{ord}(A_i)$ with σ_{o_i-j} having the standard homogeneity and growth conditions for the symbol class S^{o_i-j} [7]. The order of A is then the maximum of the o_j. It is standard that each A_i, has an asymptotic expansion as $\varepsilon \to 0$ of the form

$$\mathrm{tr}_{\varepsilon}^{Q}(A_i) \sim \sum_{j=0}^{\infty} a_j(A_i, Q)\varepsilon^{\frac{j-o_i-n}{q}} + \sum_{k=0}^{\infty} b_k(A_i, Q)(\log \varepsilon)\varepsilon^k + \sum_{\ell=0}^{\infty} c_\ell(A_i, Q)\varepsilon^\ell, \quad (4.9)$$

where $a_j(A, Q), b_k(A, Q), c_\ell(A, Q) \in \mathbb{C}$ and $n = \dim(M)$. Thus A has a similar asymptotic expansion. We set

$$\left(\mathrm{tr}_{\varepsilon}^{Q}(A_i)\right)_{\mathrm{asy}} := \sum_{j=0}^{\infty} a_j(A_i, Q)\varepsilon^{\frac{j-o_i-n}{q}} + \sum_{k=0}^{\infty} b_k(A_i, Q)(\log \varepsilon)\varepsilon^k$$

$$+ \sum_{\ell=0}^{\infty} c_\ell(A_i, Q)\varepsilon^\ell \in \mathbb{C}[\log \varepsilon][\varepsilon^{-\frac{1}{q}}, \varepsilon^{\frac{1}{q}}]],$$

and define $\left(\mathrm{tr}_{\varepsilon}^{Q}(A)\right)_{\mathrm{asy}}$ by linearity. Given a weight $Q \in \Gamma(\mathcal{C}l(\mathcal{E}))$ as above, the $\mathbb{C}$-linear morphism $\mathrm{tr}_{\varepsilon}^{Q}$ defined for fixed ε partially extends to a $\mathbb{C}[[\varepsilon]]$-morphism

$$\mathrm{tr}_{\varepsilon}^{Q} : \Gamma(\mathcal{C}l(\mathcal{E})[[\varepsilon]]) \to \mathbb{C}[\log \varepsilon][\varepsilon^{-\frac{1}{q}}, \varepsilon^{\frac{1}{q}}]], \quad A = \sum_{k=0}^{\infty} A_k \varepsilon^k \mapsto \sum_{k=0}^{\infty} \left(\mathrm{tr}_{\varepsilon}^{Q}(A_k)\right)_{\mathrm{asy}} \varepsilon^k.$$

$$(4.10)$$

In the last term in (4.10), we formally rearrange the sum to produce an element of $\mathbb{C}[\log \varepsilon][\varepsilon^{-\frac{1}{q}}, \varepsilon^{\frac{1}{q}}]]$, provided the number of terms contributing to each ε^ℓ and $(\log \varepsilon)\varepsilon^\ell$ is finite.

It is not hard to give conditions that guarantee that $\mathrm{tr}_{\varepsilon}^{Q}\left(\sum_{k=0}^{\infty} A_k \varepsilon^k\right)$ exists in this formally rearranged sense:

Lemma 4.2. *If the* $a_i := \mathrm{ord}(A_i)$ *satisfy* $\lim_{i \to \infty} qi - a_i = \infty$, *then* $\mathrm{tr}_{\varepsilon}^{Q}\left(\sum_{i=0}^{\infty} A_i \varepsilon^i\right)$ *exists as a rearranged sum.*

Proof. We may assume that each A_i is classical polyhomogeneous, as replacing A_i by a finite sum of such operators does not affect the proof. For fixed i, j, the term $a_j(A_i, Q)$ in (4.9) appears in $\mathrm{tr}_{\varepsilon}^{Q}\left(\sum_{i=0}^{\infty} A_i \varepsilon^i\right)$ as a coefficient of $\varepsilon^{\frac{j-a_i-n+qi}{q}}$. The hypothesis guarantees that only a finite number of i, j can contribute to the coefficient of a fixed ε^{k_0}. Similar arguments apply to $b_k(A_i, Q)$, $c_\ell(A_i, Q)$. $\square$

Motivated by Lemma 4.1, we introduce a $C\ell(\mathcal{E})[[\varepsilon]]$-valued connection ∇_ε^Q defined in terms of the connection $\nabla^{C\ell(\mathcal{E})} = [\nabla, \cdot] = \nabla^{\text{Ad } P^{\mathcal{E}}}$ on $C\ell(\mathcal{E})$ and the weight Q:

$$\nabla_\varepsilon^Q \alpha := \nabla^{C\ell(\mathcal{E})} \alpha - \sum_{j=1}^{\infty} \frac{\varepsilon^j}{j!} [[\nabla, Q]]_Q^{j-1} \alpha, \quad \alpha \in \Omega^*(B, C\ell(\mathcal{E})). \qquad (4.11)$$

We now show that ∇_ε^Q has the key property of commuting with the weighted trace tr_ε^Q:

Lemma 4.3. *Let ∇ be a $C\ell_{\leq 0}$-connection on $\mathcal{E}$, and let Q be a weight on $\mathcal{E}$ with scalar leading symbol. For $\alpha \in \Omega^*(B, C\ell(\mathcal{E}))$, we have*

$$d \circ \text{tr}_\varepsilon^Q \alpha = \text{tr}_\varepsilon^Q \circ \nabla_\varepsilon^Q \alpha.$$

Proof. By Lemma 4.1, we have

$$d \circ \text{tr}_\varepsilon^Q \alpha - \text{tr}_\varepsilon^Q \circ \nabla_\varepsilon^Q \alpha = [\nabla_\varepsilon^Q, \text{tr}_\varepsilon^Q](\alpha)$$

$$= [\nabla, \text{tr}_\varepsilon^Q](\alpha) - \sum_{j=1}^{\infty} \frac{(-\varepsilon)^j}{j!} \text{tr}_\varepsilon^Q \left([[\nabla, Q]]_Q^{j-1} \alpha \right) = 0,$$

provided we show that tr_ε^Q can be applied to $\nabla_\varepsilon^Q \alpha$. In fact, if $d := \text{ord}[\nabla, Q] \leq q$, then the order of $[[\nabla, Q]]_Q^{j-1}$ is $a_j \leq d + (j-1)(q-1)$, since Q has scalar leading symbol. Thus the hypothesis of Lemma 4.2 is satisfied. $\qquad\square$

Remarks. (i) The preceding proof assumes that Lemma 4.1 extends to formal power series of operators. This justification, while not difficult, is somewhat lengthy and is omitted.

(ii) If ∇_ε^Q were induced from a connection on $\mathcal{E}$, Lemma 4.3 would guarantee a Chern–Weil theory for the curvature Ω_ε^Q: each coefficient in $\text{tr}_\varepsilon^Q(\Omega_\varepsilon^Q)$ would be a closed from independent of the connection. However, we will see in Corollary 4.7 that for loop groups, the leading order coefficient is the Kähler form $\text{Tr}^Q(\Omega)$ for the $H^{1/2}$ Levi-Civita connection. The corresponding non-zero Kähler class is certainly not independent of connection, since TLG is trivial. Theorem 4.6 gives a more refined analysis of this example.

Despite the last remark, we can use Lemma 4.3 to produce a Chern–Weil theory under additional hypotheses.

Theorem 4.4. *Let ∇ be a $C\ell_{\leq 0}$-connection and Q a weight on $\mathcal{E}$ of order q and with scalar leading symbol. Let d be the order of the $C\ell(\mathcal{E})$-valued form $[\nabla, Q]$, and set $r := q - d$.*

(i) *For $\alpha \in \Omega^*(B, C\ell(\mathcal{E}))$ of constant order a, we have*

$$d\text{tr}_\varepsilon^Q(\alpha) = \text{tr}_\varepsilon^Q(\nabla^{C\ell(\mathcal{E})}\alpha) + o\left(\varepsilon^{\frac{-a-n+r-\eta}{q}}\right),$$

for all $\eta > 0$.

(ii) *Let Ω, the curvature of ∇, have constant order a. Then the coefficient of $\varepsilon^{\frac{\gamma}{q}}$ in the asymptotic expansion of $\mathrm{tr}_\varepsilon^Q(\Omega^k)$ is closed, for all $\gamma < -ka - n + r$. In particular, if $r > 0$, the coefficient of the leading order term $\varepsilon^{\frac{-ka-n}{q}}$ is closed. The coefficients of $\log \varepsilon \cdot \varepsilon^\ell$ are closed for all $\ell < \frac{-ka-n+r}{q}$.*

(iii) *Let Ω have constant order a. The coefficient of $\log \varepsilon$ in the asymptotic expansion of $\mathrm{tr}_\varepsilon^Q(\Omega^k)$ is closed.*

Note that part (ii) of the theorem only applies if $r > 0$, which occurs e.g. if the leading order symbol of Q is independent of $b \in B$. We need $r > ka + n$ to obtain information about the $\log \varepsilon \cdot \varepsilon^\ell$ terms with $\ell > 0$.

Proof. (i) By Lemma 4.3, we have

$$d\mathrm{tr}_\varepsilon^Q(\alpha) = \mathrm{tr}_\varepsilon^Q(\nabla_\varepsilon^Q \alpha) = \mathrm{tr}_\varepsilon^Q(\nabla^{Cl(\mathcal{E})}\alpha) - \sum_{j=1}^\infty \frac{\varepsilon^j}{j!}\mathrm{tr}_\varepsilon^Q([[\nabla, Q]]_Q^{j-1}\alpha).$$

We want to show that for $\eta > 0$,

$$\lim_{\varepsilon \to 0} \varepsilon^{\frac{n+a-r+\eta}{q}}\left(\sum_{j=1}^\infty \frac{\varepsilon^j}{j!}\mathrm{tr}_\varepsilon^Q([[\nabla, Q]]_Q^{j-1}\alpha)\right) = 0.$$

Since the infinite sum is a rearrangeable formal power series, we mean that each exponent k_0 of the rearranged series satisfies $\frac{n+a-r+\eta}{q} + k_0 > 0$. Since the leading asymptotic term $a_j\varepsilon^{\gamma_j}$ of $\mathrm{tr}_\varepsilon^Q([[\nabla, Q]]_Q^{j-1}\alpha)$ contributes the exponent $j+\gamma_j$, it suffices to show that

$$\frac{n+a-r+\eta}{q} + j + \gamma_j > 0, \tag{4.12}$$

for all $j \in \mathbb{N}$. (A similar argument treats the case where the leading asymptotic term contains $\log \varepsilon$.) As in Lemma 4.3, $d_j := \mathrm{ord}([[\nabla, Q]]_Q^{j-1})$ satisfies $d_j \le d + (j-1)(q-1)$ for $d := \mathrm{ord}[\nabla, Q]$. Thus

$$\gamma_j \ge \frac{-d_j - a - n}{q} \ge \frac{r+j-1-a-n-jq}{q},$$

which implies (4.12).

(ii) Since $\nabla^{Cl(\mathcal{E})}\Omega^k = 0$ and $\mathrm{ord}(\Omega^k) = ka$, it follows from (i) that

$$d\mathrm{tr}_\varepsilon^Q(\Omega^k) = -\sum_{j=1}^\infty \frac{\varepsilon^j}{j!}\mathrm{tr}_\varepsilon^Q([[\nabla, Q]]_Q^{j-1}\Omega^k) = o\left(\varepsilon^{\frac{-ka-n+r-\eta}{q}}\right), \tag{4.13}$$

for all $\eta > 0$. Thus all coefficients of powers ε^γ with $\gamma < -ka - n + r$ are closed. A similar argument handles the $\log \varepsilon$ terms.

Finally, since the coefficient of the $\log \varepsilon$ term is proportional to the Wodzicki residue $\text{res}_W(\Omega)$, (iii) follows from the discussion in §3.1.   □

This clarifies the non-closed weighted traces of [17].

Corollary 4.5. *Let* ∇ *be a* $Cl_{\leq 0}$ *connection,* Q *a weight on* $\mathcal{E}$ *with scalar leading symbol,* $\Omega = \nabla^2$ *the curvature of* ∇, *and* $\text{tr}^Q(\Omega^k)$ *its* Q-*weighted trace, i.e.* $\text{tr}^Q(\Omega^k)$ *is the finite part of* $\text{tr}_\varepsilon^Q(\Omega^k)$ *as* $\varepsilon \to 0$. *Then* $d\text{tr}^Q(\Omega^k)$ *is an explicit finite linear combination of coefficients in the asymptotic expansion of* $\text{tr}_\varepsilon^Q\big([[\nabla, Q]]_Q^{j-1}\Omega^k\big)$, $j \in \mathbb{N}$.

Proof. This follows from (4.13) and the fact that $\sum_{j=1}^{\infty} \frac{\varepsilon^j}{j!}\text{tr}_\varepsilon^Q\big([[\nabla, Q]]_Q^{j-1}\Omega^k\big)$ is rearrangeable. In particular, the coefficient of ε^0 is constructed as stated.   □

We now discuss the independence of the closed forms in Theorem 4.4 on the choice of connection. Note that the hypotheses are more stringent than in Theorem 4.4.

Theorem 4.6. (i) *Let* Q *be a weight on* $\mathcal{E}$ *with scalar leading symbol, and let* $\{\nabla_t : t \in [0, 1]\}$ *be a smooth family of* $Cl_{\leq 0}(\mathcal{E})$ *connections such that* $\dot{\nabla}_t \in \Omega^1(B, Cl_{\leq -s}(\mathcal{E}))$, *for some* $s \geq 0$, *for all* t. *Then*

$$\frac{d}{dt}\text{tr}_\varepsilon^Q(\Omega_t^k) = d\text{tr}_\varepsilon^Q\left(\sum_{j=1}^{k}\Omega_t^{k-j}\dot{\nabla}_t\Omega_t^{j-1}\right) + o\left(\varepsilon^{\frac{-n+s-(k-1)a-\eta}{q}}\right),$$

for all $k \in \mathbb{N}$ *and for all* $\eta > 0$.

(ii) *Let* $a = \text{ord}(\Omega_t)$ *be independent of* t. *Let* $q = \text{ord}(Q)$, *and set* $r := q - d$, *where we assume that* $d := \text{ord}([\nabla_t, Q])$ *is independent of* t. *If* $s > r - a$, *then the cohomology class of the coefficient of* $\varepsilon^{\frac{\gamma}{q}}$ *in the asymptotic expansion of* $\text{tr}_\varepsilon^Q(\Omega_t^k)$ *is independent of* t *for all* $\gamma < -ka - n + r$. *The cohomology class of the coefficient of* $\log \varepsilon \cdot \varepsilon^\ell$ *is independent of* t *for all* $0 < \ell < \frac{-ka-n+r}{q}$.

(iii) *Let* $a = \text{ord}(\Omega_t)$ *be independent of* t. *The cohomology class of the coefficient of* $\log \varepsilon$ *in the asymptotic expansion of* $\text{tr}_\varepsilon^Q(\Omega_t^k)$ *is independent of* t.

As with Theorem 4.4, this theorem is only meaningful if $s > 0$.

Proof. Mimicking the finite dimensional proof, we have

$$\frac{d}{dt}\text{tr}_\varepsilon^Q(\Omega_t^k) = \text{tr}_\varepsilon^Q\left(\sum_{j=1}^{k}\Omega_t^{k-j}[\nabla_t, \dot{\nabla}_t]\Omega_t^{j-1}\right)$$

$$= \text{tr}_\varepsilon^Q\left(\sum_{j=1}^{k}\Omega_t^{k-j}(\nabla_t^{Cl(\mathcal{E})}\dot{\nabla}_t)\Omega_t^{j-1}\right)$$

$$= \text{tr}_\varepsilon^Q\left(\nabla_t^{Cl(\mathcal{E})}\sum_{j=1}^{k}\Omega_t^{k-j}\dot{\nabla}_t\Omega_t^{j-1}\right)$$

$$= \mathrm{tr}_\varepsilon^Q \Big(\nabla_{\varepsilon,t}^Q \sum_{j=1}^k \Omega_t^{k-j} \dot{\nabla}_t \Omega_t^{j-1} \Big) + \sum_{j=1}^\infty \frac{\varepsilon^j}{j!} \mathrm{tr}_\varepsilon^Q \Big([[\nabla_t, Q]]_Q^{j-1} \sum_{j=1}^k \Omega_t^{k-j} \dot{\nabla} \Omega_t^{j-1} \Big)$$

$$= d\mathrm{tr}_\varepsilon^Q \Big(\sum_{j=1}^k \Omega_t^{k-j} \dot{\nabla}_t \Omega_t^{j-1} \Big) + \sum_{j=1}^\infty \frac{\varepsilon^j}{j!} \mathrm{tr}_\varepsilon^Q \Big([[\nabla_t, Q]]_Q^{j-1} \sum_{j=1}^k \Omega_t^{k-j} \dot{\nabla} \Omega_t^{j-1} \Big).$$

The leading term in the asymptotics of $\frac{\varepsilon^j}{j!} \mathrm{tr}_\varepsilon^Q \Big([[\nabla_t, Q]]_Q^{j-1} \sum_{j=1}^k \Omega_t^{k-j} \dot{\nabla} \Omega_t^{j-1} \Big)$ is of the form $a_j \varepsilon^{\gamma_j}$, with

$$\gamma_j > \frac{-n + jq - d - (j-1)(q-1) + s - (k-1)a}{q} \geq \frac{-n + s - (k-1)a}{q}.$$

This proves (i). For (ii), we note that this last fraction will be greater than $(-n - ka + r)/q$ provided $s > r - a$. As in the previous theorem, the proof of (iii) follows from properties of the Wodzicki–Chern class of §3.1. $\qquad \square$

With the previous two theorems, we have developed a theory of characteristic forms that explains why Freed's conditional first Chern form is closed and why its cohomology class cannot be connection independent.

Corollary 4.7. *Let* $\Omega = \Omega^{(\frac{1}{2})}$ *be the curvature of the Levi-Civita connection for the* $H^{\frac{1}{2}}$ *metric on the loop group* LG. *Then Freed's conditional first Chern form coincides with the weighted first Chern form* $\mathrm{tr}^Q(\Omega)$ *for any left invariant scalar weight* Q *on* LG, *and hence is closed.*

Proof. The conditional trace of the Levi-Civita curvature in [6] is $\mathrm{tr}(\mathrm{tr}_{\mathrm{Lie}}(\Omega))$, where $\mathrm{tr}_{\mathrm{Lie}}$ denotes the trace with respect to the Killing form in the Lie algebra of G, and the outer trace is the ordinary operator trace. In particular, $\mathrm{tr}_{\mathrm{Lie}}(\Omega)$ is a trace class $\Psi\mathrm{DO}$ on the trivial $\mathbb{C}$ bundle over S^1. As in [5], for any left invariant scalar weight Q we have

$$\mathrm{tr}\,(\mathrm{tr}_{\mathrm{Lie}}(\Omega)) = \lim_{\varepsilon \to 0} \mathrm{tr}\big(\mathrm{tr}_{\mathrm{Lie}}(\Omega) e^{-\varepsilon Q}\big) = \lim_{\varepsilon \to 0} \mathrm{tr}_\varepsilon^Q(\Omega) = \mathrm{tr}^Q(\Omega),$$

so Freed's conditional first Chern form is the weighted trace $\mathrm{tr}^Q(\Omega)$. Recall that the curvature two-form is a $\Psi\mathrm{DO}$ of order $a = -1$. Since Q is left invariant and scalar, $[\nabla, Q] = dQ + [\theta, Q] = [\theta, Q]$ has order $r = q - 1 = 1$. Theorem 4.4 with $n = 1, q = 2, r = 1, a = -1$ shows that the constant term $a_0(\Omega, Q)$ in the asymptotic expansion of $\mathrm{tr}_\varepsilon^Q(\Omega)$ is closed. Since this constant term equals $\lim_{\varepsilon \to 0} \mathrm{tr}_\varepsilon^Q(\Omega) = \mathrm{tr}^Q(\Omega)$, it follows that $\mathrm{tr}^Q(\Omega)$ is closed for loop groups. $\qquad \square$

Remark. Theorem 4.6 does not apply to Freed's conditional first Chern form. In particular, if we shrink the connection one-form θ to zero using the family $t\theta$, we cannot apply Theorem 4.6, since for this family we have $s = 0$.

References

[1]   H. Araki, Expansional in Banach algebras, *Ann. Sci. École Norm. Sup.* 6 (1973), 67–84.

[2]   M. F. Atiyah and R. Bott, The Yang-Mills equations over Riemann surfaces, *Philos. Trans. Roy. Soc. London Ser. A* 308 (1983), 523–615.

[3]   N. Berline, E. Getzler, and M. Vergne, *Heat Kernels and Dirac Operators*, Grundlehren Math. Wiss. 298, Springer-Verlag, Berlin 1992.

[4]   J.-M. Bismut, The Atiyah-Singer index theorem for families of Dirac operators: two heat equation proofs, *Invent. Math.* 83 (1985), 91–151.

[5]   A. Cardona, C. Ducourtioux, J.-P. Magnot, and S. Paycha, Weighted traces on algebras of pseudo-differential operators and geometry of loop groups, *Infin. Dimens. Anal. Quantum Probab. Relat. Top.* 5 (4) (2002), 503–540.

[6]   D. Freed, The geometry of loop groups, *J. Differential Geometry* 28 (1988), 223–276.

[7]   G. Grubb and R. Seeley, Weakly parametric pseudodifferential opeartors and Atiyah-Patodi-Singer boundary problems, *Invent. Math.* 121 (1995), 481–529.

[8]   N. Jacobson, *Lie Algebras*, Interscience Publishers, New York 1962.

[9]   A. Jaffe, A. Lesniewski, and K. Osterwalder, Quantum K-theory, *Commun. Math. Phys.* 118 (1988), 1–14.

[10] C. Kassel, Le résidu non commutatif [d'aprés Wodzicki], *Astérisque* 177–178 (1989), 199–229.

[11] A. Kriegl and P. Michor, *The Convenient Setting in Global Analysis*, Math. Surveys Monogr. 53, Amer. Math. Soc., Providence, RI, 1997.

[12] M. Lesch, On the non-commutative residue of pseudo-differential operators with log-polyhomogeneous symbols, *Ann. Global Anal. Geom.* 17 (1998), 151–197.

[13] Y. Maeda, S. Rosenberg, and F. Torres-Ardila, The geometry of loop spaces, in preparation.

[14] J.-P. Magnot, Sur la géométrie d'espaces de lacets, Thèse, Université Blaise Pascal (Clermont II), 2002.

[15] B. Maissen, Über Topologien im Endomorphismenraum eines topologischen Vektorraumes, *Math. Ann.* 151 (1963), 283–285.

[16] K. Okikiolu, The Campbell-Hausdorff theorem for elliptic operators and a related trace formula, *Duke Math. J.* 79 (1995), 687–722.

[17] S. Paycha and S. Rosenberg, Curvature of determinant bundles and first Chern forms. *J. Geom. Phys.* 45 (2003), 393–429.

[18] S. Paycha and S. Rosenberg. Chern-Weil constructions on ΨDO-bundles. mathDG/0301185.

New classical limits of quantum theories

S. G. Rajeev

Department of Physics and Astronomy,
University of Rochester
Rochester, New York 14627, U.S.A.
email: rajeev@pas.rochester.edu

Abstract. Quantum fluctuations of some systems vanish not only in the limit $\hbar \to 0$, but also as some other parameters (such as $\frac{1}{N}$, the inverse of the number of 'colors' of a Yang–Mills theory) vanish. These lead to new classical limits that are often much better approximations to the quantum theory. We describe two examples: the familiar Hartree–Fock–Thomas–Fermi methods of atomic physics as well as the limit of large spatial dimension. Then we present an approach of the Hecke operators on modular forms inspired by these ideas of quantum mechanics. It explains in a simple way why the spectra of these operators tend to the spectrum of random matrices for large weight for the modular forms.

2000 Mathematics Subject Classification: 81V45; 11F03.

Contents

1 Introduction

It is well-known that a classical mechanical system has many possible quantizations. The classical theory is the limiting case as $\hbar \to 0$, so it is not surprising that there would be many quantum theories that in this limit reduce to the same classical theory. In this, largely expository, paper I will discuss the opposite phenomenon: how the same quantum theory can be obtained by quantizing radically different classical systems. Viewed another way, a quantum theory could depend on two parameters, say $\hbar, \lambda$ and the quantum fluctuations of some class of observables are of order $\hbar\lambda$. Then both the limits $\hbar \to 0$ and $\lambda \to 0$ are classical theories. These classical theories could be entirely different. In an example from atomic physics, the conventional classical limit has a finite number of degrees of freedom, while the new one has an infinite number. We will refer to the new limits (obtained by taking parameters other than $\hbar$ to zero) as 'neo-classical limits'.

This phenomenon is physically interesting because one of the new classical limits may be a better approximation to the quantum theory than the naive classical limit. These ideas came into the high energy physics literature from the work of 't Hooft and Witten on the large N limit of gauge theories. Witten [1] in particular worked out several simpler cases to popularize the notion that even case $N = 3$ may be well approximated by the large N limit. But historically, the various mean field theories of condensed matter physics [2] (the spherical model for example) and even the theory of Nuclear Magnetic Resonance can be thought of as precursors of these ideas.

For example, in atomic physics, in the usual classical limit $\hbar \to 0$ there is no ground state: the hamiltonian is not bounded from below. However, the neo-classical limit (in this case a version of the Hartree–Fock approximation) has a ground state. Moreover, it gives an excellent first approximation to the ground state energy of the atom. The neo-semiclassical expansion gives a systematic way of calculating corrections to arbitrary accuracy, although the complexity grows rapidly with the desired accuracy.

Another important example, discussed below, is also from atomic physics: the quantum fluctuations in electron distances become small in the limit as the dimension of space becomes large. We will derive a simple effective potential that explains the stability of the atom. For another approach to this see the work of Hershbach [3].

Some important corners of mathematics are also illuminated by this phenomenon. The theory of modular forms can be viewed as the quantization of a classical mechanical system whose phase space is the upper half plane. The limit as the weight of the modular form goes to infinity corresponds to a classical limit. But there is also another classical limit corresponding to letting the level (the area of the fundamental domain) go to infinity. These limits lead to interesting new approaches to the problem of determining the spectrum of the Hecke operators on modular forms.

The mathematical formulation of a classical dynamical system has expanded steadily in generality throughout history: as new physical theories are discovered we are led to enlarge the formalism to incorporate the new developments. In the progression from ordinary differential equations to Hamilton–Jacobi theory, symplectic geometry and the currently fashionable Poisson algebra formulation, we learn to deal with increasingly sophisticated systems and symmetries. Neo-classical limits produce classical systems of even greater generality, often with non-local action principles and no simple hamiltonian description [4]. We have described some examples of this before. There are classical limits of quantum field theories that retain asymptotically freedom and require a renormalization [5]. The challenge of finding the right mathematical description of these new kinds of classical systems remains.

Much of the story told in this paper is, of course, well-known. I hope that organizing them in this way will help to understand common themes in apparently distant subjects.

2 Hartree–Fock theory of atoms

2.1 The classical limit of the atom

We start with a basic problem of quantum mechanics that cannot be solved exactly: an atom (or ion) with more than one electron. We usually start with the classical hamiltonian

$$H = \sum_{i=1}^{m} \frac{p_i^2}{2\mu} - \sum_{i=1}^{m} \frac{Ze^2}{|r_i|} + \sum_{1 \leq i < j \leq m} \frac{e^2}{|r_i - r_j|}. \tag{1}$$

As usual, Z is the atomic number of the nucleus, e the charge of the electron and μ its mass. For each $1 \leq i \leq m$, the position r_i and momentum p_i of the electron are vectors in the Euclidean space R^3. Then we pass to the quantum theory whose hamiltonian is

$$\hat{H} = -\hbar^2 \sum_{i=1}^{m} \frac{\nabla_i^2}{2\mu} - \sum_{i=1}^{m} \frac{Ze^2}{|r_i|} + \sum_{1 \leq i < j \leq m} \frac{e^2}{|r_i - r_j|}. \tag{2}$$

This is an operator on the complex Hilbert space of anti-symmetric wavefunctions $\mathcal{F}_m = \Lambda^m(\mathcal{H})$. The anti-symmetry incorporates the Pauli exclusion principle. The space of single particle wavefunctions is $\mathcal{H} = L^2(R^3, C^2)$; the wavefunction of each electron takes values in C^2, since it can exist in two spin states [1].

The limit $\hbar \to 0$ is the usual classical limit of the atom. It is well-known that this is a spectacularly bad approximation to the quantum theory of the atom. The quantum hamiltonian is self-adjoint and bounded below and hence has a well-defined ground state. Indeed the central problem of atomic physics is the determination of this ground state wavefunction and the corresponding eigenvalue. The hamiltonian of the classical limit on the other hand, has no ground state: we can let position of an electron approach the nucleus, $r_a \to 0$, thus decreasing the energy down to $-\infty$.

What is missing here is the uncertainty principle: in the quantum theory it is not possible to make the position of the electron close to the nucleus without making its kinetic energy large. This might suggest that there is no way to produce a classical approximation to the atom with a stable ground state.

We will now produce a completely different classical system (with an infinite number of degrees of freedom in fact) whose quantization yields exactly the above quantum theory of the atom. Moreover, it has a ground state which is even a good approximation to the quantum ground state. What we will describe is just a reformulation of the standard Hartree–Fock approximation in atomic physics. This reformulation allows a generalization to relativistic many-fermion systems which we have described elsewhere [6].

2.2 The neo-classical theory of the atom

There are many standard texts that discuss the material in this section, although often without the geometric interpretation in terms of the Grassmannian. See for example [7]. Let $\mathcal{H} = L^2(R^3, C^2)$ be the familiar complex Hilbert space. We define the *Grassmannian* $\mathrm{Gr}_m(\mathcal{H})$ to be the space of linear self-adjoint trace-class[2] projection operators of rank m:

$$\mathrm{Gr}_m(\mathcal{H}) = \{\rho : \mathcal{H} \to \mathcal{H} | \rho^\dagger = \rho; \ \rho^2 = \rho; \ \mathrm{tr}\,\rho = m\}. \tag{3}$$

It is clear that an eigenvalue of ρ is equal either to zero or to one. Corresponding to each such projection operator there is a subspace of $\mathcal{H}$ of dimension m: the eigenspace of ρ with eigenvalue one. Conversely, each such m-dimensional subspace V defines an orthogonal decomposition $\mathcal{H} = V \oplus V^\perp$; then we can construct ρ as the hermitean projection operator to V. Thus we can see that $\mathrm{Gr}_m(\mathcal{H})$ is really the set of all m-dimensional subspaces of $\mathcal{H}$. Thus we have an infinite dimensional (but finite rank) generalization of the usual definition of the Grassmannian [8]. $\mathrm{Gr}_m(\mathcal{H})$ is an

---

[1] We will, for simplicity, ignore relativistic and spin-dependent terms in the hamiltonian.

[2] If a projection operator is trace class, its trace must be an integer, the dimension of the vector space it projects.

infinite dimensional manifold, whose tangent space is the space of rank m self-adjoint operators on $\mathcal{H}$.

A classical dynamical system is specified by (i) a manifold which will be its phase space, (ii) a symplectic form on this phase space which will determine the Poisson brackets, and (iii) a real function on the phase space which is its hamiltonian.

In our theory, $\text{Gr}_m(\mathcal{H})$ is the phase space. The symplectic form on it generalizes the standard symplectic form on finite-dimensional Grassmannians. The Poisson brackets that it implies for a pair of functions is:

$$\{f, g\} = \text{tr } \rho[df, dg].\tag{4}$$

Here df is the infinitesimal variation of f which can be thought of as a linear operator on $\mathcal{H}$. If we represent the operator ρ by its integral kernel $\rho_b^a(x, y)$ (where $a = 1, 2$ labels spin) we can write this Poisson bracket as

$$\{\rho_b^a(x, y), \rho_d^c(z, u)\} = \delta_b^c \delta(y, z)\rho_d^a(x, u) - \delta_d^a \delta(x, u)\rho_b^c(z, y).\tag{5}$$

The last piece of information is the hamiltonian, which we postulate to be

$$H_1 = \int \frac{p^2}{2\mu} \tilde{\rho}(x, p)d^3x \frac{d^3p}{(2\pi\hbar)^3} - \int \frac{Ze^2}{|x|}\rho_a^a(x, x)\, d^3x$$
$$+\frac{1}{2}\int \frac{e^2}{|x - y|}\left[\rho_a^a(x, x)\rho_b^b(y, y) - \rho_b^a(x, y)\rho_a^b(y, x)\right]d^3xd^3y.$$

Here, $\tilde{\rho}$ is the *symbol* of the operator ρ:

$$\tilde{\rho}_b^a(x, p) = \int \rho_b^a\left(x + \frac{u}{2}, x - \frac{u}{2}\right)e^{-\frac{i}{\hbar}p\cdot u}\, d^3u,$$
$$\rho_b^a(x, y) = \int \tilde{\rho}_b^a\left(\frac{x + y}{2}, p\right)e^{\frac{i}{\hbar}p\cdot(x-y)}\frac{d^3p}{(2\pi\hbar)^3}.\tag{6}$$

Clearly, $\rho_a^a(x, x) = \int \tilde{\rho}_a^a(x, p)\frac{d^3p}{(2\pi\hbar)^3}$.

So far it is clear that this system depends on the parameters μ, e, Z, m of the of the atom. Although the dynamical variables are operators, and $\hbar$ appears in the formula for the hamiltonian, it is a bona fide classical dynamical system.

We will show that a quantization of this system is exactly the quantum theory of the atom. The physical meaning of the operator ρ is that it is the 'density matrix' of the electrons. Indeed $\rho_a^a(x, x)$ is the number density of the electrons at the point x; $\text{tr } \rho = \int \rho_a^a(x, x)d^3x = m$ is the total number of electrons. More generally, $\tilde{\rho}_a^a(x, p)$ is the density of electrons of momentum p and position x. The Pauli exclusion principle which allows for at most one electron per single particle state, becomes the condition that this density matrix be a projection operator, so that its eigenvalues can only be zero or one. Thus this classical dynamical system realizes many of the facts we usually associate with quantum theory.

The first term represents the kinetic energy and the second term the potential energy due to the nucleus. We can combine these 'single-particle' terms in the hamiltonian into the form

$$\text{tr } \rho K, \quad K = -\frac{\hbar^2}{2\mu}\nabla^2 - \frac{Ze^2}{|x|}. \tag{7}$$

Now, K is bounded below by $E_1 = -\frac{1}{2}\mu\frac{Z^2 e^4}{\hbar^2}$, as we know from the elementary theory of an ion with one electron. Since ρ is positive and $\text{tr } \rho = m$, we see that $\text{tr } \rho K \geq m E_1$. (A stricter bound can be obtained using the fact that ρ is a projection. But we don't need it.) This way, the system avoids the catastrophe of the conventional classical limit.[3]

The interaction of the electrons induces two kinds of terms. The first is obvious, the Coulomb energy of a charge cloud of density $e\rho_a^a(x, x)$. The last term is not so obvious–it is the 'exchange energy'. It is needed to get back the correct quantum theory (see below). By a version of the Schwarz inequality it should be possible to see that

$$\int \left[\rho_a^a(x, x)\rho_b^b(y, y) - \rho_b^a(x, y)\rho_a^b(y, x) \right] \frac{e^2}{|x - y|} d^3x d^3y \geq 0. \tag{8}$$

This expresses the physical fact that the electron-electron interaction is repulsive. Thus the total hamiltonian is bounded below by at least $m E_1$.

To actually find the ground state of this classical system, we must vary the hamiltonian subject to the constraints on ρ. Such a variation of ρ is always of the form $\delta\rho = -i[\rho, u]$ for some hermitean operator u. The condition for an extremum is then

$$[\rho, dH_1] = 0. \tag{9}$$

Here $dH_1 = \frac{\partial H_1}{\partial \rho}$ is a linear operator

$$dH_1 = K + \mathcal{U} + \mathcal{W}. \tag{10}$$

Here, K is as defined above and $\mathcal{U}$ is the multiplication by the 'mean field'

$$\mathcal{U}_b^a(x) = \delta_b^a \int \frac{Ze^2}{|x - y|} \rho_a^a(y, y) d^3y. \tag{11}$$

The 'exchange energy' contributes an operator W whose integral kernel is

$$\mathcal{W}_b^a(x, y) = -\frac{Ze^2}{|x - y|}\rho_b^a(x, y). \tag{12}$$

---

[3]Strictly speaking H_1 exists only on some dense domain of $\text{Gr}_m(\mathcal{H})$. This domain should be some class of pseudo-differential operators, and is the true phase space of the system. The correct statement is that the hamiltonian is bounded below within this domain. This is a technical project that I am unable to complete. It would be interesting to produce a functional analytic realization of the physical ideas that are described here.

Thus we can find the extremum by solving a non-linear eigenvalue problem self-consistently. We have,

$$\rho = \sum_{a=1}^{m} \psi_a \otimes \psi_a^\dagger \qquad (13)$$

where each of the vectors $\psi_a \in \mathcal{H}$ is an eigenstate of dH_1.

This is exactly the Hartree–Fock approximation to the atomic ground state. We find the wave-functions that are eigenstates of some single particle hamiltonian; the potential in this hamiltonian is self-consistently determined by postulating that m of these are occupied by electrons. Our description avoids the usual Slater determinants for the wavefunction-the hamiltonian only depends on the density matrix of the electrons and not the wavefunction itself. We have shown that this way of formulating the Hartree–Fock theory allows for generalization to systems containing an infinite number of fermions such as relativistic theories [6, 5].

Our point in this paper is that this theory can be thought of as minimizing the hamiltonian of a classical system on $\mathrm{Gr}_m(\mathcal{H})$. This extends to the time evolution as well: the Hamilton equations of our system are the usual equations of time-dependent Hartree–Fock theory.

2.3 Back to the quantum theory

How do we quantize a system whose phase space is a Grassmannian? It is not possible to cover the Grassmannian by a single co-ordinate system, so it is inconvenient to look for canonical variables. However, the Grassmannian is a Kähler manifold, and we can apply the ideas of geometric (or Berezin–Toeplitz) quantization [9]. In an earlier paper (the appendix of [6]) we used the representation theory of the unitary group to quantize this theory.

Recall the situation in the case of finite dimensional Grassmannians: let $\mathcal{V}$ be a finite dimensional vector space, and $\mathrm{Gr}_m(\mathcal{V})$ the set of its m-dimensional subspaces. $\mathrm{Gr}_m(\mathcal{V})$ is a compact Kähler manifold. Its canonical line bundle $\mathcal{L}$ admits a hermitean metric and a connection whose curvature is just the symplectic form. (A line bundle that admits such a metric and connection is said to be quantizable [9].) The holomorphic sections of the dual of this line bundle, $\mathrm{Hol}(\mathcal{L}^*)$ form a finite dimensional vector space isomorphic to $\Lambda^m(\mathcal{V})$. This space of holomorphic sections is a subspace of the Hilbert space of square-integrable sections with a projection operator $\Pi : L^2(\mathcal{L}) \to \mathrm{Hol}(\mathcal{L}^*)$. These geometric facts can be used to construct a quantization [9] of the dynamical system whose phase space is $\mathrm{Gr}_m(\mathcal{V})$.

From any function $f : \mathrm{Gr}_m(\mathcal{V}) \to R$ we will construct an operator $\hat{f} : \mathrm{Hol}(\mathcal{L}^*) \to \mathrm{Hol}(\mathcal{L}^*)$ by the formula

$$\hat{f} = \Pi f. \qquad (14)$$

That is, we multiply a holomorphic section by the function to get a section of $\mathcal{L}^*$ that may not be holomorphic; then we simply project out the holomorphic part. The operator we construct this way is self-adjoint (it is just a finite dimensional hermitean matrix in fact).

In what sense is it a quantization of the dynamical system on Gr_m $(\mathcal{V})$? How will we recover the classical limit? The idea [9] is that it is merely a special case of a one-parameter family of quantum theories where $\mathcal{L}$ above is replaced by $\mathcal{L}^N$. As long as N is a positive integer the above ideas go through: there is still a projection $\Pi_N : L^2(\mathcal{L}^N) \to \mathrm{Hol}(\mathcal{L}^{*N})$ to a finite dimensional space of holomorphic sections. [4] Also in the limit $N \to \infty$ the operator algebra tends to the Poisson algebra of functions in the sense that

$$||\Pi_N f \ \Pi_N g - \Pi_N(fg)|| = O\left(\frac{1}{N}\right), \tag{15}$$

and moreover

$$||iN[\Pi_N f, \ \Pi_N g] - \Pi_N(\{f, g\})|| = O\left(\frac{1}{N}\right). \tag{16}$$

Also the operator norm of $T_N f$ approaches the sup norm of the function f.

These ideas also extend to Gr_m $(\mathcal{H})$ when $\mathcal{V}$ is replaced by the infinite dimensional vector space $\mathcal{H}$. The technical aspects are simpler than in [10, 11], since we need only finite rank projections. We give only a very brief outline here. Any subspace of dimension m can be brought to some standard subspace whose orthogonal complement is $\mathcal{H}_\perp$; hence Gr_m $(\mathcal{H})$ is a coset space [5] Gr_m $(\mathcal{H}) = \mathrm{U}_0(\mathcal{H})/\mathrm{U}(\mathcal{H}_\perp) \times \mathrm{U}(m)$. Using the trivial representation of $\mathrm{U}(\mathcal{H}_\perp)$ and the determinant representation of $\mathrm{U}(m)$, we can construct a line bundle

$$\mathcal{L} = (\mathrm{U}_0(\mathcal{H}) \times C)/\mathrm{U}(\mathcal{H}_\perp) \times \mathrm{U}(m). \tag{17}$$

The holomorphic sections of this bundle can now be constructed and shown to form $\Lambda^m(\mathcal{H})$. It is thus clear that $\Lambda^m(\mathcal{H})$ is the Hilbert space of at least one way of quantizing our system on Gr_m $(\mathcal{H})$. Indeed, the hamiltonian of the system when worked out in this way is exactly the quantum hamiltonian of the atom we had earlier.

This quantum hamiltonian is the special case as $N = 1$ of a one-parameter family of theories. For $N > 1$ these describe fermions that carry a 'color' quantum number, except that only observables that are invariant under $\mathrm{U}(N)$ are realized in the Hilbert space $\mathrm{Hol}(\mathcal{L}^{*N})$. The Hartree–Fock method thus approximates the theory for $N = 1$ by the neoclassical limit as $N \to \infty$. In effect $\frac{1}{N}$ measures the size of the quantum corrections.

---

[4]The space $\mathrm{Hol}(\mathcal{L}^{*N})$ carries a representation of the unitary group $\mathrm{U}(\mathcal{V})$ given by the Young diagram of height n and width N, generalizing the completely anti-symmetric tensor representation of the case $N = 1$ above.

[5]$\mathrm{U}_0(\mathcal{H})$ is the group of unitary transformations that mixes the standard m-dimensional subspace with its orthogonal complement only by a finite rank operator. It is analogous to the 'restricted Grassmannian' of Sato [12] except that it is modelled on finite rank operators rather than compact ones.

A relativistic generalization of this theory is described in [6]. There I developed an approach to two-dimensional QCD where the large N limit was realized as a classical theory. The story above appears in the appendix to that paper. Later I found that many other problems in physics and mathematics can be thought of in a unifying way as different classical limits of the same quantum theory.

3 The Thomas–Fermi approximation

The problem of minimizing the energy on the Grassmannian is still a hard problem. Further approximations are needed. It turns out that there is a way to consider the limit[6] $\hbar \to 0$ (a kind of semi-classical approximation) which yields a simpler theory. The ideas go back to the Thomas-Fermi approximation of early atomic physics [13] and have seen several revivals. There seems to be a connection with the density functional [14] method as well. Our point of view is based on symbol calculus and was in part inspired by the work of Lieb, Thirring [15] and others on the stability of matter. We will work out explicitly the leading terms but indicate how higher order terms can be calculated systematically if needed.

It is possible to develop the theory in a more general context than in the last section without much additional work[7]. We therefore consider a system of m fermions with the hamiltonian

$$H = \sum_{i=1}^{m} [T(-i\partial_i) + U(r_i)] + \sum_{i<j} G(r_i, r_j). \tag{18}$$

The configuration space of each fermion is R^n. We will allow the fermions to carry a 'spin' quantum number $\sigma = 1, \ldots, N_f$. The above hamiltonian is assumed to be independent of this quantum number. (We will usually suppress the spin index). Here, $T(p)$ is the 'dispersion relation'; i.e., the dependence of kinetic energy on momentum. $T(p)$ is usually spherically symmetric. $U(x)$ is the external potential that all the fermions are subject to and $G(x, y) = G(y, x)$ is the two body potential.

In the last section we had the following special case: the dimension of space $n = 3$, the spin takes two values, $N_f = 2$, the kinetic energy is $T(p) = \frac{p^2}{2\mu}$ and the inter-electron potential is the Coulomb potential $G(x, y) = \frac{e^2}{|x-y|}$. The cases $n = 1, 2, 3$ for the dimension of space are also of interest in other contexts.

---

[6]It is important to take $\hbar \to 0$, after the theory has been formulated on the Grassmannian as above; if we took $\hbar \to 0$ we would get a theory without a ground state, as explained above.

[7]In fact I worked this theory in this way (in 1992) to find a relativistic generalization of the Thomas-Fermi method. It remains unpublished.

We have then the Hartree–Fock energy

$$H_1(\rho) = \text{Tr}(T + U)\rho$$
$$+ \frac{1}{2} \int dx dy \, G(x, y)[\, \text{tr} \, \rho(x, x) \, \text{tr} \, \rho(y, y) - \, \text{tr} \, \rho(x, y)\rho(y, x)]$$

to be minimized over all operators satisfying $\rho^\dagger = \rho$, $\rho^2 = \rho$, $\text{Tr}\rho = m$. We denote by tr the trace over flavor while Tr includes the integral over position as well. As before, the first term represents the single-particle kinetic energy and potential energy, the second term the direct interaction and the last term the exchange interaction.

The minimization problem above leads to the variational equations

$$[\rho, dH_1] = 0 \tag{19}$$

where the Hartree–Fock self consistent hamiltonian itself depends on ρ. A self-consistent solution is clearly[8] $\rho = \Theta(E_F - H_1)$ where the 'Fermi energy' E_F is determined by the condition $\text{Tr}\rho = m$. It is often too hard to solve this problem, so yet another approximation is needed. We could minimize over some smaller set of operators ρ thereby obtaining a variational bound that is simpler to calculate. Or we could calculate the function $H_1(\rho)$ in a semi-classical approximation.

The essence of the Thomas–Fermi approximation (a modern version is the density functional method [14]) is a combination of these two ideas:

1. use the variational ansatz $\rho = \Theta(-h)$ where $h = t(-i\partial) + v(x)$ is a separable hamiltonian (i.e., a function of p alone plus a function of x alone);

2. expand the energy function

$$H_1(\Theta(-t - v)) \tag{20}$$

semiclassically;

3. minimize the leading term in this expansion $H_{TF}(t, v)$ with respect to the variational parameters t and v. (In many treatments, however, t is chosen to be the same as $T(p)$ and only $v(x)$ is varied.)

The semiclassical expansion will amount to an expansion in powers of the derivatives of v. The above ansatz for h is motivated by the form of the Hartree–Fock hamiltonian. If the two body potential G is absent, the first step is automatic, since H_1 is already in this form. Even for interacting fermions, the direct interaction is already of the separable form. The indirect energy may not be separable in general, but as long as it is a monotonic function of momentum, the projection operator $\Theta(E_F - H_1)$ will agree with that of some separable hamiltonian. Thus one expects this separable ansatz to be a good approximation.

---

[8]The theta-function of a self-adjoint operator $\Theta(A)$ is defined to be the projection operator to the subspace on which A is positive.

The projection operator $\Theta(h - E)$ can be expressed in terms the resolvent operator, $\frac{1}{h-E}$, since

$$\Theta(-x) = \int_D \frac{dE}{2\pi i} \frac{1}{x - E} \tag{21}$$

where D is a contour that surrounds the negative real axis in a counterclockwise direction. There is a semi-classical expansion for the resolvent which can be used to derive one for the projection operator $\Theta(h - E)$.

3.1 The semiclassical expansion of the resolvent

To do the semiclassical expansion, it is convenient to restate the problem in terms of (matrix-valued) functions on the phase space $R^n \oplus R^n$ rather than operators on the Hilbert space $L^2(R^n, C^{N_f})$. There is a systematic theory of this procedure (symbol calculus) described in detail in, for example, [16]. The main idea is to use Weyl ordering to set up a one-one correspondence between functions on the phase space and operators on the Hilbert space. From a function $\tilde{A}(x, p)$ we construct the operator A whose kernel is

$$A(x, y) = \int \tilde{A}\left(\frac{x + y}{2}, p\right) e^{\frac{i}{\hbar} p \cdot (x - y)} [dp]. \tag{22}$$

We use the abbreviation $[dp] = \frac{dp}{(2\pi\hbar)^n}$. If we apply this to simple functions such as polynomials we can check that this definition corresponds to Weyl ordering. For example, the function xp becomes the operator $-i\hbar\partial x + x(-i\hbar\partial)$.

Conversely, given an operator, we define its symbol to be the function

$$\tilde{A}(x, p) = \int A\left(x + \frac{z}{2}, x - \frac{z}{2}\right) e^{-\frac{i}{\hbar} p \cdot z} dz. \tag{23}$$

The idea is that p is the momentum conjugate to the *relative* coordinate of the operator kernel. The operator multiplication can now be translated into the multiplication of symbols. The result can be expressed in closed form:

$$\tilde{A} \circ \tilde{B}(x, p) = \left\{ e^{\frac{-i\hbar}{2}\left(\frac{\partial}{\partial x^i} \frac{\partial}{\partial p_i'} - \frac{\partial}{\partial p_i} \frac{\partial}{\partial x^{i'}}\right)} \tilde{A}(x, p)\tilde{B}(x', p') \right\}_{x=x'; p=p'}. \tag{24}$$

The trace of operators becomes an integral in phase space

$$\mathrm{Tr}A = \mathrm{tr} \int A(x, x)dx = \mathrm{tr} \int dx[dp]\tilde{A}(x, p). \tag{25}$$

We emphasize that the algebra of symbols under the multiplication law is exactly the same (isomorphic) to the algebra of operators on a Hilbert space; i.e., no approximation is involved in replacing an operator by its symbol.

We see that to the leading order the above multiplication law is just the pointwise multiplication of the classical theory. In the next order there is a correction proportional to the Poisson bracket. If we expand the exponential,

$$\tilde{A} \circ \tilde{B}(x, p) = \sum_{n=0}^{\infty} \left(\frac{-i\hbar}{2} \right)^n \frac{1}{n!} \{\tilde{A}, \tilde{B}\}_{(n)}. \tag{26}$$

Here we see a sequence of generalized Poisson brackets

$$\{\tilde{A}, \tilde{B}\}_{(n)} = \sum_{r=0}^{n} (-1)^r \tilde{A}^{j_1...j_r}_{i_1...i_{n-r}} \tilde{B}^{i_1...i_{n-r}}_{j_1...j_r} \tag{27}$$

where $\tilde{A}^i = \frac{\partial \tilde{A}}{\partial p_i}$ and $\tilde{A}_i = \frac{\partial \tilde{A}}{\partial x^i}$ etc. $n = 1$ corresponds to the usual Poisson bracket. If $\tilde{A}$ and $\tilde{B}$ commute as matrices on spin, the odd brackets are antisymmetric and the even ones are symmetric. Otherwise, there is no particular symmetry property.

Consider now the resolvent operator of a hamiltonian h,

$$r(E) = \frac{1}{h - E}. \tag{28}$$

We will now derive a semiclassical expansion for the symbol $\tilde{r}(E)$ of this operator. The resolvent symbol satisfies

$$\tilde{r}(E) \circ (\tilde{h} - E) = 1. \tag{29}$$

Expand $r(E)$ in power series in $\hbar$ and put into the expansion of the above equation to get,

$$\tilde{r}(E) = \sum_{k=0}^{\infty} \tilde{r}_{(k)}(E) \hbar^k, \tag{30}$$

$$\sum_{n=0}^{\infty} \sum_{k=0}^{\infty} \left(\frac{-i\hbar}{2} \right)^n \frac{1}{n!} \hbar^k \{\tilde{r}_{(k)}(E), \tilde{h} - E\}_{(n)} = 1. \tag{31}$$

Equating the powers of $\hbar$ on both sides of this equation, we get a set of recursion relations

$$\tilde{r}_{(0)}(E) = (\tilde{h} - E)^{-1}, \tag{32}$$

$$\tilde{r}_{(m)}(E) = -\sum_{n=1}^{m} \left(\frac{-i}{2} \right)^n \frac{1}{n!} \{\tilde{r}_{(n-m)}(E), \tilde{h}\}_{(n)} (\tilde{h} - E)^{-1}. \tag{33}$$

If $\tilde{h}$ is diagonal in flavor space, the odd terms $\tilde{r}_{(2m+1)}$ vanish. The above expansion can be used to derive the usual WKB quantization conditions as well as higher order corrections to it.

Of particular interest to us is the case where h is a separable operator:

$$\tilde{h}(x, p) = t(p) + v(x). \tag{34}$$

In this case the mixed derivatives in the generalized Poisson brackets vanish and we get

$$\{\tilde{r}_{(k)}, \tilde{h}\}_{(n)} = \tilde{r}_{(k)i_1\ldots i_n} t^{i_1\ldots i_n} + (-1)^n \tilde{r}_{(k)}^{i_1\ldots i_n} v_{i_1\ldots i_n}. \tag{35}$$

If[9] $t(p) = p_i p_i$ as in nonrelativistic quantum mechanics, there is only one term for $n > 2$,

$$\{\tilde{r}_{(k)}, \tilde{h}\}_{(n)} = (-1)^n \tilde{r}_{(k)}^{i_1\ldots i_n} v_{i_1\ldots i_n} \tag{36}$$

while

$$\{\tilde{r}_{(k)}, \tilde{h}\}_{(1)} = 2 p_i \tilde{r}_{(k)i} - \tilde{r}_{(k)}^i v_i \tag{37}$$

and

$$\{\tilde{r}_{(k)}, \tilde{h}\}_{(2)} = 2 \tilde{r}_{(k)ii} + \tilde{r}_{(k)}^{ij} v_{ij}. \tag{38}$$

If moreover, v is diagonal in flavor space, $\tilde{r}_{(1)} = 0$ and

$$\tilde{r}_{(2)}(E) = \frac{1}{2(\tilde{h} - E)^2} \left[\frac{v_i v_i}{\tilde{h} - E} + 2(p_i v_i)^2 - v_{ii} \right]. \tag{39}$$

3.2 Derivative expansion of energy function

Now we can rewrite the Hartree–Fock energy in terms of the symbol $\tilde{\rho}(x, p)$ of the projection operator ρ.

$$\begin{aligned}
H_1(\tilde{\rho}) = \ &\mathrm{tr} \int \tilde{\rho}(x, p) T(p) \, dx[dp] + \mathrm{tr} \int U(x) \rho(x) \, dx \\
&+ \frac{1}{2} \int G(x, y) \, \mathrm{tr}\, \rho(x) \, \mathrm{tr}\, \rho(y) \, dx dy \\
&- \frac{1}{8} \int dx \int [dp][dp'] \tilde{G}(x, p - p') \, \mathrm{tr}\, \tilde{\rho}(x, p) \tilde{\rho}(x, p')
\end{aligned} \tag{40}$$

Here,

$$\rho(x) = \rho(x, x) = \int [dp] \tilde{\rho}(x, p). \tag{41}$$

We must minimize this subject to the constraints

$$\tilde{\rho} \circ \tilde{\rho}(x, p) = \tilde{\rho}(x, p); \quad \mathrm{tr} \int \tilde{\rho}(x, p) \, dx[dp] = m. \tag{42}$$

---

[9]We use units here such that $2\mu = 1$, to simplify the formulas.

We reiterate that although the problem has been formulated on the classical phase space, no approximation has been made yet. All the complications are in the multiplication law of the functions (hence in the quadratic constraint on $\tilde{\rho}$).

Now we put in the separable ansatz (which satisfies the constraint automatically) and expand in powers of $\hbar$. We will have

$$\tilde{\rho}(x, p) = \sum_{k=0}^{\infty} \tilde{\rho}_{(k)}(x, p)\hbar^k \tag{43}$$

Using the integral representation in terms of the resolvent symbol,

$$\tilde{\rho} = \int_D \frac{dE}{2\pi i} \tilde{r}(E) \tag{44}$$

we get

$$\tilde{\rho}_{(k)} = \int_D \frac{dE}{2\pi i} \tilde{r}_{(k)}(E). \tag{45}$$

These terms in the expansion of $\tilde{\rho}$ are distributions on the phase space involving the delta function and its derivatives, although the $\tilde{r}_{(k)}$ are ordinary functions. In the same way, we have expansions for the number density,

$$\rho(x) = \sum_{k=0}^{\infty} \tilde{\rho}_{(k)}(x)\hbar^k, \tag{46}$$

$$\tilde{\rho}_{(k)}(x) = \int_D \frac{dE}{2\pi i} \int [dp]\tilde{r}_{(k)}(E, x, p) \tag{47}$$

and the kinetic energy $K = \text{tr} \int T(p)\tilde{\rho}(x, p)\, dx[dp]$,

$$K = \sum_{k=0}^{\infty} K_{(k)}\hbar^k. \tag{48}$$

Also,

$$\tilde{q}_{(k)} = \int_D \frac{dE}{2\pi i} \text{ tr} \int dx[dp]\tilde{r}_{(k)}(E, x, p)T(p). \tag{49}$$

The direct energy can be written in terms of the density function $\rho(x)$. The exchange integral is more complicated, being quadratic in $\tilde{\rho}$; however, in most cases it is quite small and explicit calculation in higher orders is not necessary.

It is now straightforward to calculate the Thomas–Fermi energy to lowest order in the case of nonrelativistic quantum mechanics with a potential v that is diagonal in flavor space. We get upon evaluating the integrals, (it is convenient to introduce a new variable by $v(x) = -\phi^2(x)$),

$$\rho_{(0)}(x) = \int [dp]\Theta(-p^2 + \phi(x)) = \omega_n' \frac{\phi^n(x)}{n},$$

$$K_{(0)} = \text{tr} \int \omega'_n \frac{\phi^{n+2}(x)}{n+2} \, dx. \tag{50}$$

Here

$$\omega'_n = \frac{\omega_n}{(2\pi)^n} = 2 \left[\frac{1}{4\pi} \right]^{\frac{n}{2}} \frac{1}{\Gamma(\frac{n}{2})} \tag{51}$$

is the area of a sphere of unit radius in momentum space. The exchange integral is, to lowest order,

$$I_{(0)} = \frac{1}{2} \int [dp \, dp'] \Theta(\phi^2(x) - p^2) \Theta(\phi^2(x) - p'^2) \tilde{G}(p - p'). \tag{52}$$

With $\tilde{G}(p) = \frac{e^2}{p^2}$ as for the Coulomb interaction, we can evaluate this more explicitly by introducing spherical polar coordinates in momentum space. We get

$$I_{(0)} = \frac{1}{2} \alpha \frac{\omega'_n \omega_{n-1}}{(2\pi)^n} C_n \int \phi^{2n-2}(x) \, dx \tag{53}$$

where

$$C_n = \int_0^1 dy \int_0^1 dy' (yy')^{(n-1)} \int_0^\pi d\theta \frac{\sin^{n-2}\theta}{y^2 + y'^2 - 2yy'\cos\theta}. \tag{54}$$

This leads to

$$E_{TF}(\phi) = \text{tr} \int \left[\omega'_n \frac{\phi^{n+2}}{n+2}(x) - \frac{e^2}{2} \frac{\omega'_n \omega_{n-1}}{(2\pi)^n} C_n \phi^{2n-2}(x) + \omega'_n U(x) \frac{\phi^n(x)}{n} \right] dx$$
$$+ \frac{1}{2} \frac{\omega'^2_n}{n^2} \int G(x, y) \, \text{tr} \, \phi(x)^n \, \text{tr} \, \phi^n(y) \, dx \, dy. \tag{55}$$

In our expansion the exchange term appears in the lowest order. However, in atomic physics it is as small as the terms involving derivatives of ϕ, so it is often ignored in the lowest order treatments. Also, in many discussions, the energy is expressed as a function of the density $\rho(x)$, but one can make the change of variable from the Fermi momentum $\phi(x)$ to $\rho(x)$ easily.

Now we can vary this w.r.t. to ϕ to get an integral equation that determines the ground state in this approximation. Actually a more convenient variable to use is the mean field induced by this electron density: in terms of it we get a differential equation instead. If the distribution is spherically symmetric, as for an atom, this becomes a second order non-linear ordinary differential equation, the celebrated Thomas-Fermi differential equation [13].

Thus it is indeed possible to take the limit as $\hbar \to 0$ on systems such as the atom and get a sensible approximation to the ground state. However, this leads to a density function in the classical phase space and not conventional classical mechanics. Moreover, it has to be derived through an intermediary that is a bona-fide classical mechanical system but of infinite dimensions.

4 Atoms in the limit of large dimension

As another example of a neoclassical limit, again in atomic physics, we consider the limit of large spatial dimension. This idea originates in an observation of Witten that in this limit the quantum fluctuations in the rotation invariant quantities will become small. Let r_{ai} for $a = 1, \ldots, m$ and $i = 1, \ldots, n$ be the positions of m electrons in an atom (or ion) of atomic number Z. Although the physically interesting case is $n = 3$ we can, as a mathematical device, extend the system to n spatial dimensions. The problem of determining the ground state becomes that of minimizing

$$\int \left[\frac{\hbar^2}{2\mu} \frac{\partial \psi^*}{\partial r_{ai}} \frac{\partial \psi}{\partial r_{ai}} + \left(-\sum_a \frac{Ze^2}{|r_a|} + \sum_{1 \leq a < b \leq m} \frac{e^2}{|r_a - r_b|} \right) |\psi(r)|^2 \right] \prod_{ai} dr_{ai} \quad (56)$$

subject to the condition that

$$\int |\psi(r)|^2 \prod_{ai} dr_{ai} = 1. \quad (57)$$

Here $\psi \in \Lambda^m \left(L^2(R^n, C^{N_f}) \right)$.

Now, the hamiltonian is invariant under the rotation group $O(n) \times U(N_f)$. We now take the limit as n and N_f tend to infinity, and recover a classical theory. When N_f is large, we can assume that the wavefunction is completely anti-symmetric in the 'spin' indices; the position dependent part of the wavefunction is then symmetric. Indeed we can assume that this part is rotation invariant, [10] so that it depends only the invariant quantities $q_{ab} = \frac{1}{n} r_{ai} r_{bi}$.

These inner products form a positive $m \times m$ matrix. A complete set of $O(n)$ invariants are given by the remaining bilinears $\hat{L}_a^b = \frac{1}{2n}[r_{aj}, \frac{\hbar}{i} \frac{\partial}{\partial r_{aj}}]_+$, $\hat{P}^{ab} = -\frac{\hbar^2}{n} \frac{\partial^2}{\partial r_{ai} \partial r_{bi}}$. They form a representation of the symplectic Lie algebra $Sp(2n)$:

$$[q_{ab}, q_{cd}] = 0 = \left[\hat{P}^{ab}, \hat{P}^{cd} \right]$$

$$[\hat{L}_b^a, \hat{L}_d^c] = \frac{i\hbar}{n} \left(\delta_b^c \hat{L}_d^a - \delta_d^a \hat{L}_b^c \right)$$

$$[\hat{L}_b^a, \hat{P}^{cd}] = \frac{i\hbar}{n} \left(\delta_b^c \hat{P}^{ad} + \delta_b^d \hat{P}^{ac} \right) \quad (58)$$

$$[\hat{L}_b^a, q_{cd}] = -\frac{i\hbar}{n} \left(\delta_c^a q_{bd} + \delta_d^a q_{bc} \right).$$

These commutators are proportional to $\frac{\hbar}{n}$. Hence, there are two limits where the quantum fluctuations vanish: the conventional classical limit where we let $\hbar \to 0$ keeping n fixed (at the value 3 for example), or the neo-classical limit where we let $n \to \infty$ keeping $\hbar$ fixed. In this neo-classical limit, the quantum observables tend to

---

[10]In the real world, the ground state wavefunction is invariant under $O(n)$ at least for the noble gases: all the shells are filled.

classical ones satisfying the Poisson brackets of the symplectic Lie algebra:

$$\{q_{ab}, q_{cd}\} = 0 = \{P^{ab}, P^{cd}\}$$
$$\{L_b^a, L_d^c\} = \hbar\left(\delta_b^c L_d^a - \delta_d^a L_b^c\right)$$
$$\{L_b^a, P^{cd}\} = \hbar\left(\delta_b^c P^{ad} + \delta_b^d P^{ac}\right)$$
$$\{L_b^a, q_{cd}\} = -\hbar\left(\delta_c^a q_{bd} + \delta_d^a q_{bc}\right).$$

(59)

These Poisson brackets will determine the neo-classical equations of motion, once the hamiltonian is determined.

There are some subtleties in determining the hamiltonian of this neo-classical theory: there is a new term in the potential arising from the change of the measure of integration. (It is possible to interpret this as a kind of 'Fischer information' while the measure determines a kind of 'entropy' [17]. But we don't need this idea here.). Once the correct hamiltonian has been determined, this classical theory gives a relatively simple minimization problem for the ground state energy. Here we will consider only the static limit (time independent solution) that determines the ground state of theory.

4.1 The change of variables

Let us return to the variational problem of determining the ground state. This will reduce to the minimization of an effective potential that depends only on q_{ab}. First of all, we need to determine the measure of integration $\mu(q)\prod_{c\leq d} dq_{cd} := \mu(q)dq$ determined by the change of variables $q_{ab} = \frac{1}{n}r_{ai}r_{bi}$ on the Lebesgue measure $\prod_{ai} dr_{ai}$. This can be done by evaluating the following integral in two different ways:

$$Z(J) = \int_{q\geq 0} e^{-q_{ab}J^{ab}}\mu(q)\prod_{c\leq d} dq_{cd} = \int e^{-\frac{1}{n}r_{ai}r_{bi}J^{ab}}\prod_{cj} dr_{cj}. \quad (60)$$

(Here, J is a positive matrix.) On the r.h.s. we have a standard Gaussian integral yielding

$$\int_{q\geq 0} e^{-q_{ab}J^{ab}}\mu(q)\prod_{c\leq d} dq_{cd} = k(n, m) (\det J)^{-\frac{n}{2}} \quad (61)$$

where $k(n, m) = (\pi n)^{-\frac{nm}{2}}$ is independent of J.

Thus $Z(J)$ depends on J only through its determinant. It follows[11] that $\mu(q)$ can only depend on q through $\det q$, prompting the ansatz $\mu(q) = \tilde{k}[\det q]^\nu$. To determine ν we note that under the transformation $q \to SqS^T$, the measure of integration

---

[11]The space of positive matrices is a homogenous space of the general linear group, since any such matrix can be mapped to the identity by the transformation $q \mapsto SqS^T$ with J transforming dually. The transformation law of $Z(J)$ under this transformation completely determines that of $\mu(q)$ as well.

transforms as $dq \mapsto [\det S]^{m+1} dq$. Thus

$$Z(J) = \int_{q \geq 0} e^{-\operatorname{tr} q(S^T J S)} [\det q]^\nu [\det S]^{2\nu+m+1} \, dq = Z(S^T J S)[\det S]^{2\nu+m+1}$$

(62)

which determines $\nu = \frac{n-m-1}{2}$.

Thus we have

$$\|\psi\|^2 = \tilde{k} \int_{q \geq 0} |\psi(q)|^2 [\det q]^{\frac{n-m-1}{2}} \, dq.$$

(63)

It is thus tempting to define a new wavefunction absorbing the determinant of q:

$$\chi(q) = \sqrt{[\mu(q)]} \psi(q), \quad \|\psi\|^2 = \int_{q \geq 0} |\chi(q)|^2 \, dq.$$

(64)

This $\chi(q)$ is a kind of 'radial wavefunction'.

4.2 The effective potential

Now we must express the hamiltonian in terms of this χ. The only calculation we need is for the gradient of the wavefunction:

$$\int \frac{\hbar^2}{2\mu n^2} \frac{\partial \psi^*}{\partial r_{ai}} \frac{\partial \psi}{\partial r_{ai}} \prod_{bj} dr_{bj} = \int \frac{\hbar^2}{2\mu} g^{ab \, cd} \mu^{-\frac{1}{2}} \frac{\partial(\mu^{\frac{1}{2}} \chi^*)}{\partial q^{ab}} \mu^{-\frac{1}{2}} \frac{\partial(\mu^{\frac{1}{2}} \chi)}{\partial q^{cd}} \, dq$$

$$= \int \frac{\hbar^2}{2\mu} g^{ab \, cd} \left[\frac{\partial \chi^*}{\partial q^{ab}} + \frac{n-m-1}{4} \frac{\partial \log \det q}{\partial q^{ab}} \chi^*(q) \right]$$

$$\left[\frac{\partial \chi}{\partial q^{cd}} + \frac{n-m-1}{4} \frac{\partial \log \det q}{\partial q^{cd}} \chi(q) \right] dq$$

(65)

where

$$g^{ab \, cd} = n^2 \frac{\partial q^{ab}}{\partial r_{ej}} \frac{\partial q^{cd}}{\partial r_{ej}} = \delta^{ac} q^{bd} + \delta^{ad} q^{bc} + \delta^{bc} q^{ad} + \delta^{bd} q^{ac}$$

(66)

is an induced metric on the new configuration space. Moreover we know from elementary matrix theory that

$$\frac{\partial \log \det q}{\partial q^{ab}} = q_{ab}^{-1}.$$

(67)

The terms proportional to $\chi^* \chi$ become a correction to the potential; the terms involving one derivative of the wavefunction combine to give a total derivative that can be dropped. Those that involve the square of the derivative of χ become a new kinetic energy term.

Thus the variational problem is now to minimize

$$\int_{q \geq 0} \left[\frac{\hbar^2}{2\mu n^2} g^{abcd} \frac{\partial \chi^*}{\partial q^{ab}} \frac{\partial \chi}{\partial q^{cd}} + \left(V_{eff}(q) + V(q) \right) |\chi(q)|^2 \right] dq \qquad (68)$$

subject to the constraint

$$\int_{q \geq 0} |\chi(q)|^2 dq = 1. \qquad (69)$$

Here,

$$V_{eff} = \frac{\hbar^2}{2\mu} \frac{(n-m-1)^2}{4n^2} \operatorname{tr} q^{-1}. \qquad (70)$$

Also, $U(q)$ is the potential energy of the electron expressed in terms of the new variables:

$$U(q) = -\sum_{a=1}^{m} \frac{Z\alpha}{\sqrt{[q^{aa}]}} + \sum_{1 \leq a < b \leq n} \frac{\alpha}{\sqrt{[q^{aa} + q^{bb} - 2q^{ab}]}} \qquad (71)$$

where $\alpha = \frac{e^2}{\sqrt{n}}$.

We are now ready to take the limit as $n \to \infty$, holding $\frac{e^2}{\sqrt{n}} = \alpha$ (not e^2 itself!) fixed. As expected, in the new variables, the kinetic energy of the ground state wavefunction will be of order $\frac{1}{n^2}$. It is very important that there is now a new term in the potential energy (arising from the kinetic energy of the old picture) which makes it bounded below:

$$V(q) = \frac{\hbar^2}{8\mu} \operatorname{tr} q^{-1} + U(q). \qquad (72)$$

In the end the correction to the potential is quite simple!

The ground state energy in our neoclassical approximation is the minimum of this function over all positive q. The condition for this is an algebraic equation for q.

The case of a hydrogenic ion is of course simplest: when $m = 1$, q is just a positive number and there is no repulsive Coulomb interaction:

$$V(q) = \frac{\hbar^2}{8\mu} q^{-1} - \frac{Z\alpha}{\sqrt{q}}. \qquad (73)$$

The minimum is

$$-2\mu \frac{Z^2\alpha^2}{\hbar^2} = -\frac{2}{n} \mu \frac{Z^2 e^4}{\hbar^2}. \qquad (74)$$

This is to be compared with the exact answer (for $n = 3$) of $-\frac{1}{2}\mu \frac{Z^2 e^4}{\hbar^2}$. Thus we get roughly the correct answer: the relative error is about $\frac{1}{n}$. It should be possible to improve on this by semi-classical methods.

More generally, it is reasonable to expect (but not guaranteed) that the minimum will respect the permutation symmetry of the problem. Then we can put the ansatz that all the diagonal elements are equal (say, $q_{aa} = \rho^2, \forall a$) and that all the off-diagonal elements are also equal, (put $q_{ab} = \rho^2 u, \forall a \neq b$). Then $|u| \leq 1$ by Schwarz inequality. The potential becomes in these new variables, (it is convenient to choose a kind of atomic units during such explicit calculations, $2\mu = \hbar = \alpha = 1$):

$$V(\rho, u) = \frac{1}{4\rho^2} f(u) - \frac{1}{\rho} g(u). \tag{75}$$

Here,

$$g(u) = mZ - \frac{m(m-1)}{2} \frac{1}{\sqrt{[2(1-u)]}}. \tag{76}$$

Moreover,

$$f(u) = \operatorname{tr} \tilde{q}^{-1}, \quad \tilde{q} = (1-u) + uC \tag{77}$$

and C is the $m \times m$ matrix all of whose matrix elements are equal to one. The spectrum of C is quite simple: it has an eigenvalue equal to zero with degeneracy $m-1$ and the remaining eigenvalue is just m. Thus we can determine the spectra of $\tilde{q}$ and $\tilde{q}^{-1}$ and hence its trace:

$$f(u) = \frac{m-1}{1-u} + \frac{1}{1+(m-1)u}. \tag{78}$$

It is simple to minimize in ρ to reduce the problem to minimizing in u of $-\frac{g^2(u)}{f(u)}$. If we change variables yet again to

$$v = \frac{1}{\sqrt{[2(1-u)]}}, \quad \frac{1}{2} \leq v, \tag{79}$$

our approximation of the ground state energy becomes the minimum of the rational function

$$\tilde{V}(v) = -\left[mZ - \frac{m(m-1)}{2} v \right]^2 \frac{2mv^2 - (m-1)}{2mv^2 \left[2(m-1)v^2 - (m-2) \right]}. \tag{80}$$

This minimum can in fact be found in closed form as an algebraic function of m and Z. But the formula (obtained by an algebraic computation program such as Mathematica) is quite complicated. But this formula is fit very well[12] in the case of a neutral atom (i.e., $m = Z$) by the polynomial

$$E(Z) = -\left[0.00152507 + 0.0871987Z - 0.920957Z + 1.83211Z^2 \right] \frac{2\mu e^4}{n\hbar^2}$$

We have restored the original units to make comparison with other methods easier.

---

[12]The fit is good to a relative error of 0.01% over the range $1 \leq Z \leq 100$

The point of this method is that it gives an exactly solvable and reasonably accurate picture for the ground state of the atom without having to deal with complicated non-linear differential equations. The answers are reasonable considering the simplicity of the calculations.

5 A physicist's view of modular forms

Next we will consider an example from mathematics: the theory of modular forms. I don't claim to have solved any deep problem in this area (of which there are many). But perhaps the point of view described will suggest new methods.

5.1 The modular group and its subgroups

We will give only a foretaste of the theory of modular forms. See reference [18] for most of the proofs and precise statements of the results. Also see reference [19] for relations to other areas of mathematics and physics.

The group of two by two matrices with integer entries and determinant one is called $SL_2(Z)$; its quotient by the center, $\Gamma(1) = SL_2(Z)/Z_2$, is the *modular group*. It is conventional to denote elements of $\Gamma(1)$ as matrices $\begin{pmatrix} a & b \\ c & d \end{pmatrix}$, their pre-images in $SL_2(Z)$.

$\Gamma(1)$ acts on the upper half of the complex plane U through the fractional linear transformations

$$z \mapsto \frac{az + b}{cz + d}. \tag{81}$$

A fundamental region for this action is,

$$D = \left\{ z \mid |z| > 1, |\mathrm{Re}\, z| < \frac{1}{2} \right\}. \tag{82}$$

This region is a spherical triangle with vertices at $i\infty$, $\pm\frac{1}{2} + \frac{\sqrt{3}}{2}i$. The point is that translations can be used to bring any point inside the strip $|\mathrm{Re}\, z| < \frac{1}{2}$; and under inversion any point is equivalent to one outside the unit circle.

The modular group is generated by $S : z \mapsto -\frac{1}{z}$ and $P : z \mapsto -\frac{1}{z+1}$. It is obvious that $S^2 = 1$, $P^3 = 1$. Indeed it can be shown that $\Gamma(1) = Z_2 * Z_3$ is the free product generated by these two elements-there are no other relations among these generators.

Many interesting groups appear as subgroups of the modular group. For example, the commutator subgroup of $\Gamma(1)$ is the free group on two generators; it is a normal subgroup of index[13] 6. Thus the modular group is both non-abelian and infinite in an essential way: free groups are the ultimate examples of such groups.

The *principal congruence subgroup* $\Gamma(n)$ *of level n* consists of all elements that are equal to the identity matrix modulo n. Now we see why the modular group is called $\Gamma(1)$; its elements are of the form $\begin{pmatrix} a & b \\ c & d \end{pmatrix}$ with $a = d = 1 \bmod n$, $c = b = 0 \bmod n$. Any subgroup Γ in between, $\Gamma(n) \subset \Gamma \subset \Gamma(1)$, is called a *congruence subgroup of level n*. The congruence subgroups are all of finite index. It is possible to show by a counting argument [18] that $[\Gamma(1) : \Gamma(n)] = n^3 \prod_{p|n}[1 - p^{-3}]$. Of particular importance [14] is the subgroup $\Gamma_0(n) = \left\{ \begin{pmatrix} a & b \\ c & d \end{pmatrix} \in \Gamma(1), c = 0 \bmod n \right\}$. The index can be shown to be [18] $[\Gamma(1) : \Gamma_0(n)] = n \prod_{p|n}[1 + p^{-1}]$.

5.2 Modular forms

A *entire modular form* of integer *weight k* associated to a subgroup $\Gamma \subseteq \Gamma(1)$ is a holomorphic function on the upper half plane (including the point at [15] $i\infty$) satisfying

$$(cz + d)^{-k} f\left(\frac{az + b}{cz + d}\right) = f(z), \quad \text{for} \quad \begin{pmatrix} a & b \\ c & d \end{pmatrix} \in \Gamma. \tag{83}$$

It is called a *cusp form* if $f(i\infty) = 0$.

If the weight is even, we can think of a modular form as a covariant tensor of order $k/2$ ('form') on U/Γ: the condition above is the statement that $f(z) [dz]^{\frac{k}{2}}$ is invariant under Γ.

The holomorphic sections of the canonical line bundle on U (in this case the cotangent bundle) $\mathcal{L}$ are entire functions $f(z)$ on U such that $f(z)dz$ is invariant under Γ. Thus the modular forms of weight k are simply holomorphic sections of $\mathcal{L}^{\frac{k}{2}}$. If k is odd these correspond to some 'spinors' on U/Γ.

An example of a modular form[16] of weight $2k$ is the *Eisenstein series*

$$G_{2k}(z) = \sum_{(m,n) \neq (0,0)} \frac{1}{(m + nz)^{2k}}. \tag{84}$$

It does not vanish at $i\infty$: $G_{2k}(i\infty) = 2\zeta(2k)$.

The most famous cusp form is

$$\Delta(z) = (2\pi)^{12} e^{i\pi z} \prod_{n=1}^{\infty} [1 - e^{2\pi inz}]^{12}. \tag{85}$$

---

[13]The *index* $[G : H]$ of a subgroup H of a group G is the number of elements in the coset G/H; alternatively, it is the number of copies of the fundamental region of G that is needed to form a fundamental region of H.

[14]It is the modular forms of weight two with respect to this subgroup that appear in the Shimura–Taniyama conjecture.

[15]A function f is holomorphic at $i\infty$ if it has a convergent Fourier expansion $f(z) = \sum_0^\infty f_n e^{2\pi inz}$. Moreover, $f(i\infty) = f_0$.

[16]If we don't specify Γ, we will be speaking of the modular group itself.

It is of weight 12. It is nonzero everywhere except for a simple zero at $i\infty$. It appears in Ramanujan's theory of partitions of numbers. If we expand the product of the twelfth root of Δ (which is called the Dedekind η-function) we can see that it is a generating function for partitions. The partitions of large numbers is given by the asymptotic behavior as $\text{Im } z \to 0$; this is an essential singularity of the function so at first this looks hopeless. However, the modular invariance relates the value of Δ at z to its value at $-\frac{1}{z}$; thus the behavior at $i\infty$ (which is trivial to determine) gives the behavior as $\text{Im } z \to 0$. Hardy and Ramanujan turned this rough stone of an idea into an exquisite jewel (further polished by Rademacher), deriving an asymptotic formula for partitions of large numbers.

Any modular form is a periodic function hence can be expanded in a Fourier series. These Fourier coefficients are of great interest. An example is the Ramanujan τ-function, which are the Fourier coefficients of the modular form above,

$$\Delta(z) = \sum_1^\infty \tau(n) e^{2\pi i n z}. \tag{86}$$

A deep conjecture of Ramanujan (proved eventually by Deligne following ideas of Grothendieck) was that

$$|\tau(p)| \le 2p^{\frac{11}{2}}. \tag{87}$$

In the theory of partitions, this inequality gives a bound on the error term to the Hardy-Ramanujan asymptotic formula for partitions of large numbers. These error terms seem to oscillate erratically yet a bound on their magnitude follows from the above inequality.

These erratic oscillations are related to yet another interesting phenomenon: if we define $2p^{\frac{11}{2}} \cos\theta(p) = \tau(p)$, the angles $\theta(p)$ seem to be distributed randomly according to the circular ensemble of random matrix theory. Indeed spectra of random matrices appear in many places in the theory of modular forms and related Dirichlet series (see the recent books [19]). We will seek a clarification of this phenomenon using ideas from quantum mechanics in the theory of Hecke operators.

Let M_k be the vector space of entire modular forms of weight k and $\mathcal{S}_k$ that of cusp forms. It is clear that $M_k M_l \subset M_{k+l}$. For $k \ge 12$, multiplication by Δ gives a linear map $M_{k-12} \to \mathcal{S}_k$. Moreover, $\dim \mathcal{S}_k = \dim M_k - 1$ since there is just one condition on the Fourier coefficients of a cusp form: that the zeroth one vanishes. It is possible to reduce the determination of the dimension of M_k to small values of k using these facts; see [18] for details.

With our definition of weight, there are no entire modular forms of odd weight. For $k = 0$ there is just one entire modular form, the constant. There are none for $k = 2$. For $k = 4, 6, 8, 10$ the only entire modular forms are multiples of the Eisenstein series.

For $k \geq 12$ and even,

$$\dim \mathcal{S}_k = \begin{cases} \left[\frac{k}{12}\right] & \text{if } k \neq 2 \bmod 12 \\ \left[\frac{k}{12}\right] - 1 & \text{if } k = 2 \bmod 12. \end{cases} \tag{88}$$

Thus for large k, the dimension grows linearly with weight. A way of understanding this is that the elements of $\mathcal{S}_k$ are holomorphic sections of the line bundle $\mathcal{L}^{\frac{k}{2}}$; as k grows this line bundle has greater Chern character allowing for more sections: it approaches a kind of classical limit.

5.3 Modular forms as wavefunctions

To a physicist, the above theory of modular forms is very reminiscent of quantum mechanics.

We can regard the upper half plane as the phase space of some classical mechanical system. The symplectic form is the Poincarè form:

$$\omega = \frac{dx \wedge dy}{y^2}, \quad z = x + iy. \tag{89}$$

The modular group (or one of its finite index subgroups) can be thought of a discrete gauge group, so that points related by such a transformation represent the same classical state. A wavefunction would be a holomorphic function on the upper half plane; more precisely it would be a holomorphic section of a line bundle $\mathcal{L}^{\frac{k}{2}}$ on U/Γ. Thus $\frac{k}{2}$ is analogous to the parameter N in our earlier discussion of compact Kähler manifolds. The base U/Γ is not usually a compact manifold because of the cusps (points at infinity and points where the stability group is finite). Nevertheless U/Γ has finite area hence most of the theory ought to generalize.

What is the dynamical system whose phase space is $U/\Gamma(1)$? We can imagine it as a model of quantum gravity in two dimensions. There are many such models that illustrate various aspects of gravity. Here, we think of space-time as a torus. Our model of gravity is conformally invariant (not crazy since two is the critical dimension for gravity). Thus the set of metrics modulo diffeomorphisms and conformal (Weyl) transformations is the phase space of gravity. Using a diffeomorphism that is connected to the identity and a Weyl transformation we can bring any metric to the form $ds^2 = |d\theta_1 + zd\theta_2|^2$, where $0 \leq \theta_1, \theta_2 \leq 2\pi$ are standard co-ordinates on the torus. Also we can choose $\text{Im} z > 0$. Now if we also allow for diffeomorphisms that are not connected to the identity, which are

$$\begin{pmatrix} \theta_1 \\ \theta_2 \end{pmatrix} \mapsto \begin{pmatrix} a & b \\ c & d \end{pmatrix} \begin{pmatrix} \theta_1 \\ \theta_2 \end{pmatrix}, \quad \begin{pmatrix} a & b \\ c & d \end{pmatrix} \in SL_2(\mathbb{Z}) \tag{90}$$

then the true phase space would be $U/\Gamma(1)$: the action of $\Gamma(1)$ on z is exactly the above fractional linear transformation.

What would be the meaning of a gauge group that is only a subgroup of $\Gamma(1)$? We might have some additional geometric object on the torus that has to be invariant as well (like a spin structure) that would reduce the size of the gauge group.

What would be the hamiltonian of our theory? A closed cosmology like a torus would at first not seem to have any meaningful time evolution. An asymptotically flat space-time would have a time at infinity with respect to which we can evolve its wavefunction. In closed universes in four dimensions, the Wheeler–DeWitt equation gives a 'time evolution' where the conformal factor of the metric itself is a kind of time variable. But we have given this up by postulating that continuous rescalings are part of the gauge group, so that the wavefunction is invariant under them.

However we can still regard time evolution as a *discrete* rescaling ('expansion') of the universe. While rescalings connected to the identity are part of the confor-mal group, there are certain discrete rescalings that for example double the size of the fundamental region. There are many different ways of rescaling such a region (e.g., double just one leg of the fundamental parallelogram) which individually vio-late modular invariance. Only by averaging over all of them would we recover modular invariance.

What we describe above is an interpretation of the Hecke operators on modular forms: they are rescalings averaged over the modular group. Because the evolution is discrete we cannot find a generator for infinitesimal transformations. The closest we get to are the prime rescalings, which cannot be decomposed as compositions of others. They yield a family of commuting hermitean operators which together play the role of the hamiltonian. In the next section we give a more detailed description of Hecke operators.

The central problem of the Hecke theory of modular forms, that of determining the simultaneous eigenvectors of the Hecke operators, is just like the central problem in quantum mechanics: finding the eigenfunctions of the hamiltonian. Quantum mechanics suggests some strategies to attack this Hecke problem. The limits of large weight or large level are like neo-classical and classical limits. For example, the number of linearly independent modular forms is $O(\nu k)$ in the limit of either large index ν or large weight k. The number of independent states of a quantum mechanical system with a two-dimensional phase space is of order of the area of the phase space divided by $\hbar$. The fundamental region D of the modular group is not compact but still has finite area with respect to the Poincaré metric:

$$A(D) = \int_{-\frac{1}{2}}^{\frac{1}{2}} dx \int_{\sqrt{[1-x^2]}}^{\infty} \frac{dy}{y^2} = \frac{\pi}{3}. \tag{91}$$

The fundamental region of a subgroup of index ν is then just $\nu A(D)$. Since the number of linearly independent modular forms is $\frac{k}{6}$ for large k, we see that the analogue of $\hbar$ in our theory is essentially $6A(D)/k = \frac{2\pi}{k}$. For example for the subgroup $\Gamma_0(p)$ which has index $\sim p$ (for prime p) there should be, according to this interpretation,

~ pk linearly independent modular forms. This is indeed known in the traditional theory of modular forms.

Thus we get simpler classical analogues of the theory of modular forms in these limits; we can then hope to understand the general case by asymptotic expansions in inverse powers of k or n. In the limit of large n, the subgroup

$$\Gamma_0(n) = \left\{ \begin{pmatrix} a & b \\ c & d \end{pmatrix} \middle| c = 0 \bmod n; \; ad - bc = 1; \; a, b, c, d \in Z \right\} \tag{92}$$

becomes essentially the group of translations

$$\begin{pmatrix} 1 & b \\ 0 & 1 \end{pmatrix}; \quad b \in Z, \; z \mapsto z + b. \tag{93}$$

This is a huge simplification: the invariance group becomes more 'abelian' as $n \to \infty$. In ordinary gauge theories (such as Yang–Mills theories) the limit as the theory becomes abelian and the limit of small quantum corrections are intimately related. (Perturbation theory is essentially the same as the loop expansion.) Thus one elementary strategy to understand modular forms is to study first this easy limiting case where modular forms reduce to periodic functions on the upper half plane. We will see then that this 'perturbative' limit is also a 'semi-classical limit' of large k.

The theory of modular forms should be viewed as a gauge theory with a non-abelian and non-compact gauge group- the modular group. It has all the essential features of the gauge theories of physics but in a much simpler mathematical setting: the group is only countably infinite instead of being an infinite dimensional Lie group. Thus there is no need for renormalization. By studying modular forms we are studying the gauge principle in its purest form without contamination by the other complications of quantum field theories. The number theory of the last century bears witness to the claim that even this simplest of all non-abelian gauge theories is very deep: some of the deepest problems of number theory could be solved if we understood the spectrum of the Hecke operators.

In this paper we develop only the analogue of lowest order perturbation theory ('abelian approximation'). Our other ideas on non-abelian gauge theories ('summing planar diagrams') also should have analogues here and should lead to deep results in the future. I hope an enterprising reader will take up this challenge.

5.4 Hecke operators

We now return to the exposition of the classic theory of modular forms due to Hecke. Hecke was motivated by the work of Mordell who in turn was trying to understand the Ramanujan conjecture on the τ-function.

A lattice on the complex plane is a set of points

$$w \sim w + r\omega_1 + s\omega_2, \quad r, s \in Z; \tag{94}$$

the fundamental region is a parallelogram with side z. A linear change of basis with integer coefficients and determinant one[17] $ad - bc = 1$,

$$\begin{pmatrix} \omega_1 \\ \omega_2 \end{pmatrix} \mapsto \begin{pmatrix} a & b \\ c & d \end{pmatrix} \begin{pmatrix} \omega_1 \\ \omega_2 \end{pmatrix} \tag{95}$$

does not change the lattice. By identifying the opposite sides of this parallelogram we get a torus. By a rotating our co-ordinate system and choosing an appropriate unit of length we can choose $\omega_2 = 1$. Also, we can reflect around this axis if needed to make ω_1 lie in the upper half plane. Thus only the ratio $z = \frac{\omega_1}{\omega_2}$ is needed to specify a lattice. A modular transformation

$$z \mapsto \frac{az + b}{cz + d}, \quad ad - bc = 1 \tag{96}$$

is just the effect of a change of basis on this ratio. We call this lattice L_z.
    A fractional linear transformation

$$z \mapsto \frac{az + b}{cz + d} \tag{97}$$

with integer coefficients will map L_z to a sublattice of index[18] $n = ad - bc$. Each sublattice of index n corresponds to an orbit of $\Gamma(1)$ on the set

$$\Gamma_n = \left\{ \begin{pmatrix} a & b \\ c & d \end{pmatrix} \middle| a, b, c, d \in Z, ad - bc = n \right\} \tag{98}$$

since an action by $\Gamma(1)$ would not have changed the original lattice L_z. (For $n \neq 1$, Γ_n is not a group; $\Gamma_m \Gamma_n \subset \Gamma_{mn}$).
    We now define the action of the *Hecke operator* $T(n)$ on a modular form f as a sum over all the sublattices of index[19] n:

$$[T(n)f](z) = n^{\frac{k}{2} - 1} \sum_{h \in \Gamma_n / \Gamma} f(h(z)) \left[\frac{dh(z)}{dz} \right]^{k/2}. \tag{99}$$

Using the fact an action by $\Gamma(1)$ merely permutes the terms of this sum (the left action of $\Gamma(1)$ on the coset $\Gamma_n / \Gamma(1)$) we can show that $T(n)f$ is also a modular form of weight k.
    By a right action of $\Gamma(1)$ we can bring any element of Γ_n to the upper triangular form ; actually we can enumerate the elements of the coset $\Gamma_n / \Gamma(1)$ by $\begin{pmatrix} a & b \\ 0 & d \end{pmatrix}$ with $ad = n$, $b = 0, 1, \ldots, d - 1$. For proofs see [18]. A more explicit formula for the

---

[17]If the determinant is not one we change the area of the fundamental domain. See below.
[18]This means that there are n fundamental domains of L_z in one fundamental domain of the sublattice.
[19]We denote $h(z) = \frac{az + b}{cz + d}$ for $h = \begin{pmatrix} a & b \\ c & d \end{pmatrix}$

Hecke operator is thus,

$$[T(n)f](z) = \frac{1}{n}\sum_{ad=n} a^k \sum_{b=0}^{d-1} f\left(\frac{az+b}{d}\right). \tag{100}$$

In particular, for prime p,

$$[T(p)f](z) = p^{k-1}f(pz) + \frac{1}{p}\sum_{b=0}^{p-1} f\left(\frac{z+b}{p}\right). \tag{101}$$

In terms of Fourier coefficients:

$$f(z) = \sum_0^\infty f_m e^{2\pi i m z}, \quad [T(n)f](z) = \sum_0^\infty \gamma_n(m)e^{2\pi i m z}, \tag{102}$$

where,

$$\gamma_n(m) = \sum_{d|(n,m)} d^{k-1} f_{\frac{mn}{d^2}}. \tag{103}$$

In particular[20],

$$[T(p)f]_m = f_{pm} + \delta(p|m)p^{k-1}f_{\frac{m}{p}} \tag{104}$$

It follows then that $T(mn) = T(m)T(n)$ if m and n are coprime. More generally we can show

$$T(m)T(n) = \sum_{d|(m,n)} d^{k-1} T\left(\frac{mn}{d^2}\right). \tag{105}$$

In particular, the Hecke operators commute with each other. It is also useful that for prime powers we have a recursion relation,

$$T(p^{r+1}) = T(p)T(p^r) - p^{k-1}T(p^{r-1}) \tag{106}$$

which can be solved in terms of Tchebycheff polynomials[21]:

$$T(p^r) = p^{\frac{r(k-1)}{2}} U_r\left(\frac{1}{2}p^{\frac{k-1}{2}}T(p)\right). \tag{108}$$

Thus the $T(p)$ for prime p determine all the $T(n)$.

---

[20] $\delta(p|m) = 1$ if p divides m and zero otherwise. We use $\sum_{b=0}^{p-1} e^{\frac{2\pi i m b}{p}} = p\delta(p|m)$ to derive this formula.

[21] The Tchebycheff polynomials are defined by

$$U_0(x) = 1, \quad U_1(x) = 2x, \quad U_{r+1}(x) = 2xU_r(x) - U_{r-1}(x). \tag{107}$$

It is not difficult to establish an inner product on $\mathscr{S}_k$ with respect to which $T(n)$ are hermitean:

$$\langle f_1, f_2 \rangle >= \int_D f_1^*(z) f_2(z)[\text{Im } z]^{k-2} d^2z, \tag{109}$$

where,

$$D = \left\{ z = x + iy \mid -\frac{1}{2} \leq x \leq \frac{1}{2}, \; x^2 + y^2 \geq 1 \right\}. \tag{110}$$

The point here is that $y^k f_1^*(z) f_2(z)$ is a modular invariant function, so that it can be integrated after multiplying by the modular invariant volume form $\frac{dx \wedge dy}{y^2}$ to get a modular invariant quantity. Of course since each tile contributes the same amount we must restrict the integral to one fundamental domain.

Thus we have a set of commuting hermitean matrices, there is an orthogonal basis of simultaneous eigenvectors, with real eigenvalues: the *Hecke forms*.

Suppose we have as simultaneous eigenvector a cuspform satisfying,

$$T(n) f(z) = \lambda_n f(z), \quad \forall n. \tag{111}$$

The convention is to normalize an eigenvector by setting the first Fourier coefficient $f_1 = 1$. Taking Fourier coefficients of both sides, we get $f_n = \lambda_n$: *the Fourier coefficients of a simultaneous eigenvector of the Hecke operators are the eigenvalues.* It follows that these coefficients satisfy a multiplicative identity

$$f_m f_n = \sum_{d|(n,m)} d^{k-1} f_{\frac{mn}{d^2}}. \tag{112}$$

There is a Dirichlet series associated to any modular form

$$f(z) = f_0 + \sum_1^\infty f_n e^{2\pi i n z}, \quad \phi(s) = \sum_1^\infty \frac{f_n}{n^s} \tag{113}$$

or,

$$\phi(s) = \frac{(2\pi)^s}{\Gamma(s)} \int_0^\infty y^{s-1}[f(iy) - f_0] dy. \tag{114}$$

The modularity of f implies a functional equation[22] for ϕ:

$$(2\pi)^{-s} \Gamma(s) \phi(s) = (-1)^{\frac{k}{2}} (2\pi)^{s-k} \Gamma(k-s) \phi(k-s) \tag{116}$$

---

[22]To see this, just take the Mellin transform of the condition for invariance under inversion:

$$f\left(-\frac{1}{z}\right) z^k = f(z). \tag{115}$$

and conversely. The multiplicative property of the coefficients of the Hecke eigenforms yields a product formula:

$$\phi(s) = \prod_p \frac{1}{1 - f_p p^{-s} + p^{k-1} p^{-2s}}. \tag{117}$$

This is reminiscent of the Riemann zeta function $\zeta(s)$. Indeed the theta function, of which $\zeta(s)$ is the Mellin transform, is a modular form of a congruence subgroup of level 2.

It is of much interest to understand the behavior of the eigenvalues of the Hecke operators. They have been related to the zeros of the zeta function of algebraic varieties over finite fields (Eichler, Sato, Deligne). There are many deep conjectures about the behavior of the eigenvalues $\lambda(n)$ for large n. The simplest case is when the dimension of the space of cusp forms is one: when $k = 12$ the only cusp form is the function $\Delta(z)$ we introduced earlier. In this case the Hecke operators are 1×1 matrices: just numbers. From the above it is clear that these numbers are just the Fourier coefficients of the function $\Delta(z)$. In other words, for the case $k = 12$, the Hecke operators reduce to the Ramanujan τ-function: $T(n) = \tau(n)$. In fact Hecke discovered these operators by generalizing some ideas of Mordell on the modular form $\Delta(z)$ to the case of higher weight.

From our earlier discussion, we are led to consider the limit of large level where the invariance group becomes abelian. We now present a simple analogue of the Hecke problem for this case of periodic functions.

5.5 Hecke operators on periodic functions

Any modular form is periodic so we can expand it in a Fourier series $f(z) = \sum_{n=1} f_n e_n(z)$. Thus it is tempting to think of the space of modular forms as a subspace of the space of periodic functions V. To make this idea precise, we would like to have a norm on V.

The inner product on modular forms given above can be written as

$$\langle f, \tilde{f} \rangle = \int_{-\frac{1}{2}}^{\frac{1}{2}} dx \int_{x^2+y^2 \geq 1} dy\, y^{k-2} f^*(z) \tilde{f}(z) = \sum_{m,n=1}^{\infty} f_n^* \tilde{f}_m g_{nm} \tag{118}$$

where

$$g_{nm} = \langle e_n, e_m \rangle = \int_{-\frac{1}{2}}^{\frac{1}{2}} dx\, e^{2\pi i (m-n)x} \int_0^{\infty} y^{k-2} e^{-2\pi(m+n)y} \theta\left(y \leq \sqrt{(1-x^2)}\right) dy \tag{119}$$

is a positive sesquilinear form. We can extend the integral to the fundamental region of the translation group to get an inner product on V:

$$(\psi, \tilde{\psi}) = \int_{-\frac{1}{2}}^{\frac{1}{2}} dx \int_0^\infty y^{k-2} \psi^* \tilde{\psi} dy = \sum_{n,m=1}^\infty \psi_n^* \tilde{\psi}_m h_{nm}. \qquad (120)$$

In terms of the basis e_m, we have h_{nm} as a diagonal sesquilinear form

$$h_{nm} = (e_n, e_m) = (k-2)![4\pi n]^{1-k} \delta_{n,m}. \qquad (121)$$

The Hecke operators are given by quite simple formulae in terms of the Fourier components. We can work out easily their dual action on the basis $e_m(z)$:

$$T(p)e_m(z) = p^{k-1} e_{mp}(z) + \delta(p|m)e_{\frac{m}{p}}(z) \qquad (122)$$

for prime p. This can then be used extend to them as operators on V. The spectral problem for $T(p)$ in the infinite dimensional space V is much simpler than its counterpart in $\mathcal{S}_k$. We will solve this simpler problem and see that it has close connections to the theory of random matrices.

How will we recover the Hecke operators on modular forms? We could try thinking of modular forms as a subspace of the space of periodic functions. But under the above inner product on V they will not be square integrable. The reason is precisely modular invariance: each fundamental region contributes an equal amount to the integral, and the region $-\frac{1}{2} \le x \le \frac{1}{2}$ contains an infinite number of such regions. Thus, the expansion of a modular form in the basis $e_m(z)$ is not convergent in the above norm $(.,.)$. On the other hand, the integral for $\langle .,.,.\rangle$ corresponding to the sesquilinear form g_{nm} also can be extended to an inner product on V. It is convergent on modular forms but is a degenerate sesquilinear form in V: the integral is restricted to the region $x^2 + y^2 \ge 1$. We can quotient V by the null space of g_{nm} to get a finite dimensional space $\tilde{\mathcal{S}}_k$ that is a 'gauge fixed' version of $\mathcal{S}_k$. That is, instead of thinking of modular forms as a subspace of V, we think of them as a quotient of V by the null space of g_{nm}. The 'gauge fixing' amounts to the choice of one particular fundamental region among the infinite number as the domain of the integral. Thus the spectral problem of the Hecke operators on modular forms can be replaced by that on the space $\tilde{\mathcal{S}}_k$.

As k grows the dimension of $\mathcal{S}_k$ grows linearly with k. We will see that in a certain sense the two inner products g_{nm} and h_{nm} approach each other. Thus the discrete eigenvalues of $T(p)$ are so close together as $k \to \infty$ that they merge to form the continuous spectrum of $T(p)$ on V.

5.6 Toeplitz operators

We now solve the spectral problem for the Hecke operators on periodic functions. It will be more transparent to transform to the orthonormal basis

$$|m> = \left[(4\pi m)^{1-k}(k-2)!\right]^{-\frac{1}{2}} e_m.$$

We then have

$$T(p)|m> = p^{\frac{k-1}{2}}\left[|mp> + \delta(p|m)|\frac{m}{p}>\right].$$ (123)

There is then a simple description in terms of Toeplitz operators. The operators

$$A^{\dagger}(p)|m> = |pm>, \quad A(p)|m> = \delta(p|m)|\frac{m}{p}>$$ (124)

are adjoints of each other and satisfy

$$A(p)A^{\dagger}(p) = 1, \quad A(p)A(p') = A(p)A(p'), \quad A(p)A^{\dagger}(p') = A^{\dagger}(p')A(p)$$ (125)

for $p \neq p'$. That is, they form a commuting set of Toeplitz operators labelled by the prime numbers. Being isometries $|A| = |A^{\dagger}| = 1$. It follows easily that $T(p)$ are hermitean and that

$$|T(p)| \leq 2p^{\frac{k-1}{2}}.$$ (126)

The analogue of this inequality on the space of modular invariant functions (rather than periodic functions) is a much deeper statement.

5.7 Connection to random matrices

The simultaneous eigenfunctions of the $T(p)$ can now be obtained in terms of Tchebychev polynomials. The spectrum is connected with the Wigner distribution for random matrices.

It is enough to study each $T(p)$ separately: the theory 'localizes' completely. To see this, represent each number m in terms of its prime decomposition:

$$m = \prod_p p^{\nu_p}.$$ (127)

The product is over the set of all primes; $\nu_p = 0, 1, \ldots$ with only a finite number of them being non-zero. Then

$$A^{\dagger}(p)|\nu_2, \nu_3, \cdots > = |\nu_2, \ldots, \nu_p + 1, \cdots >,$$ (128)

$$A(p)|\nu_2, \nu_3, \cdots > = \delta(\nu_p \neq 0)|\nu_2, \ldots, \nu_p - 1, \cdots > .$$ (129)

Thus $A(p)$, $A^{\dagger}(p)$ act only on the p-th entry.

The Toeplitz algebra is the associative algebra generated by a pair of elements satisfying the relation

$$AA^{\dagger} = 1.$$ (130)

The standard representation in terms of an orthonormal basis $|\nu>, \nu = 0, 1 \ldots$ is

$$A^{\dagger}|\nu> = |\nu + 1>, \quad A|\nu> = \delta(\nu \neq 0)|\nu - 1> .$$ (131)

This is precisely the representation that we have.

Voiculescu [20] has found a remarkable connection between the theory of random matrices and the Toeplitz algebra. Given any polynomial $f : R \rightarrow R$, define

$$\langle f \rangle_N = \frac{\int \operatorname{tr} f(X) e^{-\frac{1}{2} \operatorname{tr} X^2} dX}{\int e^{-\operatorname{tr} X^2} dX} \tag{132}$$

the integral being over all hermitean $N \times N$ matrices. Thus X is a hermitean matrix whose matrix elements are independent random variables. Then, Voiculescu shows that

$$\lim_{N \to \infty} \langle f \rangle_N = \langle 0 | f(A + A^\dagger) | 0 \rangle. \tag{133}$$

There is a probability distribution on R, the Wigner semi-circle distribution, such that

$$\lim_{N \to \infty} \langle f \rangle_N = \int_R f(x) \rho(x) \, dx. \tag{134}$$

Explicitly,

$$\rho(x) = \theta(|x| < 2) \frac{1}{2\pi} \sqrt{[4 - x^2]}. \tag{135}$$

Thus we see that the Hecke operators $T(p)$ on periodic functions are hermitean operators whose spectrum is the interval $\left[-2p^{\frac{k-1}{2}}, 2p^{\frac{k-2}{2}} \right]$. The (generalized) eigenfunctions are given by Tchebycheff polynomials. The Wigner distribution gives the spectral density. Thus each $T(p)$ behaves like a hermitean random matrix; the different Hecke operators for different prime p commute with each other, so they are not 'free' in the sense of Voiculescu; instead they are statistically independent in the more conventional sense.

5.8 The limit of large weight

Recall that the main difference between the exactly solvable model above and the theory of modular forms is that we replaced the sesquilinear form g_{nm} by the simpler one h_{nm}. We now show that in the limit of large weight k this is a small correction so that what we obtained above is the asymptotic behavior as $k \to \infty$.

Note that we can split

$$h_{nm} = g_{nm} + q_{nm} \tag{136}$$

where k is the integral over the complimentary region

$$q_{nm} = \int_{-\frac{1}{2}}^{\frac{1}{2}} dx \, e^{2\pi i (m-n)x} \int_0^{\sqrt{(1-x^2)}} y^{k-2} e^{-2\pi(m+n)y} \, dy. \tag{137}$$

If we rewrite this in the orthonormal basis of h_{nm}, we will get

$$\delta_{nm} = \tilde{g}_{nm} + \tilde{q}_{nm} \tag{138}$$

where $\tilde{q}_{nm} = \frac{q_{nm}}{\sqrt{(h_{nn}h_{mm})}}$ etc . We will show that $|\tilde{q}_{nm}|$ tends to zero as $k \to \infty$.

Now, evaluating the y-integral,

$$\tilde{q}_{nm} = \left[\frac{\frac{m+n}{2}}{\sqrt{(mn)}} \right]^{1-k}$$

$$\int_{-\frac{1}{2}}^{\frac{1}{2}} dx e^{2\pi i (m-n)x} \frac{1}{(k-2)!} \Gamma\left(k-1, 2\pi(m+n)\sqrt{(1-x^2)}\right)$$

where the incomplete Gamma function is defined by

$$\Gamma(s, u) = \int_0^u t^{s-1} e^{-t} dt. \tag{139}$$

We will study this limit as $k \to \infty$ keeping m, n fixed[23]. Now recall that as $s - 1 > u$, the maximum value of the integrand is attained at its upper limit in this case, so that $\Gamma(s, u) < u^s e^{-u}$. Then

$$|\tilde{q}_{nm}| \le \left[\frac{\frac{m+n}{2}}{\sqrt{(mn)}} \right]^{1-k} \int_{-\frac{1}{2}}^{\frac{1}{2}} \frac{1}{(k-2)!}$$

$$\left[2\pi(m+n)\sqrt{(1-x^2)}\right]^{k-1} e^{-2\pi(m+n)\sqrt{(1-x^2)}} dx. \tag{140}$$

Again, replacing the integrand by its largest value (which is attained at $x = 0$), we get

$$|\tilde{q}_{nm}| \le \frac{\left[4\pi\sqrt{(mn)}\right]^{k-1}}{(k-2)!}. \tag{141}$$

The growth of the factorial beats the exponential growth for fixed m and n.

Thus we can see why the distribution of eigenvalues of Hecke operators resemble those of random matrices by our semi-classical approximation method. Moreover we see why the spectrum of $T(p)$ is in the interval $\left[-2p^{\frac{k-1}{2}}, 2p^{\frac{k-1}{2}} \right]$.

5.9 The Poisson algebra of modular invariant functions

We should expect that in the limit of large weight, the theory of modular forms is well-approximated by a classical theory. More precisely the algebra of matrices on $\mathcal{S}_k$ should tend to the Poisson algebra of functions on $U/\Gamma(1)$. In particular there will be functions on the upper half plane which are classical approximations to the Hecke operators. These Hecke functions will have vanishing Poisson brackets relative to each other. Their range will give a classical approximation to the Hecke eigenvalue

---

[23]We should really be estimating the operator norm of $\tilde{q}$. I hope that the arguments here will motivate a more rigorous analysis.

problem. This is another, (manifestly gauge invariant) of studying the limit of large weight.

The upper half plane is a symplectic manifold with symplectic form

$$\omega = \frac{dx \wedge dy}{y^2}. \tag{142}$$

This means that x and $\frac{1}{y}$ are canonical conjugates:

$$\{y^{-1}, x\} = 1. \tag{143}$$

The set of functions on the upper halfplane form a Poisson algebra with the bracket

$$\{u, v\} = y^2 \left(\frac{\partial u}{\partial x} \frac{\partial v}{\partial y} - \frac{\partial u}{\partial y} \frac{\partial v}{\partial x} \right). \tag{144}$$

The modular transformations

$$x \mapsto x + 1, \qquad y \mapsto y \tag{145}$$

and

$$x \mapsto -\frac{x}{x^2 + y^2}, \qquad y \mapsto \frac{y}{x^2 + y^2} \tag{146}$$

are canonical transformations. Thus the space of modular invariant functions is a sub-Poisson algebra. The quotient of this by its center is the algebra of 'gauge invariant observables' if we regard the modular group as a 'gauge group'. We can construct such observables from smooth functions of the upper half plane (vanishing sufficiently fast at infinity) by averaging over orbits. The Maas forms provide nice examples of such 'observables'.

In the limit of large k, the Hecke operators should tend to certain modular invariant functions (which are not holomorphic) that have zero Poisson brackets relative to each other. The range of these functions is the large k-limit of the Hecke spectrum. We should also be able to derive a systematic semi-classical expansion in powers of $\frac{1}{k}$. But this paper is getting long already; I hope to return to these questions in a later publication.

Acknowledgement. I thank Teoman Turgut and Edwin Langmann warmly for reading through various versions of this paper. Also, Ersan Demilrap brought references [3, 7, 14] to my attention. Thanks are due also to A. Agarwal, L. Akant and G. Krishnaswami for many discussions. The gracious hospitality of the Erwin Schrödinger Institute (Vienna) and the Feza Gürsey Institute (Istanbul), where this paper was written, is also acknowledged.

248   S. G. Rajeev

References

[1]   E. Witten, *Nucl. Phys.* B 160 (1979), 57.

[2]   E. Brezin and S. Wadia (ed.), *The Large N Expansion in Quantum Field Theory and Statistical Physics: From Spin Systems to 2-Dimensional Gravity*, World Scientific, Singapore 1993.

[3]   D. R. Hershbach, *Int. J. Quant. Chem.* 57 (1996), 295.

[4]   S. Guruswamy, S. G. Rajeev and P. Vitale, *Nucl. Phys.* B 438 (1995), 491 [arXiv:hep-th/9406010].

[5]   R. J. Henderson and S. G. Rajeev, *Internat. J. Modern Phys.* A 10 (1995), 3765. [arXiv:hep-th/9501080].

[6]   S. G. Rajeev, *Internat. J. Modern Phys.* A 9 (1994), 5583. [arXiv:hep-th/9401115].

[7]   R. G. Parr and W. T. Yao, *Density-Functional Theory of Atoms and Molecules*, Oxford Science Publications, 1989.

[8]   S. S. Chern, *Complex Manifolds without Potential Theory*, Springer-Verlag, 1979.

[9]   A. Huckleberry and T. Wurzbacher (ed.), *Infinite-dimensional Kaehler manifolds*, DMV-Seminar 31, Birkhäuser Verlag, 2001; M. Schlichenmaier, in *Conference Moshe Flato* (G. Dito and D. Sternheimer, eds.), Kluwer, 2000 [math.QA/9910137]; and other references given in these books.

[10] J. Mickelsson and S. G. Rajeev, *Commun. Math. Phys.* 116 (1988), 365.

[11] S. G. Rajeev and O. T. Turgut, *Commun. Math. Phys.* 192 (1998), 493.

[12] A. Pressley and G. Segal, *Loop Groups*, Oxford Math. Monogr., The Clarendon Press, Oxford, New York 1986.

[13] L. D. Landau and E. M. Lifshitz, *Quantum Mechanics*, Butterworth-Heinemann, 1997.

[14] D. P. Chong (ed.), *Recent Advances in Density Functional Methods*, World Scientific, Singapore 1995.

[15] W. Thirring (ed.), *The Stability of Matter: From Atoms to Stars*, Selecta of E. Lieb, Springer-Verlag, 1997.

[16] L. Hormander, Comm. Pure. Appl. Math. 32, 359 (1979); P. Folland, *Harmonic Analysis in Phase Space*, Princeton University Press, 1989.

[17] L. Akant, G. S. Krishnaswami and S. G. Rajeev, *Internat. J. Modern Phys.* A 17 (2002), 2413. [arXiv:hep-th/0111263]; A. Agarwal, L. Akant, G. S. Krishnaswami and S. G. Rajeev, arXiv:hep-th/0207200, to appear in *Internat. J. Modern Phys.*

[18] T. Apostol, *Modular Functions and Dirichlet Series in Number Theory*, Springer-Verlag (1990).

[19] P. Sarnak, *Some Applications of Modular Forms*, Cambridge University Press, 1990; P. Sarnak and N. M. Katz, *Random Matrices, Frobenius Eigenvalues and Monodromy*, Amer. Math. Soc., Providence, R.I., 1999.

[20] M. L. Mehta, *Random Matrices*, Academic Press, New York 1967; D. Voiculescu, K. Dykema and A. Nica, *Free Random Variables*, Amer. Math. Soc., Providence, R.I., 1992.

www.ingramcontent.com/pod-product-compliance
Lightning Source LLC
Chambersburg PA
CBHW081059220326
41598CB00038B/7158